AF180890

W. Herzog H.-C. Deter
W. Vandereycken (Eds.)

The Course of Eating Disorders

Long-Term Follow-up Studies of Anorexia and Bulimia Nervosa

With 29 Figures and 80 Tables

Springer-Verlag
Berlin Heidelberg New York
London Paris Tokyo
Hong Kong Barcelona
Budapest

Dr. Wolfgang Herzog
Abteilung für Allgemeine Klinische und
Psychosomatische Medizin, Medizinische Universitätsklinik
Bergheimer Straße 58, W-6900 Heidelberg, FRG

Prof. Dr. Hans-Christian Deter
Abteilung für Psychosomatik und Psychotherapie
Universitätsklinikum Steglitz
Hindenburgdamm 30, 1000 Berlin 45, FRG

Prof. Dr. Walter Vandereycken
Department of Psychology, University of Leuven
Leuvenselaan 85, 3300 Tienen, Belgium

ISBN-13: 978-3-642-76636-7 e-ISBN-13: 978-3-642-76634-3
DOI: 10.1007/978-3-642-76634-3

Library of Congress Cataloging-in-Publication Data
The Course of eating disorders: long-term follow-up studies of anorexia and bulimia nervosa / W. Herzog, H.-C. Deter, W. Vandereycken (eds.).
p. cm. Includes bibliographical references and index.

1. Anorexia nervosa-Longitudinal studies. 2. Bulimia-Longitudinal studies. I. Herzog, W. (Wolfgang) II. Deter, Hans-Christian.
III. Vandereycken, Walter, 1949– .
[DNLM: 1. Anorexia Nervosa. 2. Bulimia. 3. Follow-Up Studies. WM 175 C861] RC552.A5C68 1992 616.85'26-dc20 DNLM/DLC 91-5233

This work is subject to copyright. All rights are reserved, whether the whole or part of the material is concerned, specifically the rights of translation, reprinting, reuse of illustrations, recitation, broadcasting, reproduction on microfilm or in any other way, and storage in data banks. Duplication of this publication or parts thereof is permitted only under the provisions of the German Copyright Law of September 9, 1965, in its current version, and permission for use must always be obtained from Springer-Verlag. Violations are liable for prosecution under the German Copyright Law.

© Springer-Verlag Berlin Heidelberg 1992
Softcover reprint of the hardcover 1st edition 1992

The use of general descriptive names, registered names, trademarks, etc. in this publication does not imply, even in the absence of a specific statement, that such names are exempt from the relevant protective laws and regulations and therefore free for general use.

Product liability: The publishers cannot guarantee the accuracy of any information about dosage and application contained in this book. In every individual case the user must check such information by consulting the relevant literature.

Typesetting: Best-set Typesetter Ltd., Hong Kong
19/3130-5 4 3 2 1 0 – Printed on acid-free paper

Preface

Clinicians and scientists are becoming more and more interested in anorexia nervosa and bulimia nervosa, as evidenced by the increasing number of publications on the subject. In addition to studies on symptomatology, diagnosis, etiology, and treatment, studies on the course of disease are gaining in significance. More and more clinicians and researchers have now realized that anorexia nervosa and bulimia nervosa are often difficult to treat and that short-term outcome can be very misleading. In many cases these eating disorders follow a long-term course and may have serious consequences for the patient's life in the future.

The purpose of this book is to provide a detailed overview of the latest developments on long-term outcome in patients suffering from anorexia and bulimia nervosa. The subject is dealt with in such a way as to develop a more comprehensive perspective on the conditions, whereby various methodological approaches to studying outcome have been adopted – biomedical, psychological, sociocultural, and even historical.

In the first part "General Aspects of Follow-up Research" those topics, issues, and problems are treated which confront every clinician who wants to get an idea of long-term outcome of anorexia and bulimia nervosa and every scientist who is planning such a study. As well as presenting preconsiderations on the significance of follow-up studies, the book describes methodological prerequisites (Vandereycken and Meermann), reviews the literature on the long-term course of anorexia nervosa (Herzog et al.), discusses the particular problem of outcome assessment in eating disorders (Morgan et al., Deter), and deals separately with the peculiarities of eating disorders in male patients (Andersen).[1]

---

[1] As regards the characteristics in younger patients see the chapter by Steinhausen et al.

In the second part of the book "Follow-up Studies of Anorexia and Bulimia Nervosa," ten extensive long-term follow-up studies are presented. The data from these studies, which for the most part are being published for the first time, reflect the current state of research and make it possible to compare the results of long-term follow-up studies of anorexia from specialized psychiatric clinics (Russell, Theander), child and adolescent psychiatric institutions (Steinhausen, Tolstrup), and medical centers (Deter et al., Herzog et al., Engel et al.). Since a diagnostic distinction of bulimia was not made until about 10 years ago, there are no results of observations of bulimia nervosa patients over longer periods of time which are comparable to the data on anorexia patients. However, data from methodologically sophisticated studies in large numbers of patients over a follow-up time of 2–4 years have been reported (Fichter and Quadflieg, Lacey, Mitchell and Pyle). As in the Leuven study (Vandereycken and Pieters), prospective study designs were put to use. The last three chapters of this part present not only the results of the most well-known "ultralong-term studies," including those with observation times of over 30 years (Russell, Theander, Tolstrup). Based on their wealth of experience as clinicians and researchers, the authors have also summarized their thoughts on prognosis, on a new concept of chronicity, and also on the question of what can be learned from the long-term study of patients with eating disorders.

The third part of the book deals with "Special Issues." A broad spectrum of methods have been used to analyze various aspects of the subject of long-term follow-up in eating disorders. A biomedical approach was taken in twin studies (Schepank), and a thorough analysis of menstrual disorders and hormonal disturbances is also presented (Beumont). Eating behavior has been studied epidemiologically, too, in healthy individuals and in those with eating disorders (Rathner). The recovery process is presented through the eyes of the women affected (Noordenboos) and is followed by studies of psychosocial conceptions such as "social support" in the long-term follow-up of anorexia (Manz et al.) or the significance of ego strength, object relations, and life events as predictors of outcome (Sohlberg and Norring), as well as a historical follow-up of what happened to the fasting girls (Van Deth and Vandereycken). In the last chapter an experienced therapist describes his personal view of the developments and changes in the treatment of eating disorders over the past two decades (Petzold). Here, the question which from the start has been the main motivation for studying long-term follow-up arises again: How can we use and integrate our knowledge of the outcome of anorexia and bulimia nervosa so as to develop and offer optimal treatment?

December 1991

W. Herzog
H.-C. Deter
W. Vandereycken

Contents

List of Contributors

Andersen, A.E., M.D., Professor
Department of Psychiatry, The University of Iowa
500 Newton Road, Iowa City, IA 52242, USA

Beumont, P.J.V., M.D., Professor and Sub-dean
Department of Psychiatry, University of Sydney
Head of Royal Prince Alfred Hospital and Lynton Hospital
Camperdown, N.S.W. 2050, Australia

Deter, H.-C., Prof. Dr.
Abteilung für Psychosomatik und Psychotherapie
Universitätsklinikum Steglitz
Hindenburgdamm 30, 1000 Berlin 45, FRG

Engel, K., Priv.-Doz. Dr.
Westfälische Klinik für Psychiatrie, Universität Bochum
Marsbruchstraße 179, W-4600 Dortmund, FRG

Fichter, M.M., Prof. Dr.
Psychosomatische Klinik Roseneck
Am Roseneck 6, W-8210 Prien, FRG

Hentze, M., Dr.
Benkiser GmbH
Achenbachstraße 58, W-6700 Ludwigshafen, FRG

Herzog, W., Dr.
Abteilung für Allgemeine Klinische und
Psychosomatische Medizin, Medizinische Universitätsklinik
Bergheimer Straße 58, W-6900 Heidelberg, FRG

Kröger, F., Dr.
Psychosomatische Klinik
Universität Aachen, RWTH
Pauwelsstraße, W-5100 Aachen, FRG

Lacey, J.H., M.D., Professor
Adult Psychiatry Section, St. George's Hospital
Medical School, University of London
Department of Psychiatry, Jenner Wing
Cranmer Terrace, Tooting, London SW17 ORE
United Kingdom

Macquire-Samson, I.M.P.J., M.B., B.S., M.R.C. Psych.,
Senior Registrar, Department of Psychiatry, Barrow Hospital
41 St. Michael's Hill, Bristol BS2 8DZ, United Kingdom

Manz, R. Dipl. Psych.
Psychosomatische Klinik, Universität Heidelberg
Thibautstraße 2, W-6900 Heidelberg, FRG

Mayer, H.
Streß-Forschungsgruppe, Universität Heidelberg
Bergheimer Straße 58, W-6900 Heidelberg, FRG

Meyer, A.E., Prof. Dr.
Abteilung für Psychosomatische Medizin
Medizinische Universitätsklinik
Martinistraße 52, W-2000 Hamburg 20, FRG

Meermann R., Dr. Dipl. Psych., Priv.-Doz.
Psychosomatische Fachklinik
Bombergallee 10, W-3280 Bad Pyrmont, FRG

Minne, H., Prof. Dr.
Abteilung für Endokrinologie
Medizinische Universitätsklinik
Bergheimer Straße 58, W-6900 Heidelberg, FRG

Mitchell, J.E., M.D., Associate Professor
Department of Psychiatry, University of Minnesota
University Hospitals
Mayo Box 393, 420 Delaware Street S.E.
Minneapolis, MN 55455, USA

Morgan, H.G., M.A., M.D., Professor
Department of Mental Health, University of Bristol
41 St. Michael's Hill, Bristol BS2 8DZ, United Kingdom

Noordenbos, G., Ph.D.
Department of Women's Studies, State University of Leiden
Wassenaarseweg 52, 2333 AK Leiden, The Netherlands

Norring, C., M.D.
Department of Clinical Psychology, Uppsala University
P.O. Box 1225, 751 42 Uppsala, Sweden

Petzold, E., Prof. Dr.
Psychosomatische Klinik
Universität Aachen, RWTH
Pauwelsstraße, W-5100 Aachen, FRG

Pieters, G., M.D.
University Psychiatric Center
3070 Kortenberg, Belgium

Purgold, J., A.A.P.S.W., Research Associate (deceased)
Department of Mental Health, University of Bristol
41 St. Michael's Hill, Bristol BS2 8DZ, United Kingdom

Pyle, R.L., M.D., Associate Professor
Department of Psychiatry, University of Minnesota
Mayo Box 393, 420 Delaware Street S.E.L.
Minneapolis, MN 55455, USA

Quadflieg, N., Dipl. Psych.
Abteilung für Psychiatrie, Universität München
W-8000 München, FRG

Rathner, G., Dr.
Psychotherapeutische Abteilung
Universitätsklinik für Kinderheilkunde
Universität Innsbruck
Anichstraße 35, 6020 Innsbruck, Austria

Russell, G.F.M., M.D., Professor
Department of Psychiatry, University of London
Bethlehem Royal Hospital and Maudsley Hospital
Institute of Psychiatry, De Crespigny Park
Denmark Hill, London SE5 8AF, United Kingdom

Rief, W., Dr.
Psychosomatische Klinik Roseneck
Am Roseneck 6, W-8210 Prien, FRG

Sarembe, E., Dr.
Abteilung für Endokrinologie
Medizinische Universitätsklinik
Bergheimer Straße 58, W-6900 Heidelberg, FRG

Schellberg, D., Dipl. Psych.
Psychosomatische Klinik
Zentralinstitut für Seelische Gesundheit
J5, W-6800 Mannheim 1, FRG

Schepank, H., Prof. Dr.
Psychosomatische Klinik
Zentralinstitut für Seelische Gesundheit
J5, W-6800 Mannheim 1,.FRG

Seidel, R., Dr.
Psychiatrische Universitäts-Poliklinik für Kinder
und Jugendliche, Universität Zürich
Freiestraße 15, 8028 Zürich, Switzerland

Seilkopf, S., Dr.
Abteilung für Allgemeine Klinische und
Psychosomatische Medizin, Medizinische Universitätsklinik
Bergheimer Straße 58, W-6900 Heidelberg, FRG

Sohlberg, S., Ph.D.
Department of Clinical Psychology, Uppsala University
P.O. Box 1225, 75142 Uppsala, Sweden

Steinhausen, H.-C., Prof. Dr.
Psychiatrische Universitäts-Poliklinik für Kinder
und Jugendliche, Universität Zürich
Freiestraße 15, 8028 Zürich, Switzerland

Theander, S., M.D., Docent
Department of Psychiatry, University Hospital
22185 Lund, Sweden

Tolstrup, K., M.D., Professor (em.)
Department of Child Psychiatry, University of Copenhagen
Osterbrogade 212, 2100 Copenhagen, Denmark

Vandereycken, W., M.D., Ph.D., Professor
Department of Psychology, University of Leuven
Leuvenselaan 85, 3300 Tienen, Belgium

van Deth, R., Dipl. Psych.
Gans 17, 2381 KA Zoeterwoudel, The Netherlands

Vassilas C.A., M.D., Ch.B., M.Phil., M.R.C. Psych.
Lecturer in Mental Health and Honorary Senior Registrar
University of Bristol
41 St. Michael's Hill, Bristol BS2 8DZ, United Kingdom

Wittern, M., Dr.
Schemannstraße 5b, W-2000 Hamburg 67, FRG

Part I
General Aspects of Follow-up Research

The Significance of Follow-up Investigations

W. Vandereycken and R. Meermann

The questions raised by follow-up research are of a conceptual, methodological, practical, and ethical nature (Vandereycken and Meermann 1984; Garner 1985). We will discuss the most important issues within a more general framework, but with special emphasis on the aspects related to the subject of this book: the long-term course of eating disorders with or without treatment.

Conceptual Issues

While discussing the many difficulties associated with long-term assessment in behavior therapy, Mash and Terdal (1980) found "a remarkable consensus on two issues: first, the importance of conducting follow-up evaluations; and second, the lack of adequate follow-up information across a wide range of treatments and problems. This discrepancy between the perceived importance of follow-up and the nonavailability of meaningful information highlights the issues and disagreements surrounding the conduct of follow-up investigations in psychotherapy generally and behavior therapy in particular" (p. 102). The extent to which a particular treatment procedure produces *lasting* effects continues to be an important concern for researchers and practitioners alike. Although it is conceivable that in some situations even "short-term relief" may be a worthwhile outcome, treatments with short-lived results appear to be insufficient in meeting the needs of both the patient and the society at large.

In debates on the effectiveness of psychotherapy, clinicians have been almost "brainwashed" with the ever-recurring warning that the absence of follow-up data in clinical research makes the long-term effects of any treatment method a matter of pure speculation. One has argued that *durability* rather than short-term outcome seems to be the great equalizer of treatments: as time following therapy increases, the magnitude of the treatment effect decreases. Thus the progressive diminishment of therapeutic effects with time, both within and across treatments, stresses the importance of examining changes over long time intervals (Mash and Terdal 1980). This should be a fortiori the case in phenomena such as eating disorders which are known for their periodic (partial) remissions and relapses. Although Nicholson and Berman (1983) have questioned the necessity of follow-up in evaluating psychotherapy, they suggest a selective use of follow-up designs "when examining diagnoses found to have high relapse rates or where the specific focus is on maintenance of treatment gains" (p. 275). But, besides some methodological and practical problems (to be discussed further on), the researcher should also bear in mind some conceptual pitfalls.

Research guided by the axiom "the longer, the better" might derail into a *follow-up fallacy*: "Without an adequate consideration of purposes and method, the conclusions to be drawn from follow-up investigations of any length will necessarily be limited" (Mash and Terdal 1980 p. 102). When we study, for instance, a consecutive series of patients, the differences in time span between first presentation and follow-up assessment constitute a complex issue. The longer this time interval, the more likely subsequent therapeutic interventions and/or life events might have influenced the evolution and contributed to the patient's actual state. When evaluating patients of different ages after a similar time period, it could be that we are just assessing the "normal" development of an individual or the "natural" course of a disorder (Vandereycken and Pierloot 1983b). Furthermore, the longer the time span, the greater the likelihood that shifts will occur in the normative standards used in evaluating and reporting outcome (Mash and Terdal 1980).

The main purpose for conducting follow-up studies is to extend our clinical knowledge such that we would be able to put findings at a given moment, summarized as a diagnosis, in a time perspective satisfying man's eternal wish to predict the future. Scientists nowadays are highly valued prophets and in matters of life and death medicine stands under great pressure to provide reliable prophecies. Hence, a meaningful diagnosis is expected to imply some form of forecast of the probable course of an illness. *Prognosis* – to know beforehand – constitutes the alpha and omega of follow-up research. The crucial issue here is one of generalization. For several diseases, especially those with a chronic course, the long-term prognosis is often subject to disagreement among clinicians, between clinicians and researchers, and also among researchers. Some of these discrepancies arise from the difference in professional roles vis-à-vis patients, but a major source of disagreement may be attributed to a difference in the populations they sample: "In sum, the clinician's illusion is the attribution of the characteristics and course of those patients who are currently ill to the entire population contracting the illness" (Cohen and Cohen 1984 p. 1180).

This problem of sampling bias will be discussed in the paragraph on methodological issues with regard to the problem of comparability. But another major issue should be stressed here: even if patient populations sampled by clinicians are comparable, we still do not know what happens to those individuals with eating disorders who stay outside the health-care system. What determines a benign evolution of the disorder (protective factors) or when and how do spontaneous recoveries occur? Therapists must accept that many patients can do without them or may be even better off without any treatment at all (Vandereycken and Meermann 1984). In view of the popularity of the self-help movement in eating disorders, it is astonishing how little research has been done on (long-term) outcome of self-help groups. They might constitute a most interesting control group for professional therapies! Although in recent years the epidemiological interest in eating disorders increased tremendously, we still know little or nothing about the *natural course*. Instead of simply counting the cases, epidemiological studies should adopt a longitudinal perspective. An example is the study by Yager et al. (1987, 1988) who followed up after an average of 20 months a large group of females with eating disorders who had responded to a solicitation in a women's magazine. Another example can be found in this book (see Rathner, this volume).

A final conceptual remark: besides the purpose of evaluation, follow-up research may have interesting *side effects*. It may yield important diagnostic information, e.g., the early detection of warning signals that attract our attention to a possible deterioration of the patient's status. This could be useful as to secondary prevention by inducing the need for a new therapeutic intervention or the reformulation of current treatment. Follow-up assessments may also evoke unintentional reactivity: some behavioral or attitudinal changes may be induced by the assessment context itself and/or the influence of the assessor, particularly when there was a previous therapeutic relationship between (ex-)patient and assessor. Finally, follow-up contacts may be intentionally treatment oriented in their purpose: the assessment itself is then presumed to have some value (a kind of booster) in enhancing the persistence of treatment effects (Mash and Terdal 1980). The possible negative side effects of follow-up research will be discussed in the section on ethical issues.

Methodological Issues

The literature on anorexia nervosa, and more recently also bulimia nervosa, shows a fortunate trend toward more attention to follow-up investigations. A very thorough review of the world literature (including publications in English, German and French which described at least 10 patients followed up for at least 1 year) revealed that between 1930 and 1980 more than 70 reports had been published on treatment outcome, course of illness, or prognosis in anorexia nervosa patients. In the last decade at least as many similar reports have appeared as in the previous half century: a comparable review of the literature from 1980 to 1989 ended up with a total of 65 publications (both reviews were the subject of two Dutch dissertations, by Magda Duerinkx in 1981 and Chantal Van Elsuwege in 1990, at the Department of Psychology of the Catholic University of Leuven, under supervision of the first author WV). The majority of studies after 1980 were methodologically much better than the previous ones. But if judged according to strict research standards, many recent studies suffer from basic shortcomings and still are difficult to interpret and compare. We will now summarize the most important methodological issues encountered in follow-up studies of eating disorder patients. This summary is based on our own analysis of the literature as well as on a series of publications on the same subject (Deter 1981; Garfinkel and Garner 1982; Herzog et al. 1988, 1990; Hsu 1980, 1987, 1988, Rohrmeier 1982; Schwartz and Thompson 1981; Steinhausen and Glanville 1983; Swift 1982; Theander 1985; Thompson and Gans 1985; Vandereycken and Meermann 1984; Vandereycken and Pierloot 1983b).

Incomplete Information on Patient Characteristics

Although physicians may prefer some flexibility in the use of diagnostic criteria within a therapeutic context, a diagnosis for research purposes should be made according to standardized criteria and preferably on the basis of consensus between at least two clinicians. With the worldwide acceptance of DSM-III criteria a big step has been set towards a common diagnostic language, regardless of the weaknesses of this classification system. But for long-term studies revisions of the system (now DSM-III-R and tomorrow DSM-IV) pose a lot of problems: for example, the DSM-III bulimia diagnosis differs substantially from the DSM-III-R delineation. An adaptation of a previously diagnosed sample to the new criteria risks becoming a rather questionable manipulation of the data since it is always based on retrospective information.

The description of the eating disorder should also include systematic assessment of important behaviors and attitudes known to be related to the syndrome (see "Deficiencies in Follow-up Assessment"). This would allow distinguishing between possible subtypes and give an impression of degree of severity. The latter may also be reflected by information on duration (or age at onset and at assessment) and previous course (including treatment attempts) of the eating disorder. It most probably makes a difference when additional diagnoses are present such as mood disorder, substance abuse, and personality disorder.

Patient Selection or Sampling Bias

Patient selection is one of the major problems in follow-up studies (see "Practical Issues"). We may distinguish four types of patient selection (Vandereycken and Pierloot 1983b):

1. *Referral selection*: A specialized center known for treating eating disorders may attract a particular subset of patients and/or the selection depends on the source of referral (general practitioners, psychiatrists, other centers, announcement in local newspaper or widely read magazine, etc.). In our experience, especially cases with long, complex or intractable problems (i.e., negative selection) tend to be referred to specialized (research or academic) centers.
2. *Criteria for admission*: Although seldom explicitly mentioned in the literature, we informally learned that many treatment centers select the patients they like to admit; for instance, they exclude the older ones, the males, those with a complicated illness history, a dangerous physical status, lack of motivation (compulsory admissions) or those living too far away or without a collaborating family, etc.
3. *Dropout during treatment*: This third selection factor is also scarcely commented on in the reports; in spite of the frustration they evoke in the therapists, these cases must always be included in an outome study particularly so because we have to deal very often with patients known for their reluctance to engage in treatment (Vandereycken and Pierloot 1983a).
4. *Subject attrition during follow-up*: This may be due to a failure to trace patients or to the subjects' refusal to participate in follow-up assessment, investigators should not only specify the proportion of sample attrition but also ascertain whether these drop-outs differ in any essential aspect from the rest of the sample (the question of representativity).

Poorly Planned Follow-up

All too many follow-up studies seem to arise from a clinician's wish at a certain moment "to know how my patients are doing." Hence, the majority of studies are designed in essence in a retrospective way and will be inevitably limited by the quality and amount of information on the admission sample gathered long before the follow-up was planned. Hence many studies lack some clear guiding hypotheses or they are fabricated post hoc. Long-term measurements should be planned early in the program, preferably with repeated assessments after fixed time intervals. One single endpoint observation after whatever time period has limited value. Prospective repeated observations, however, allow a standardization across time (temporal congruence of measurements), enhance the comparability of results, can establish certain patterns or trends in the course, and avoid the error that a random fluctuation in the patient's state be taken for systematic change.

Follow-up Interval as Empty Space

The choice of follow-up interval(s) will depend on the purpose of the investigation: The more one is interested in finding treatment effects, the shorter the time span (e.g., up to 1 year after treatment). Of course in disorders like anorexia and bulimia nervosa, which have a strong tendency to show an undulating course with periodic remissions and relapses or exacerbations, repeated measurements over several years are required. But we have warned already against the follow-up fallacy of "the longer, the better." Moreover, it is striking how many investigators seem to be interested only in comparing two cross sections in time: What lies in between the first and the final assessment is considered a black box or "empty space." But the longer the time interval, the more likely subsequent therapeutic interventions and/or major life events (if not just landmarks in the normal growth of young individuals) have influenced the evolution. So we need to know what has happened to the patient in the meantime, and not just with regard to stressful events (see, e.g., Strober 1984), but also concerning available social support. Sometimes, however, this information may be difficult to interpret: for example, if a patient appears to do fairly well but is still in regular treatment, should we judge this as a weakness or a strength?

Deficiencies in Follow-up Assessment

Follow-up assessment should make use of a direct method to be reliable (in rank order of diminishing quality): face-to-face interview, telephone interview, questionnaires sent by mail. Less direct methods are more subject to an individual's distortions or denial and are far more likely to yield inaccurate, incomplete, contradicting, or ambiguous information. The assessment should be multidimensional (see, e.g., Wilson 1987) using standardized procedures with sufficient validity and reliability (see Morgan et al., this volume). It is recommended that information be obtained from other sources than just the patient as only information from the patient might present too subjective a picture. Also, other feedback will enlarge the scope of the assessment onto an interpersonal level, especially as to family or marital interactions (see Vandereycken et al. 1989).

Lack of Clear Outcome Criteria

What clinicians mean by normal, cured or recovered is often only clear to them and represents the conclusion of an impressionistic appraisal. In eating disorders so many variables are included (biological, psychological, social) that it is very difficult to weigh them against one another and/or translate them into a quantifiable judgement (Hall 1985). Recommendations and warnings in this respect may be found in the chapter by Morgan et al. (this volume). May we add here that no matter how practically useful they can be, global outcome scores may obscure the findings unless such scores are accompanied by detailed accounts of the individual categories they are based upon. A final remark regarding such judgements: Follow-up assessment, especially if intent upon the evaluation of a particular treatment, has to be carried out by an independent assessor (or, even better, two assessors trying to reach a high interrater reliability).

Practical Issues

All the authors of this book as well as other researchers carrying out similar investigations are well acquainted with the frustrating difference between the beautiful research design and its concrete execution. But the investigators seldom report about the hindrances encountered during the project, although some of their frustrations can be found in the discussion paragraph of their publications where they usually "apologize" for the shortcomings. It is refreshing and encouraging at the same time when some researchers openly discuss the many practical problems involved in a follow-up study. We refer the frustrated reader to the recent discussion of the difficulties our colleagues in London encountered in the execution of a prospective treatment study of anorexia nervosa (Gowers et al. 1989).

A major difficulty all follow-up researchers struggle with is obtaining both reliable and complete data on all subjects involved, a difficulty which increases with the duration of the follow-up period. As discussed above, *sample attrition* is a crucial factor for a study to have sufficient value from a methodological viewpoint. Obtaining more than 50%–60% of the original sample after 5 years seems to require, at least for American investigators, "the skills of a detective" (Liberman 1978). Indeed, both a higher social mobility and a less well organized registration system hamper the location of former patients and account for the generally lower rate of return in American versus European studies. And even if one can trace former patients, many may refuse to cooperate. So how far should/can one go? This certainly is not just a matter of time, personnel or money: "detective-like behavior must be tempered by the priority value and ethic of confidentiality" (Liberman 1978 p. 119). We address this problem in the section "Ethical Issues."

Subjects' refusal to participate in follow-up assessment may be interpreted in different ways: it may suggest that things are going well and that the individual does not want to be bothered, or it may reflect a dissatisfaction with the former treatment (and/or therapists involved). In the latter case this might be related to a treatment failure but does not necessarily mean that the subject is now in bad shape. Since this interpretation problem is practically unsolvable, regardless of the fact that statistical procedures can be used to compensate any difference between the follow-up sample and the original sample, the study should be designed such that maximum effort is made to enhance the likelihood of participation within the limits of ethical rules. Mash and Terdal (1980) give the following recommendations in this respect:

1. The follow-up assessment should be made as important as earlier measurements.
2. The importance of long-term assessment should be explained early in the program.
3. The rationale for follow-up should be made explicit, emphasizing the diagnostic and therapeutic role of repeated measures for long periods of time.
4. Researchers themselves must give priority to follow-up assessment.
5. The practical demands of follow-up assessment should be matched as much as possible to the characteristics and circumstances of the population under study.

As with regard to treatment itself, probably the best way to detect and correct problems in a follow-up study is to regularly evaluate the cost to benefit ratio of the whole project for all parties involved.

Ethical Issues

Follow-up research presents the investigator with a difficult dilemma. As a scientist he feels obliged to carry out appropriate research in order to evaluate and consequently improve treatment methods, but as a practitioner he must also protect and promote the well-being of his (former) patients. The ethical issues involved are particularly delicate in psychiatry (see, for example, Nelson and Grunebaum 1972; Robins 1977).

Follow-up investigations raise important ethical questions since they may be viewed as an intrusion or invasion into the *patient's privacy*. The individual concerned may, for instance, not want to be considered a patient anymore or reminded of having been in treatment. Even an innocent invitation by mail or phone to participate in a follow-up assessment could be a threatening event for the person who might be put in the position of being forced, for example, to explain to people (partner, children) they are currently involved with something which belonged to a far remote past which has perhaps been kept secret from them. In order to protect patients from unwilling and potentially harmful involvement in research, the rule of *informed consent* has been given strong emphasis (Appelbaum et al. 1982). But some patients, anorectics in particular, are known for their overcompliance or willingness to please authority figures, a reflection of their low self-esteem and lack of assertiveness. Hence, some of them might be too eager to consent (Eth et al. 1984). Moreover, can we expect them to realize – at a moment of crisis and/or ill-health – what all the implications of their consent may be, also in the long run?

Follow-up research is an inevitable by-product of the pressure on clinicians to justify their work. The more time and expertise are involved in treatment, the more it costs and, hence, the more guarantees (proofs) regarding treatment efficacy (and efficiency) are required, especially when third parties (insurance companies, state health services) pay the bills. But in order to show the most favorable or clear-cut *cost to benefit ratio*, the researcher might select special aspects of the treatment and patient status, while (deliberately or not) neglecting other facets which do not demonstrate as much or are difficult to translate into quantitative terms. This issue is closely related to the endless discussions of the attempts to evaluate psychotherapy. The sometimes sharp controversies on this subject are centered around the competing value systems of *meaningfulness versus effectiveness*, i.e., experienced sense of the treatment versus demonstrable results: "The application of 'scientific' methodology to the evaluation of psychotherapy leads to an emphasis upon that which is measurable, although possibly not relevant, certain narrow parameters of effectiveness, and on ignoring of that which is most relevant, meaningfulness. (. . .) The notion of appropriateness as to what has worth in therapy will have to evolve as an outcome of both subjective report and observable data, placed into the cauldron of critical analysis" (Kisch and Kroll 1980 pp. 409–410). Applied to eating disorder patients: Should we weigh them or ask what life means to them?

In assessing outcome the researcher-clinician will be confronted with his own values and expectations. The results of his work become part of his self-evaluation. So, to what point does he want the patient to (help) evolve? Recovery and cure are primarily medical concepts, but medical treatments do not necessarily have cure as a goal. In fact, the medical syndromes for which the relevant measure of treatment is whether the patient is cured at follow-up represent only a small minority of the total. "There are *incurable illnesses* with spontaneously occurring remission and exacerbation (e.g., arthritis) where the question of follow-up is not cure or improvement but whether the treatment has increased the proportion of follow-up intervals during which the patient can function effectively. There are *self-limited illnesses* from which almost all recover (e.g., influenza) where the follow-up question is not improvement but whether the course of the disease was milder as a result of treatment" (Mash and Terdal 1980 p. 139). Applied to eating disorders: When faced with chronic cases, one may be inclined to think of anorexia or bulimia nervosa as an incurable disease; knowing of "spontaneous" or quick recoveries in young people, the eating disorder may be viewed as a self-limited disorder.

Though every therapist is hoping to cure his patients, there exist many differences in the meaning of the word *recovery*, depending on the conceptualization of the eating disorder (Vandereycken and Meermann 1984). A medically oriented clinician will be interested in the restoration of the normal nutritional status and physiological functioning of the patient as reflected, for instance, in the maintenance of a normal body weight and the resumption of regular menstruations. The systems-oriented therapist, on the other hand, will define recovery in more ecological-interactional terms, especially emphasizing healthy family functioning. But the actual follow-up research (exemplified in this book) shows the fortunate tendency to base judgements – how ever subjective they still may be – upon both the symptomatic status and the psychosocial adjustment of the patient compared to normal or healthy subjects of the same age group.

Careful follow-up research is a good antidote against rescue fantasies or feelings of omnipotence so often present in inexperienced or uncritical therapists. Selvini-Palazzoli (1974) had stressed that we can say of anorexics what Bleuler has said of schizophrenics: The more accurate the follow-up, the fewer the cures! This could lead, however, to a fatalistic attitude in the therapist. Instead, treatment failures should become a challenge to the clinicians involved. An accurate prospective follow-up study may yield important information as to the factors (including therapeutic interventions) which determine *relapse* and, hence, may improve our knowledge about the etiology of the disorder (Vandereycken and Meermann 1984). A relapse will mean something different to each individual; each patient will define it in her own terms and according to her own behavior and/or situation. Perhaps we have to abandon the term relapse altogether because of its negative connotation. Indeed, a relapse is a chance to learn more about what leads a person into a (new or old) maladaptive behavior pattern (see, for example, Mitchell et al. 1985). It allows us to gain more insight into the recovery process and the factors influencing its course. But we have to unlearn that almost incurable tendency to ascribe favorable outcomes to our own therapeutic intervention, whereas an unfavorable evolution is attributed to factors (usually outside of the therapy situation) which the therapist cannot be blamed for . . .

References

Appelbaum PS, Roth LH, Lids C (1982) The therapeutic misconception: informed consent in psychiatric research. Int J Law Psychiatry 5:319–329

Cohen P, Cohen J (1984) The clinician's illusion. Arch Gen Psychiatry 41:1178–1182

Deter H-C (1981) Zur Methodik von katamnestischen Untersuchungen bei psychosomatischen Patienten am Beispiel einer Gruppe von 31 Anorexiepatienten. Psychother Psychosom Med Psychol 31:48–52

Eth S, Eth C, Edgar H (1984) Can a research subject be too eager to consent? Hastings Cent Rep 11(4):20–21

Garfinkel PE, Garner DM (1982) Anorexia nervosa. A multidimensional perspective. Brunner/Mazel, New York

Garner DM (1985) Iatrogenesis in anorexia nervosa and bulimia nervosa. Int J Eating Dis 4:701–726

Gowers S, Norton K, Yeldham D, Bowyer C, Levett G, Heavey A, Bhat A, Crisp A (1989) The St. George's prospective treatment study of anorexia nervosa: a discussion of methodological problems. Int J Eating Dis 8:445–454

Hall A (1985) What constitutes recovery in anorexia nervosa? In: Touyz SW, Beumont PJV (eds) Eating disorders: prevalence and treatment. Williams and Wilkins, Sydney, pp 32–44

Herzog DB, Keller MB, Lavori PW (1988) Outcome in anorexia nervosa and bulimia nervosa. J Nerv Ment Dis 176:131–143

Herzog DB, Keller MB, Lavori PW, Bradburn IS, Ott IL (1990) Course and outcome of bulimia nervosa. In: Fichter MM (ed) Bulimia nervosa: basic research, diagnosis and therapy. Wiley, Chichester, pp 126–141

Hsu LKG (1980) Outcome of anorexia nervosa. A review of the literature (1954 to 1978). Arch Gen Psychiatry 37:1041–1046

Hsu LKG (1987) Outcome and treatment effects. In: Beumont PJV, Burrows GD, Casper RC (eds) Anorexia and bulimia nervosa. Elsevier, Amsterdam, pp 371–377 (Handbook of eating disorders, part 1)

Hsu LKG (1988) The outcome of anorexia nervosa: a reappraisal. Psychol Med 18: 807–812

Kisch J, Kroll J (1980) Meaningfulness versus effectiveness: paradoxical implications in the evaluation of psychotherapy. Psychother Theory Res Pract 17:401–413

Liberman BL (1978) The maintenance and persistence of change: long-term follow-up investigations of psychotherapy. In: Frank JD, Hoehn-Saric R, Imber SD, Liberman BL, Stone AR (eds) Effective ingredients of successful psychotherapy. Brunner/Mazel, New York, pp 107–129

Mash EJ, Terdal LG (1980) Follow-up assessments in behavior therapy. In: Karoly P, Steffen JJ (eds) Improving the long-term effects of psychotherapy. Gardner, New York, pp 99–147

Mitchell JE, Davis L, Goff G (1985) The process of relapse in patients with bulimia. Int J Eating Dis 4:457–463

Nelson SH, Grunebaum H (1972) Ethical issues in psychiatric follow-up studies. Am J Psychiatry 128:1358–1362

Nicholson RA, Berman JS (1983) Is follow-up necessary in evaluating psychotherapy? Psychol Bull 93:261–278

Robins LN (1977) Problems in follow-up studies. Am J Psychiatry 134:904–907

Rohrmeier F (1982) Langzeiterfolge psychosomatischer Therapien. Springer, Berlin Heidelberg New York

Schwartz, DM, Thompson MG (1981) Do anorectics get well? Current research and future needs. Am J Psychiatry 138:319–323

Selvini-Palazzoli M (1974) Self-starvation: from the intrapsychic to the transpersonal approach to anorexia nervosa. Chaucer, London

Steinhausen H-C, Glanville K (1983) Follow-up studies of anorexia nervosa: a review of research findings. Psychol Med 13:239–249

Strober M (1984) Stressful life events associated with bulimia in anorexia nervosa. Int J Eating Dis 3(2):3–16

Swift WJ (1982) The long-term outcome of early onset anorexia nervosa. A critical review. J Am Acad Child Psychiatry 16:703–715

Theander S (1985) Outcome and prognosis in anorexia nervosa: some results of previous investigations, compared with those of a Swedish long-term study. J Psychiatr Res 19:493–508

Thompson MG, Gans MT (1985) Do anorexics and bulimics get well? In: Emmett SW (ed) Theory and treatment of anorexia nervosa and bulimia. Brunner/Mazel, New York, pp 291–303

Vandereycken W, Meermann R (1984) Anorexia nervosa: a clinician's guide to treatment. De Gruyter, Berlin
Vandereycken W, Pierloot R (1983a) Drop-out during in-patient treatment of anorexia nervosa: a clinical study of 133 patients. Br J Med Psychol 56:145–156
Vandereycken W, Pierloot R (1983b) Long-term outcome research in anorexia nervosa: the problem of patient selection and follow-up duration. Int J Eating Dis 2(4):237–242
Vandereycken W, Kog E, Vanderlinden J (1989) The family approach to eating disorders: assessment and treatment of anorexia nervosa and bulimia. PMA Publ New York
Wilson GT (1987) Assessing treatment outcome in bulimia nervosa: a methodological note. Int J Eating Dis 6:339–348
Yager J, Landsverk J, Edelstein CK (1987) A 20-month follow-up study of 628 women with eating disorders. I. Course and severity of simulated diagnoses and criterion symptoms. Am J Psychiatry 144:1172–1177
Yager J, Landsverk J, Edelstein CK, Jarvik M (1988) A 20-month follow-up study of 628 women with eating disorders: II. Course of associated symptoms and related clinical features. Int J Eating Dis 7:503–513

Long-Term Course of Anorexia Nervosa: A Review of the Literature

W. Herzog, G. Rathner, and W. Vandereycken

Introduction

W. Vandereycken and R. Meermann have already elucidated the significance of follow-up studies in eating disorders for the patient and her familiy, who are personally interested in prognosis, and for clinicians and researchers, who need to be acquainted with the spectrum of possible courses and prognostic factors and who also want to be able to justify their own methods of treatment. Here, some of the fundamental results from the literature on the long-term course of anorexia nervosa will be summarized. Since no results are presently available from long-term course studies of bulimia (though some are in preparation; see, for example, Fichter et al., this volume), this presentation will be restricted to the research on the long-term course of anorexia nervosa.

In the past 15 years an increasing number[1] of studies on the course of anorexia have been published. The great differences in the goals of the individual studies, the study samples, and the study methods make a comparison of the published results extremely difficult. Even data as basic as relative weight or menstruation status are not indicated in some of the work. There are two in principle differing options, then, as to how to proceed with a review of the literature: For one, the results of all studies can be included in the analysis, irrespective of methodology and comparability, in hopes that the results of clinical course studies are as comprehensive as possible, while accepting the inexactness due to differences in quality or methods. There are a number of excellent such review articles (Hsu 1980, 1988; Steinhausen and Glanville 1983a,b; D.B. Herzog et al. 1988; Steinhausen et al. 1991). We make special reference to Steinhausen et al. (1991) whose review allows comparison of publications from the past 10 years with studies of the course of anorexia from 1950 to 1980. In the other review strategy, first, quality criteria and minimum requirements

---

[1] The number corresponds to the increasing relevance of eating disorders: Eating disorder patients make up about 15% of the psychosomatic inpatient treatments presently available in the Federal Republic of Germany.

of methodology are defined and then those studies are compared which have fulfilled these criteria. The disadvantages of this method are that the phenomenonology of the results of clinical course cannot be considered in its entirety and that any selection of qualitative and methodological criteria is always an arbitrary one. The advantage, however, lies in greater clarity and transparence. The studies which have been selected can really be compared with respect to outcome since the results were derived in a comparable way. Because of this, the latter strategy has been selected for our review of the literature.

Criteria for Long-Term Course Studies in Anorexia Nervosa

Studies of outcome in eating disorders should be prospective and longitudinal in design. Then detailed information can be expected on the course of the disease process.

The question of adequate outcome criteria (What does "good" mean?) is not a trivial one and can have far-reaching implications. Szmukler and Russell (1986) also have indicated that in the formulation of criteria of success, basic etiologic considerations are involved – at least implicitly – and have pointedly sketched three variations of such basic considerations:

1. If anorexia is understood to be a conglomeration of the symptoms weight loss, amenorrhea, and fear of losing weight, then recovery means normal weight, return of menstruation, and a change in disturbed eating habits and modification of body perception. This view does not take the structure of predisposing personality factors into account and so aspects of personality do not need to be considered in outcome research.
2. If anorexia nervosa is understood to be the expression of a "deeper" disturbance in personality, then a "real" improvement in the respective personality disturbance would be an integral aspect of the assessment of the outcome of disease. An improvement merely in weight, menstruation, and eating symptoms with continuing psychopathologic disturbance could be understood as a syndrome shift and would have to be assessed accordingly.
3. Independently of basic etiologic considerations, it might be useful to investigate the course of the psychopathologic symptoms and family and social factors along with course of core symptoms (weight, menstruation, eating habits). With this third approach a specific clinical and etiologic picture would be dispensed with (as in model 2); however, data regarding personality, family, and social integration could be acquired beyond the purely symptomatic level.

Considering only eating symptoms is obviously too limited: No one would speak of a good outcome in a former anorexia patient who has regained normal weight and is menstruating again after a few years, but who is in the hospital for an extended period because of chronic schizophrenia. The sole definition of outcome criteria in terms of strictly psychoanalytic, behavioral therapeutic, or systemic etiologic considerations (model 2) makes it impossible to compare data with those from other centers. Hence, the smallest common denominator from the various schools of psychotherapeutic thought for studying course turns out to be the investigation of eating symptoms, personality factors, family aspects, and social integration without an all-encompassing, uniform clinical model.

This view has asserted itself of late, in that the course of eating disordes is pragmatically described with respect to the core symptoms and comorbidity – meaning either psychopathologic or somatic comorbidity. Of course the construct "core symptoms versus comorbidity" contains implicit etiologic preassumptions: first the core symptoms, which are abstracted from the psychopathologic condition, then the comorbidity. From a pragmatic point of view, at present we see no other alternative other than describing the course of the condition: Course is to be described, then, in terms of a matrix with a separate axis for symptoms, psychopathology, and family and social factors.

But it would not do justice to the "classical" studies of the outcome of anorexia nervosa if the methodological requirements formulated in this book (see "Outlook") were applied to them: In these studies it was usually avid clinicians and researchers following up large groups of patients for up to 30 years. Since diagnostic criteria, treatment possibilities, and research strategies have continually changed during this time, the "classical" studies cannot meet today's methodological requirements. Though before Russell (this volume) could establish that the time for "classical" long-term follow-up studies of anorexia is over and before Theander (1985) could express disappointment about the predictor analysis of long-term follow-up they had to carry out their studies. These studies will continue to be valuable, however, since they have paved the way for new studies which will more exactly analyze the procedural aspects of follow-up.

This chapter also serves to summarize the results of the classical long-term follow-up studies of anorexia, which for the most part were retrospective and cross-sectional investigations. The minimal methodological requirements, also according to Szmukler and Russell (1986) and Hsu (1988), can be formulated as follows:

1. Inclusion of all consecutively treated patients within a defined period of time who fulfill explicit diagnostic criteria (see Table 1)
2. Minimal catamnesis time 4 years
3. Failure to trace rate less than 5%
4. Direct interview rate over 70%
5. Use of the Morgan-Russell General Outcome Categories, which have been well defined (Morgan and Russell 1975; see also Morgan et al., this volume)

Table 1. Sample description of long-term follow-up studies of anorexia nervosa

	No. of patients (Males)	Follow-up period [Mean years] (Range)	Age[a] [Mean years] (SD)	Duration of illness[a] [Mean years] (SD)	Weight [Mean % ABW] (SD)	Diagnostic criteria	Department	Inpatient treatment (%)
Burns and Crisp (1984)	27 (27)	8.0 (2–20)	21.6 (5.7)	3.8 (3.3)	68 (8.6)	Crisp (1977)	Psychiatric	74
Deter et al. (this volume) Herzog et al. (this volume)	103 (12)	12.7 (9–19)	20.8 (6.1)	2.8) (4.0)	65.7 (9.9)	Feighner (1972)	Medical and psychosomatic	86
Eckert (1990)	76 (0)	n.r. 8–10	20.1 n.r.	1.75 (1.7)	n.r.	Feighner (mod)	Psychiatric	100
Hall et al. (1984)	50 (0)	8.0 (4–12)	20.1 (5.6)	3.9	73.9	Feighner (1972)	Psychiatric and medical[b]	72
Hsu et al. (1979)	105 (0)	5.9 (4–8)	20.8 (6.2)	3.5 (4.3)	68.2[b] (8.2)	Crisp (1977)	Psychiatric	59
Morgan and Russell (1975)	41 (3)	5.0 (4–10)	21.5 (8.6)	3.7 (3.7)	63.4 (9.1)	Russell (1970)	Psychiatric	100
Morgan et al. (1983)	78 (5)	5.8 (4–8.5)	n.r.	Median 1.6	67.8[b] (8.2)	Russell (1970)	Psychiatric	54
Remschmidt et al. (1988)	103 (4)	11.7	14.5 (1.9)	1.2	n.r.	Own	Child and adolescent psychiatric[b]	100
Russell (this volume)	41 (3)	20	21.5 (8.6)	3.7 (3.7)	63.4 (9.1)	Russell (1970)	Psychiatric and others	100
Theander (1985)	94 (0)	33 (min.24)	17.2 (3.6)	n.r.	n.r.	Theander (1970)	Psychiatric and others	100

ABW, average body weight, n.r., not reported
[a] At first assessment.
[b] Minimal weight until presentation.

These criteria guarantee a reasonable minimal catamnesis interval, a low dropout rate, a sufficiently high percentage of personally examined former patients, and the application of an albeit crude, but uniform outcome criterion (Morgan-Russell General Outcome Categories). These outcome criteria are based merely on physiologic data of weight and menstruation, but can be correlated adequately with psychologic and social outcome aspects (Morgan and Hayward 1988; see also Morgan et al., this volume).

Along with the intermediate-term follow-up studies already mentioned by Hsu (1988), Morgan and Russell (1975), Hsu et al. (1979), Morgan et al. (1983), Hall et al. (1984), the long-term follow-ups by Ratnasuriya et al. (cited in Russell, this volume), Eckert (1990) and Deter et al. (this volume) will be included in the analysis. Although not all the minimal methodological requirements have been fulfilled, other studies have been included – Theander (1985; own diagnostic criteria) because of the long period of catamnesis, Remschmidt et al. (1988; own diagnostic criteria and 18% of patients not classified) for follow-up of anorexia in young patients, and Burns and Crisp (1984; catamnesis partly under 2 years) to illustrate intermediate-term follow-up of anorexia in men (see Table 1).

Results

Mortality

The follow-up mortality rates indicated in the entire literature fluctuate between 0% and 21% (Halmi et al. 1976). This also corresponds approximately to the spectrum found in the studies we analyzed: mortality ranges from 0% to 18% (Theander 1985). With the exception of Remschmidt's study, which is based on a sample of child and adolescent psychiatric patients, the very high mortality rates occur in the very long follow-ups. Theander found a rate of 18% mortality at an average catamnesis time of 33 years, and Russell (this volume) reports a 15% mortality at 20-year follow-up. Deter et al. (this volume) and Herzog et al. (this volume) found a 16% mortality rate at 12-year follow-up, whereby it must be taken into account here that the statistics come from an internal medicine – psychosomatic department and 5% of the sample died during first hospitalization after having been admitted in very critical physical condition.

The mortality rate (below 5%) at intermediate-term follow-up with an average catamnesis interval of 5–8 years is significantly lower. The optimism of Hall and coworkers (1984), who established that the mortality rates for anorexia do not exceed 5% these days, does not hold for long-term follow-up. Although Morgan and Russell (1975) reported, for example, at first follow-up a mortality rate of 5%, after 20 years of observation it was 15% (Russell, this volume).

In summary the mortality rate is 1/2% to 1% per year of observation. In the studies analyzed here the mortality ratio was not standardized: Patton (1989) found an average mortality rate of 3% in a large sample of 481 anorexia patients, which was six times higher than the mortality rate in the normal population. For patients with particularly low weight (under 35 kg) the mortality risk was 15 times higher.

The cause of death was suicide in from one half to one third of patients. After this come various medical complications, which for the most part were a result of the extreme emaciation, manipulative use of medicines (laxatives, diuretics, potassium), or pathologic eating habits (for example, electrolyte shift after vomiting). (See also Herzog et al., this volume, on somatic findings in long-term follow-up; Table 2).

Weight and Menstruation (Morgan-Russell General Outcome Categories)

If the physiologic core variables of the course of anorexia weight and menstruation are summarized (see Morgan et al., this volume), then for intermediate-term follow-up (Fig. 1) and long-term follow-up (Fig. 2) the following results can be derived:

- Good outcome: simultaneous normalization of menstruation and weight at intermediate-term follow-up of from 36% (Hall et al. 1984) to 58% (Morgan 1983). This increases up to 76% (Theander 1985) in the long term.

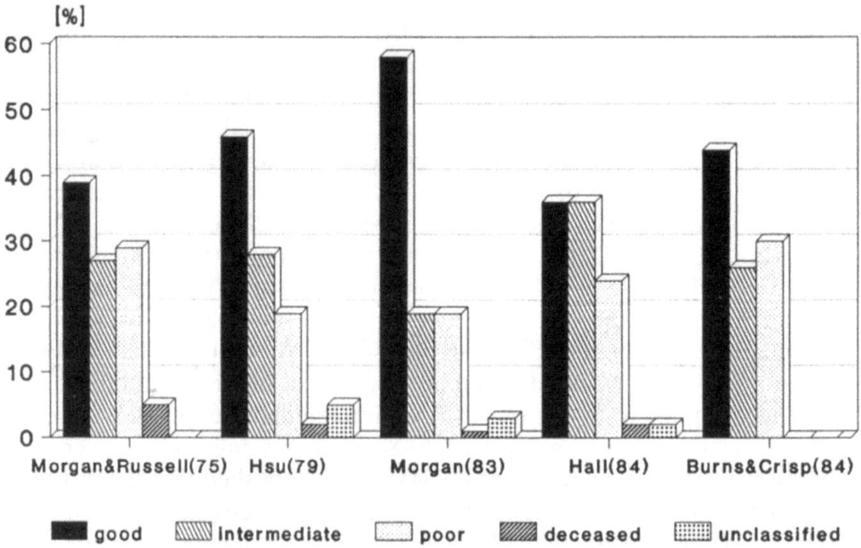

Fig. 1. Intermediate-term outcome of anorexia nervosa in terms of the Morgan-Russell General Outcome Categories in various studies

Table 2. Results of long-term follow-up studies of anorexia nervosa

	Failure to trace (%)	Direct interview (%)	Mortality rate (%)	Morgan-Russell-General-Outcome Categories (%)				Bulimia (%)	Social adjustment		Psychiatric comorbidity (%)
				Good	Intermediate	Poor	Unclassified		Marriage (%)	Children (%)	
Burns and Crisp (1984)	0	85	0	44	26	30	0	7	30	11	Drug and alcohol abuse 11
Deter et al. (this volume) Herzog et al. (this volume)	1	92	16	50	22	10	2	14	42.5	47	Substance abuse 14 Personality disorder 22 Phobic disorder 12 Major depression 8
Eckert (1990)	0	n.r.	7	51	31	11	0	25	n.r.	n.r.	n.r.
Hall et al. (1984)	0	88	2	36	36	24	2	40	42	n.r.	Major depression 7 Alcohol abuse 5 Schizophrenia 2 Mania 2 Personality disorder 2
Hsu et al. (1979)	3	73	2	46	28	19	5	20	29	14	Depressive symptoms 37 Severe disturbances 24 Obsessive symptoms 22
Morgan and Russell (1975)	0	83	5	39	27	29	0	n.r.	n.r.	n.r.	Affective symptoms 42 Obsessive symptoms 22
Morgan et al. (1983)	0	69	1	58	19	19	3	n.r.	n.r.	n.r.	n.r.
Remschmidt (1988)	7	77	3	56	9	14	18	n.r.	27	n.r.	Psychosomatic disturbances 43 Depressive symptoms 21 Obsessive symptoms 17
Russell (this volume)	0	70	15	32	29	22	2	15	n.r.	n.r.	n.r.
Theander (1985)	0	n.r.	18	76	1	5	0	n.r.	n.r.	n.r.	Schizophrenia 1

n.r., not reported.

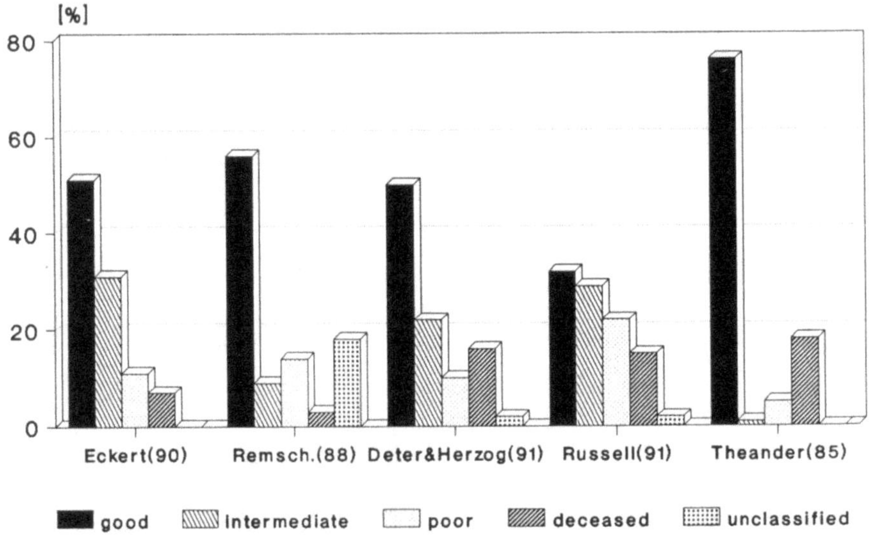

Fig. 2. Long-term outcome of anorexia nervosa in terms of the Morgan-Russell General Outcome Categories in various studies

- Poor outcome: continued existence of the anorexia criteria amenorrhea and reduced weight (<85% of average body weight) at intermediate-term follow-up of from 19% (Hsu et al. 1979; Morgan et al. 1983) to 29% (Morgan and Russell 1975). The proportion of patients with poor outcome decreases in long-term follow-up. Theander reports a minimum of only 5% anorectic patients at 33-year follow-up.

In the analysis of the follow-up studies, with increasing length of catamnesis, there is a tendency towards the following:

1. An increasing number of patients presenting with normal weight and menstruation (good outcome)
2. A steadily decreasing number of patients fulfilling the criteria for anorexia with respect to weight and menstruation status (poor outcome)
3. An increasing number of deceased patients

In the long term a polarization seems to take place, in that in terms of the Morgan-Russell Outcome Categories either the patients get well or they die.

The time dynamics of convalescence according to Morgan-Russell General Outcome Categories become clear when comparing the outcome analysis of the various follow-up times during the long-term study (Theander 1985; Russell, this volume; Herzog et al. and Deter et al., this volume). After 2 years only 15% of the patients in Theander's sample were well. In addition Theander was able to show that in his sample at over 12 years after first admission the patients may recover and improve but that this becomes more unlikely. This corresponds to Russell's (this volume) results, who found that up to 12 years after first admission, the probability of recovery in 3-year intervals lies at around 15%, but decreases to below 7% after 15 years. Here too, though, individual improvement at a later time was not excluded. Herzog et al. (this volume) report similar findings.

If the studies are compared with respect to the fact that the samples come from different departments (child and adolescent psychiatry, psychiatry, psychosomatics/internal medicine), then it seems that Remschmidt's child and adolescent psychiatric patients have a better prognosis in the long term than do the adult groups. The death rate is higher due to emergency admissions according to the results from the psychosomatic department of the internal medical clinic but otherwise findings seem to correspond to those found in the adult psychiatric department. Both intermediate- and long-term outcome results show that Russell's sample is marked by a particularly negative patient selection. The results for male anorexia patients in the intermediate term seem to correspond to those from other samples.

Bulimia

Between 7% and 40% of anorexia nervosa patients report episodes of bulimia during intermediate- to long-term follow-up. The interesting question of whether these patients come from the good outcome group, i.e., are normal weight patients with bulimia who also had normal menstruation, or come from the intermediate and poor outcome groups can only be answered by the studies of Eckert (1990) and Deter et al. and Herzog et al. (this volume).

In the long-term course of anorexia nervosa we are confronted with nosologic problems when patients develop bulimia at a later stage: then the diagnostic criteria for bulimia nervosa may well apply! In older studies this factor might have been overlooked at the initial assessment of the anorectic patients (bulimia did not become widely "known" until 1980). Although there still is discussion as to the exact diagnostic delineation, especially regarding bulimic anorectics (see Garner and Garfinkel 1988), long-term studies might yield important information as to the relationship between anorexia and bulimia nervosa. Vandereycken and Pierloot (1983b), for example, found striking evidence that anorectics who had no bulimia at first assessment but were binge eating at follow-up were characterized by features similar to those patients in whom bulimia and/or purging was already manifest at first assessment: especially those patients who develop anorexia at an older age and whose illness is of longer duration appear to be at risk of deterioration in the direction of bulimia.

Comorbidity

Knowing how difficult it is to find studies which have used similar methods of sample selection and the simple physiologic outcome parameters of weight and menstruation, it does not come as a surprise that it is almost impossible to find common assessment criteria as regards psychiatric comorbidity. Hence, only some of the existing findings will be described:

- Acute schizophrenia was diagnosed in up to 2.5% of follow-up patients. Earlier reports of rates of sometimes 10% and more (for example, Hawley 1985) could not be confirmed in our comparison of large samples and seem to be based on the effects of selection or small samples.
- Manic psychoses were reported in individual instances (1%).
- Psychotic episodes, especially in borderline personality, are reported more frequently in patient histories.
- Personality disorders are reported in 22% of patients by Deter et al. (this volume).
- The most frequently reported symptoms are marked depression, symptoms of compulsion, and substance abuse.

The uniform assessment of psychic symptoms in long-term outcome of anorexia nervosa is certainly one of the areas in which further research is needed. The question of what percentage of which psychopathologic diagnoses occur within the outcome groups of Morgan-Russell General Outcome Categories is particularly interesting. In the studies analyzed here some 40% of patients appear to have recovered in the long term (no eating disturbance, no psychiatric diagnosis). In about 20% of patients the physiologic parameters of weight and menstruation seem to have normalized but symptoms of the psychopathologic conditions which are in need of treatment persist.

Investigations of physical comorbidity seldom occur because they require a physical examination. Of the studies mentioned only Herzog et al. (this volume) in their discussion of somatic findings have undertaken detailed and extensive investigations.

Social Integration

Rough estimates as to social integration can be made from the marital status of the patients and from the proportion of patients who have their own children. The percentage of married patients amounts to slightly over 40%, that of patients with children about 47%. Even though here, as for mortaility rates, there are comparative figures from the normal population, the proportion of married anorexia patients probably lies well under that in the normal population (see also Engel et al., this volume). A more exact analysis of the social network in long-term follow-up studies of anorexia nervosa can be found in the chapter by Manz et al. (this volume).

Prognostic Factors

The search for predictors of a favorable or unfavorable course of anorexia nervorsa is one of clinicians' and researchers' main motives for conducting such extremely time-consuming follow-up investigations. It is hoped that such studies will answer some of the urgent questions regarding individual prognosis which patients and their families and those clinicians dealing with eating disorders have. Thus it appears to be all the more important before beginning with a discussion of the study results that the limitations of such predictor investigations be pointed out:

1. Even the best conducted outcome studies merely project probabilities of the outcome of the entire group of anorexia patients. The individual course may, however, be entirely unpredictable in individual cases.
2. With increasing length of catamnesis an increasing number of factors affect the lives of the patients, which also have a part in determining outcome, and these factors cannot be controlled. To reduce the differences in outcome to a set of variables which is available 5, 10, 20, or even 30 years beforehand presents quite a risk: Even if predictors could be isolated, the question remains as to the evaluability of the findings. It is difficult to exclude that these predictors may be more apt to reproduce the individual sample characteristics than predict outcome of anorexia nervosa in general.
3. The question of evaluability of determined predictors is all the more important since they have been indicated in the studies analyzed here, too, without it being indicated on which statistic models and preassumptions they have been based. Whereas correlative models are used most often in which it is not clear how many variables have been tested and whether an alpha adjustment was necessary or performed, it is not usually evident in more complex statistical models how the individual outcome factors are intercorrelated: The predicting power of individual variables can, then, derive more from specific samples which have little to do with the phenomenon of outcome of anorexia nervosa in general.

In the studies reviewed here the following factors have been found to be predictors of a less favorable outcome (Table 3).

1. Longer period of illness before first admission. In two thirds of our studies a long period of illness was found to be a predictor of a poorer outcome. This may refer to a long postponement before beginning treatment or unsuccessful previous treatment.

Table 3. Unfavorable prognostic factors in anorexia nervosa

Burns and Crisp (1984)	Low minimal weight
	Disturbed family relationship
	Absence of premorbid sexual phantasy
Deter et al. (this volume)	Duration of illness
Herzog et al. (this volume)	Low serum albumin
	Bulimia
Eckert (1990)	Previous hospitalizations
	Alcohol abuse
	Lowest weight ever
	Higher age at onset
	Vomiting
	Laxative abuse
	Abuse of diurectics
Hall et al. (1984)	Duration of illness
	Being married at presentation
Hsu et al. (1979)	Higher age
	Longer duration of illness
	Previous psychiatric treatments
	Lower social class
Morgan and Russell (1975)	Late onset
	Longer duration of illness
	Previous psychiatric treatment
	Disturbed relationship with family
Morgan et al. (1983)	Duration of illness
	Disturbed relationship with family
Remschmidt et al. (1988)	Higher age at onset
	Weight stabilization in less than 7 weeks
Russell (this volume)	Higher age at onset
	Longer duration of illness
Theander (1985)	n.r.

n.r., not reported.

2. Higher age at onset. The second most frequent factor to be reproduced in our studies was that a high age at onset corresponds to a negative outcome. Comparing long-term studies of the past 4 decades Steinhausen et al. (1991) found that higher age at onset was the most frequently mentioned negative predictor in all of the studies he investigated (before period of illness at admission).
3. Disturbed family relationships. The third most frequent negative predictor of anorexia outcome is represented by severely disturbed

relationship of the index patient to the family or within the family. This factor had third place in the study of Steinhausen et al. (1991), too.

4. Extreme weight loss. The extent of weight loss – according to the study by Patton (1989) already mentioned – is a factor which can lead to an increase in mortality. In our opinion it is remarkable that extreme loss of weight is only explicitly given twice as a negative predictor. Remschmidt's (1988) finding that a marked increase in weight within 7 weeks in his inpatient treatment program was a negative predictor might be a further indication that it makes little sense and that there is little promise of success in attaching too much importance just to weight in the assessment of long-term prognosis.
5. Other prognostic factors. Bulimia seems to be a negative prognostic factor (Eckert 1990; Herzog et al., this volume; see also Steinhausen et al. 1991). Finally, of the individual findings the significance of low serum albumin levels and low serum calcium levels must be mentioned. Herzog et al. (this volume) take this finding to be a predictor of fatal outcome of anorexia.

Summary and Perspectives

According the "classical" long-term outcome studies analyzed in this chapter anorexia nervosa is a condition which goes hand in hand with a considerable mortality rate of $\frac{1}{2}$% to 1% per observation year. In the long term about 40% of patients can count in complete recovery. Marked comorbidity in patients who have improved with respect to the physical and eating symptoms and somatic complications in patients who continue to be chronically anorectic complete the picture.

Even though the diagnosis of anorexia nervosa is made earlier nowadays and due to an increased treatment capacity the illness is more likely to be treated than was the case in the studies which have been reviewed, this will at best shift the proportions to better outcome within the outcome groups. The total spectrum of long-term outcome of anorexia nervosa with the possibility of fatal, chronic anorectic, and chronic psychiatric outcome, though, is not affected.

But even the high proportion of chronic disease courses in the studies should not lead to therapeutic resignation: Long-term anorexia can still improve after a long time, though the probability after 12 years' duration decreases.

Further methodologically refined methods (prospective studies; see Vandereycken et al., this volume; and "Outlook") will contribute to our better understanding of aspects of the disease process and healing and hence will promote the development of optimal methods of treatment which correspond to the respective phase of the illness.

References

Burns T, Crisp AH (1984) Outcome of anorexia nervosa in males. Br J Psychiatry 145:319–325

Deter HC, Petzold E (1979) Klinische Befunde und Verlaufsuntersuchungen bei Patienten mit Anorexia nervosa. Therapiewoche 29:6320–6331

Deter HC, Petzold E, Hengst-Theis R, Breiden U, Lanzinger-Rossnagel G (1983) Katamnestische Ergebnisse einer klinisch-psychosomatischen Behandlung von 103 Patienten mit Anorexia nervosa aus internistischer Sicht unter besonderer Berücksichtigung der Mortalität. Innere Med 1:3–12

Deter HC, Petzold E, Hehl FJ (1989) Differenzierung der Langzeitwirkungen einer stationären psychosomatischen Therapie von Anorexia-nervosa-Patienten. Z Psychosom Med Psychoanal 35:68–91

Eckert E (1990) Results of a 10-year follow-up. 4th International Conference on Eating Disorders, New York, 27–29 April, 1990

Garner DM, Garfinkel PE (1988) Diagnostic issues in anorexia nervosa and bulimia nervosa. Brunner/Mazel, New York

Hall A, Slim E, Hawker F, Salmond C (1984) Anorexia nervosa: long-term outcome in 50 female patients. Br J Psychiatry 145:407–413

Halmi KA, Brodland G, Rigasc (1976) A follow-up study of seventy-nine patients with anorexia nervosa. In: Winoker G, Rolff M (1976) Life history research in psychopathology, vol 4. University of Minnesota Press, Minneapolis, pp 290–300

Hawley RM (1985) The outcome of anorexia nervosa in younger subjects. Br J Psychiatry 146:657–660

Herzog DB, Keller MB, Lavori PW (1988) Outcome in anorexia nervosa and bulimia nervosa. A review of the literature. J Nerv Ment Dis 176(3):131–143

Hsu LKG (1980) Outcome of anorexia nervosa. A review of the literature (1954–1978). Arch Gen Psychiatry 37(9):1041–1046

Hsu LKG (1988) The outcome of anorexia nervosa: a reappraisal. Psychol Med 18:807–812

Hsu LKG, Crisp AH, Harding B (1979) Outcome of anorexia nervosa. Lancet I:61–65

Morgan HG, Hayward AE (1988) Clinical assessment of anorexia nervosa. The Morgan-Russell Outcome Assessment Schedule. Br J Psychiatry 152:367–371

Morgan HG, Russell GFM (1975) Value of family background and clinical features as predictors of long-term outcome in anorexia nervosa: 4-year follow-up study of 41 patients. Psychol Med 5:355–371

Morgan HG, Purgold J, Welbourne J (1983) Management and outcome in anorexia nervosa. A standardized prognostic study. Br J Psychiatry 143:282–287

Patton G (1989) The course of anorexia nervosa. Br Med J 299:139–140

Remschmidt H, Müller H (1987) Stationäre Gewichts-Ausgangsdaten und Langzeitprognose der Anorexia nervosa. Z Kinder Jugendpsychiatr 15:327–341

Remschmidt H, Wienand F, Wewetzer C (1988) Der Langzeitverlauf der Anorexia nervosa. Monatsschr Kinderheilkd 136:726–731

Steinhausen HC, Glanville K (1983a) Retrospective and prospective follow-up studies in anorexia nervosa. Int J Eat Disorders 2(4):221–235

Steinhausen HC, Glanville K (1983b) Follow-up studies of anorexia nervosa: a review of research findings. Psychol Med 13:239–249

Steinhausen HC, Rauss-Mason C, Seidel R (1991) Follow-up studies of anorexia nervosa: a review of four decades of outcome research. Psychol Med (in press)

Szmukler GI, Russell GFM (1986) Outcome and prognosis of anorexia nervosa. In: KD Brownell, JP Foreyt (eds) Handbook of eating disorders. Basic Books, New York, pp 284–300

Theander S (1970) Anorexia nervosa: a psychiatric investigation of 94 female patients Acta Psychiat Scand Suppl 214

Theander S (1983) Research on outcome and prognosis of anorexia nervosa and some results from a Swedish long-term study. Int J Eating Disorders 2:167–174
Theander S (1985) Outcome and prognosis in anorexia nervosa and bulimia: some results of previous investigations, compared with those of a Swedish long-term study. J Psychiatr Res 19:493–508
Vandereycken W, Pierloot R (1983a) Long-term outcome research in anorexia nervosa. The problem of patient selection and follow-up duration. Int J Eating Disorders 2:237–242
Vandereycken W, Pierloot R (1983b). The significance of subclassification in anorexia nervosa: a comparative study of clinical features in 141 patients. Psychol Med 13:543–549

Outcome Assessment

H.G. Morgan, I.M.P. Macquire-Samson, J. Purgold,*
and C.A. Vassilas

Introduction

This chapter is concerned with ways of assessing clinical outcome in patients with eating disorders. To begin with consideration will be given to matters of methodology relevant to follow-up studies both in general and with specific reference to eating disorders. Following this the various specialised instruments which may be of use in standardising such clinical assessment will be described and their respective roles evaluated.

Essential Methodological Issues in Follow-up Studies

Before deciding exactly how to proceed in any follow-up investigation, whether it concerns eating disorders or not, it is essential to be clear about its purpose because this will dictate the detailed design of the study. One which aims to evaluate natural history of the condition under investigation will need to extend over a considerable period of time, in order to allow for those patients whose recovery is slower than the others. It has long been agreed in the case of anorexia nervosa, for example, that assessment should be delayed for several years in order to achieve a reliable picture of long-term outcome. It may of course be quite acceptable to take shorter periods of follow-up, for example, in assessing the short-term effects of some form of therapy, but either way the precise purpose of the study needs to be defined beforehand in order to choose the appropriate follow-up period.

In order to obtain a reliable clinical assessment it is preferable to adopt direct rather than indirect forms of assessment, and contact with the maximum possible proportion of all patients should be achieved. Thus a postal questionnaire is very unsatisfactory according to both these criteria: it is indirect and even in the best of circumstances will only achieve about a 50% response rate, unless of course it is merely a preliminary to

---

* It is with great regret that we have to report the recent death of Miss Joan Purgold.

more intensive efforts to contact nonresponders. As an example, a survey of the early literature on anorexia nervosa (Morgan 1973) revealed a confusing and conflicting picture of outcome until close analysis revealed that those studies which utilised indirect methods of assessment, and which inevitably suffered from a high proportion of nonresponders, claimed a better outcome than the studies which involved direct interviews and in which the numbers of defaulters were much smaller. These findings are likely to have been related to the fact that patients with continued active anorexic illness are highly represented in those who fail to respond to the invitation to be involved in follow-up.

Ideally, data collection should be prospective and systematic, both for the initial illness and the subsequent follow-up assessment. Only in this way can it be certain that a comprehensive clinical picture is obtained and that assessment ratings are not rendered unreliable by confounding factors such as selective recall and other forms of distorted judgement.

The precise purpose of a follow-up study will also dictate the type of information that should be sought. In eating disorders any evaluation of overall outcome will need to embrace not only psychological and physical measures but complex relationship and social aspects of a patient's life situation, too. A fundamental problem revolves around the fact that those criteria which are easiest and most reliable to measure are not necessarily those which will provide the richest and most sensitive picture of outcome; for example matters concerning sexual attitudes and relationships with others are difficult and unreliable to measure yet vital indices of wellbeing in any individual. It must remain true that although questionnaires permit standardisation of data collection, their structured approach must mean that they are always at risk of missing important finer nuances of change in any individual, and a flexible, extended clinical interview remains important in order to maintain sensitive contact with the patient as a whole.

Hazards Encountered in the Clinical Assessment of Eating Disorders

Even when the most rigorous of criteria have been met, any follow-up study of eating disorders should take account of the ways in which the illness itself can limit and distort clinical measures. Denial, resentment at any enquiry and poor compliance tends to be greater the more severe the anorexic or bulimic illness; hence the need in following up any group of patients to account for all of them, and more than 95% of compliance should be the rule. This presents particular problems in patients with eating disorders as they tend to be a young and geographically mobile population.

Many of the clinical characteristics of these illnesses include behaviours that are essentially very private, besides attitudes which depend so much on cooperation for their measurement. Attempts to reduce the unreliability of clinical assessment by seeking information from other key informants may help, but they may be of little value in assessing patterns of menstrual loss, or such matters as morbid patterns of food intake and vomiting which regularly occur when the individual is alone. Secondary family tension and emotional upset in other relatives can at times be as severe as that exhibited by the individual who is ill. The unwary will interpret such family problems as causal rather than secondary to the eating disorder.

Self-report Questionnaires

Several self-report questionnaires concerning the clinical assessment of eating disorders are now available. They have the obvious advantage of ease of administration in a way that is often more acceptable and less threatening than direct interview, particularly when the latter is carried out by someone who is not acquainted with the patient. Self-report techniques are particularly useful in screening for the presence of eating disorders prior to more direct clinical assessment. The instruments to be described are noteworthy for the meticulous care that has been taken to maximise their reliability and validity.

These tests rely heavily on the assumption that subjects will describe their symptoms accurately and reliably. This is of particular concern for patients with anorexia nervosa since they often display significant denial surrounding their disorder. Several studies have shown that some subjects, though suffering from anorexia nervosa, will fill in self-report questionnaires so as to give scores in the "normal range" (Szmukler 1983; Vanderdeycken and Vanderlinden 1983). High test-retest reliability would not solve this problem, whereas involvement of collateral informants might help.

The Eating Attitudes Test

The Eating Attitudes Test (EAT-40) is a 40-item self-report questionnaire which subsequently, as a result of factor analysis, was reformulated into a 26-item version (EAT-26; Garner and Garfinkel 1979; Garner et al. 1982).

The EAT-40 represents a broad range of symptoms, behaviour and attitudes common in anorexia. Items are grouped into seven factors and each item is presented as a six-point forced choice format. It is simple to score with each extreme response in the "anorexic" direction scoring three points and adjacent alternatives scoring two points and one point, respectively. The cut-off score is 30. Being self-explanatory it is easy to administer in a postal survey, in an out-patient clinic or a general practitioner's surgery. In general, subjects seem to find it acceptable and it only takes about 10 min to complete. The EAT-40 has a high degree of internal reliability although test-retest reliabilities have not been reported. It was validated on two female groups of anorexics and controls. Seven per cent of controls who scored in the lowest anorexic range showed significant concern regarding weight at interview. The scores of obese females and normal weight males were significantly different from anorexics.

The EAT-26 produces scores that are highly correlated with the EAT-40 ($r = 0.98$). Factor analysis revealed three factors: (1) dieting, (2) bulimia and food preoccupation, and (3) oral control. The content of subscales based on these three factors appears to be related to bulimia, weight, body-image variables and psychological symptoms. Whereas bulimic and restricter variants of anorexia nervosa do not differ on the total scores of EAT-26 or EAT-40, these groups do so on factors (2) and (3) of the EAT-26. The results from the factor analysis must be viewed with caution since the number of subjects was not really sufficient for the number of items included. EAT-26 has the same six-point forced choice format as the EAT-40 with a cut-off score of 20, which correctly classifies 84% of cases.

The EAT score is not itself diagnostic, as a high score may indicate a disturbed eating pattern but not necessarily any underlying disturbed psychopathology. Although sensitive to clinical change, the predictive capabilities of EAT scores with regard to eventual outcome have not been evaluated in detail. The scores seem to lag behind restoration of body weight. The authors suggest that their most suitable use is to measure outcome in clinical groups (severity of symptoms at the beginning and during a course of treatment). They are also used as screening tests for anorexia nervosa in general population studies. Their usefulness here is limited by the low prevalence of a clinically diagnosable disorder and they should be separately validated for different population groups.

The Bulimic Investigatory Test Edinburgh

The Bulimic Investigatory Test Edinburgh (BITE) (Henderson and Freeman 1987) is a self-rating scale for bulimia nervosa, consisting of 33 items divided into two subscales. The symptom scale is intended to measure the degree of symptoms and has a maximum possible score of 30, with 20 as the cut-off or criterion score. The severity scale as defined by frequency of symptoms has a maximum score of 39; a score of five or above is considered to be clinically significant. A total score of 25 or more is indicative of a severely disordered eating pattern. Significant scorers are easily identified with scores of between 10–25, indicating less severe levels of pathology. There is an additional front sheet which provides useful demographic details relevant to the study and treatment of binge eating.

This questionnaire is self-explanatory and easy to administer. It is also simple to score and generally acceptable to subjects, who can complete it in about 10 min. As a screening test it is used to score feelings and behaviour over the past 3 months, but as a measure of response to treatment the past 1 month should be assessed.

This questionnaire includes all items in the DSM-III and Russell's 1979 definition of bulimia nervosa. It has been validated on two groups of binge eaters and controls. Interitem reliability coefficients of 0.96 for the symptom subscale and 0.62 for the severity subscale both indicate a high degree of internal consistency. High correlations have been reported between BITE scores and the diet and binge subscale scores of the EAT, as well as all subscale scores of the EDI, except those for perfectionism and interpersonal distrust. Test-retest reliabilities range between 0.68 and 0.86. Although originally intended for the detection and description of binge eating the BITE also showed sensitivity to change in a group of 27 patients undergoing treatment.

It is used as a screening tool in epidemiological studies concerned with incidence rates in various populations, as a measure of severity of symptoms at the beginning and during a course of treatment and as an adjunct to the evaluation of outcome over a defined period of time whether looking at the natural history of bulimia or the results of treatment.

The Eating Disorder Inventory

The Eating Disorder Inventory (EDI) is a 64-item self-report questionnaire (Garner et al. 1983) designed to assess psychopathology in both anorexia and bulimia. It was designed in recognition of the fact that eating disorders comprise a heterogeneous patient population. Its eight subscales are concerned with drive for thinness, bulimia, body dissatisfaction, ineffectiveness, perfectionism, interpersonal distrust, interoceptive awareness and maturity fears. The questionnaire was validated on patients with anorexia nervosa and a group of anorectics with bulimic complications. Scores for a small group of recovered anorectics resembled those of the control population. Individuals with anorexia nervosa and bulimia nervosa were satisfactorily distinguished from a control group. Assessment of the internal consistency of the EDI reveals that most of the subscales are significantly intercorrelated, indicating satisfactory reliability.

Taking about 20 min to complete, the EDI is designed to identify subtypes of anorexia nervosa in clinical populations. It also provides a clear definition of personality traits and attempts to distinguish these from illness symptoms and changes which occur as a result of treatment. Suggested uses of this instrument (preferably followed by a clinical assessment) are as a screening instrument for identifying groups in which a high proportion of subjects may have formal eating disorders and as a way of assessing the specific psychopathology of eating disorders.

Direct Assessments

Physical Assessment

The most basic evaluation of outcome in bulimia nervosa and anorexia nervosa should include the physical parameters of weight and menstruation. In anorexia nervosa, weight has obvious implications for risk no matter how well a patient claims to feel. Amenorrhoea is an essential diagnostic criterion, even though its evaluation is dependent to a great extent on the patient's indirect account. In his study of a series of patients at the Maudsley Hospital, UK, Morgan (1973) devised a pattern of outcome based on weight and menstruation (Table 1).

It is of interest that crude though these measures may be, they do seem to correlate quite well with more sensitive measures of psychological factors and relationships. This general trend was demonstrated in the original Maudsley series (Morgan and Russell 1975), but it has only recently been placed on a more firm statistical basis. Morgan and Hayward (1988) found that both for the Maudsley series and a subsequent series of patients from Bristol, categories based on body weight and menstruation correlated significantly with scores on each of the scales in the Morgan-Russell Assessment Schedule (five at $p < 0.05$ and nine at $p < 0.001$ levels of significance).

In bulimia nervosa, the maintenance of a steady weight within the normal range is an important measure of outcome. Menstruation, although usually present, is often irregular due to the chaotic eating pattern and degree of psychological disturbance. A return to regular menstruation may indicate recovery.

Body Weight Assessment. Weight is usually measured as a percentage of "expected weight". Tables giving average body weights are widely avail-

Table 1. Anorexia nervosa outcome based on body weight and menstruation

Outcome	Criteria (over previous six months)
Good	Both weight and menstruation have returned to normal throughout the 6-month period (weight ±15% of average body weight)
Intermediate	Weight always deviated from average but only at times sufficient to cause concern, with or without menstrual abnormality, or continued menstrual abnormality though weight normal
Poor	Weight always sufficiently deviant as to cause concern. Menstruation absent or erratic with scanty flow

able (Lentner 1984) but refer to a specific population, e.g. North Americans who are policy holders with insurance companies (Society of Actuaries and Association of Life Insurance Directors of America 1980), an obviously skewed sample. The Body Mass Index (BMI), equal to weight in kg/(height in metres)^2, has been developed as a way of measuring obesity without recourse to tables of average weights and independent of any reference population. More recently it has been put forward as an appropriate means of measuring weight loss in patients with eating disorders (Llewellyn-Jones and Abraham 1984). However, at present percentage expected weight is widely used by clinicians (Garner and Garfinkel 1985); it has greater face validity than the range of values suggested for the BMI, reinforced by the fact that diagnostic criteria for anorexia nervosa are expressed in terms of percentage expected weight (American Psychiatric Association 1987).

Electrolyte Measurement. There is some debate as to what proportion of patients with eating disorders have abnormal electrolytes (Wilson 1987). In those who vomit, abuse laxatives or diuretics, assessment of serum electrolytes may provide additional objective information, although it is clear that values can remain normal for some time even in the face of marked and persistent vomiting or purgation, especially when foods of high potassium content are taken.

The Morgan-Russell Assessment Schedule

The Morgan-Russell assessment takes the form of a guided interview which is concerned with clinical factors central to the syndrome of anorexia nervosa. It evaluates eating behaviour, body weight, mental state and a variety of attitudes relevant to the anorectic disorder. The responses to these specific questions are scored on nominal scales based on a combination of assessment at interview as well as status during the previous 6 months. In order to maximise the reliability of the data, the final score should preferably be the average of the scaled data obtained from both the patient and a key informant, such as a parent in the case of young patients who are still living at home.

A profile of five scores is derived by averaging the ratings on 15 subscales. The topics covered are: food intake, menstruation, mental state, psychosexual state, socioeconomic state and relationships. The resulting five point profile may itself be of value in clinical assessment. In follow-up studies the ratings have been averaged further to produce one single mean value, the Average Outcome Score (Morgan and Hayward 1988).

The Morgan-Russell Assessment Schedule was devised at a time when there was little, if any, standardisation of data in follow-up studies of anorexia nervosa. Its introduction allowed comparison of outcome between several centres (Morgan and Russell 1975; Hsu et al. 1979; Morgan et al. 1983; Hall et al. 1984) and it has since been used widely. It is easy to administer and does not require specialised training to do so, other than the basic skills inherent in clinical interview technique.

To express outcome in terms of a single score may appear an oversimplification of a complex clinical problem. In practice, however, the average outcome score has proved to be a very sensitive measure of clinical outcome in anorexia nervosa. The schedule has recently been analysed statistically in more detail and it has been shown that not only does the average outcome score correlate very significantly with all five variables, but there is also a very high intercorrelation of scores on all 15 subscales. Factor analysis showed a high degree of general relatedness between the scales, as well as a wide spread of scores from one patient to another (Morgan and Hayward 1988). It should be noted that the Morgan-Russell schedule is not designed to assess bulimia nervosa.

The simplest application is in routine clinical documentation as a way of ensuring that thorough data collection is achieved in a way that can be easily replicated. It can also be useful to monitor progress during treatment. The schedule was devised primarily as a measure of outcome, and has proved valuable in follow-up studies of anorexia nervosa, particularly in facilitating comparison of findings from one centre to another.

The Eating Disorders Interview

The Eating Disorders Interview (EDI) has been primarily designed for use in a normal population in order to elicit the behaviour and attitudes common in patients with eating disorders (Mann et al. 1983). The EDI collects sufficient information to make clinical diagnoses. It has been designed for use in epidemiological studies but not for follow-up assessment. A full description of the EDI has not been published, but its reliability is said to be satisfactory.

The Eating Disorder Examination

The Eating Disorder Examination (EDE; Cooper and Fairburn 1987) was designed to assess the specific psychopathology of eating disorders, including disturbed eating habits and concern about weight. It is a semi-structured interview taking about 1 h to administer, and a short course of training is necessary before it can be used. Terms are defined clearly for rating purposes and interviewers are free to ask further questions to obtain additional relevant information. The EDE has been designed to cover anorexia and bulimia, but it does not include questions relating to anxiety, depression and other neurotic symptoms. The five constituent subscales cover the following topics: restraint, bulimia, eating concern, weight concern, and shape concern (Cooper et al. 1989). These appear to have a satisfactory degree of internal consistency. Inter-rate reliability has been reported within the range 0.69–1.00. The individual items of the EDE appear to discriminate well between patients with eating disorders and normal controls, as well as between patients with eating disorders and persons who show concern about eating, body shape and weight (Wilson and Smith 1989).

This interview has been validated in groups of anorexics and bulimics (Cooper et al. 1989). The suggested uses are in research into the effects of treatment and to look in detail at the psychopathology of those who have eating disorders. Wilson and Smith (1989) have reported that while the EDE can successfully distinguish between bulimics and nonbulimics who are "restrained eaters" the self-report EDI failed to do so and therefore conclude that the EDE is particularly useful for identifying the core pathology of bulimia nervosa.

Conclusions

In assessing outcome following eating disorders it is important to be mindful of the need to encompass wide-based measures of psychopathology (Hsu 1980). We know that improvement of the specific anorexic and bulimic symptoms may not be paralleled by amelioration of other forms of morbidity such as affective disturbance. A recent outcome study of patients with anorexia nervosa (Rosenvinge and Mouland 1990) has reiterated this point. Several of the assessment schedules such as the EAT and BITE which are described in this chapter appear to provide reliable measures of the psychopathology which occurs as an integral and specific component of eating disorders. If these are to be used in assessment of outcome, it is advisable also to include wider measures of psychological outcome to ensure that a complete clinical picture is obtained.

References

American Psychiatric Association (1987) Diagnostic and statistical manual of mental disorders. American Psychiatric Association, Washington

Cooper Z, Fairburn C (1987) The Eating Disorder Examination: a semi-structured interview for the assessment of the specific psychopathology of eating disorders. Int J Eating Dis 6:1–8

Cooper Z, Cooper PJ, Fairburn CG (1989) The validity of the Eating Disorder Examination and its subscales. Br J Psychiatry 154:807–812

Garner DM, Garfinkel PE (1979) The Eating Attitudes Test: an index of the symptoms of anorexia nervosa. Psychol Med 9:273–280

Garner DM, Garfinkel PE (eds) (1985) Handbook of psychotherapy for anorexia and bulimia. Guilford, New York

Garner DM, Olmsted MP, Bohr Y, Garfinkel PE (1982) The Eating Attitudes Test: psychometric features and clinical correlates. Psychol Med 12:871–878
Garner DM, Olmstead MP, Polivy J (1983) Development and validation of multidimensional eating disorder inventory for anorexia nervosa and bulimia. Int J Eating Dis 2:15–34
Hall A, Slim E, Hawker F, Salmond C (1984) Anorexia nervosa: long term outcome in 50 female patients. Br J Psychiatry 145:407–414
Henderson M, Freeman CPL (1987) A self rating scale for bulimia. The BITE. Br J Psychiatry 150:18–24
Hsu LKG (1980) Outcome of anorexia nervosa: a review of the literature 1954–1978. Arch Gen Psychiatry 37:1041–1046
Hsu LKG, Crisp AH, Harding B (1979) Outcome of anorexia nervosa. Lancet 1:61–65
Lentner C (ed) (1984) Geigy Scientific Tables. Ciba Geigy, Basle
Llewellyn-Jones D, Abraham SF (1984) The Quetelet index in diagnosis of anorexia nervosa. Br Med J 288:1800
Mann AH, Wakeling A, Wood K, Monck E, Dobbs R, Szmukler C (1983) Screening for abnormal eating attitudes and psychiatric morbidity in an unselected population of 15 year old schoolgirls. Psychol Med 13:573–580
Morgan HG (1973) Anorexia nervosa: a prognostic study. MD thesis, University of Cambridge
Morgan HG, Hayward AE (1988) Clinical assessment of anorexia nervosa: the Morgan-Russell outcome assessment schedules. Br J Psychiatry 152:367–372
Morgan HG, Russell GFM (1975) Value of family background and clinical features, as predictors of long term outcome in anorexia nervosa: four year follow-up of 41 patients. Psychol Med 5:355–371
Morgan HG, Purgold J, Welbourne J (1983) Management and outcome in anorexia nervosa. A standardised prognostic study. Br J Psychiatry 149:282–287
Rosenvinge JH, Mouland SO (1990) Outcome and prognosis of anorexia nervosa. A retrospective study of 41 patients. Br J Psychiatry 156:92–97
Russell G (1979) Bulimia nervosa: an ominous variant of anorexia nervosa. Psychol Med 9:429–448
Society of Actuaries and Association of Life Insurance Directors of America (1980) Build study 1979. Chicago
Szmukler GI (1983) Weight and food preoccupation in a population of English schoolgirls. In: Bargman GJ (ed) Understanding anorexia nervosa and bulimia. 4th Ross conference on medical research. Ross Laboratories, Ohio p 21
Vandereycken W, Vanderlinden J (1983) Denial of illness and the use of self-reporting measures in anorexia nervosa patients. Int J Eating Dis 2(4):237–242
Wilson GT (1987) Assessing treatment outcome in bulimia nervosa: a methodological note. Int J Eating Dis 6:339–348
Wilson GT, Smith D (1989) Assessment of bulimia nervosa: an evaluation of the Eating Disorder Examination. Int J Eating Dis 8:173–179

The Anorexia Nervosa Symptom Score: A Multidimensional Tool for Evaluating the Course of Anorexia Nervosa

H.-C. Deter

The evaluation of the course of anorexia nervosa has been a methodological problem for years and numerous solutions have been proposed to overcome the difficulties. Especially comparisons of different studies and metaanalyses reveal that valid criteria for assessing the course of the disease, if applied at all, are not used uniformly.

Although numerous reports have been published in the literature on the outcome of anorexia nervosa terms like "healthy" or "cured," "improvement" or "chronification" are often not defined precisely, which makes it difficult to compare the results of different studies (see Herzog et al. and Morgan et al., this volume). A key problem is the description of anorexia nervosa symptoms, which is usually done by making rather general statements about the physical and psychologic findings (Vandereycken and Meermann 1984). Several investigators have tried to establish systematic criteria for evaluating (changes in) the severity of anorexia nervosa (e.g., Thomä 1967; Petzold 1979; Moldofsky and Garfinkel 1977). The test battery has thus become differentiated and now includes more standardized procedures (e.g., Fichter 1980; Fichter and Meermann 1983; Vandereycken and Meermann 1984; Steinhausen 1984).

Only the classification of severity relating to the somatic criteria of weight and amenorrhea suggested by Morgan and Russell (1975; see also chapter by Morgan et al., this volume) has gained wide acceptance in the literature. Many of the other criteria that have been discussed do not appear to yield much additional information (Russell 1980). Nevertheless, earlier systems of evaluation, such as the one established by Moldofsky and Garfinkel (1977), have not become superfluous, but a certain simplification facilitates the comparison of different studies, an effect that is highly welcome in view of the earlier lack of orientation found in the evaluation criteria of many studies. Today, however, the question arises as to whether such a broad system of criteria relating to the physical situation can satisfy the complex requirements of an evaluation, or whether it would be useful to include further independent variables in addition to the main evaluation criteria.

The question of how the course of a disease should actually be evaluated and which criterion makes it possible to obtain the best result is evaluated very differently, depending on the viewpoint of the examiner. Based on the literature and our own experience with course of disease studies, four qestions are important for the evaluation:

1. How "hard" are the Morgan-Russell criteria and can they be solely relied upon?

The patients have to be examined in person. General statements about the patient's condition or even telephone inquiries are not at all satisfactory, as we demonstrated in a small experiment carried out during an examination 12 years after the patients had first presented at our clinic. The former patients were asked to state their body weight, which was checked later on. Forty-two percent of the patients gave a false weight (the difference being between 2 and 8 kg; see Table 1), knowing that they would undergo a physical examination shortly afterwards. Presumably, the percentage would be even higher for interview patients without a physical examination (personal or by telephone).

A similar argument applies to secondary amenorrhea: Who can say whether patients, with a known attitude of denial in individual cases, have made correct statements regarding their menses? Also the question of whether there patients take contraceptives remains unanswered. Here, too, additional verification would be helpful, for example, by hormonal assessment of the female sex hormones.

2. Shouldn't the assumption that anorexia nervosa patients are a homogeneous group be questioned?

In our experience, anorexia nervosa patients have much in common but show at the same time very characteristic differences. Not only are the differences in the time of onset – either in the teens or twenties – important, but a comparison is also questionable if the various subtypes such as type of weight loss (restricter, purger), psychiatric comorbidity, family disorders, being able to work or the presence of bulimia are not taken into account (Meyer 1984).

To follow the course of the "average anorexia phantom" is ultimately unsatisfactory for the clinical examiner, even if it is useful for the re-

Table 1. Differences between patients' weight estimates and objective measurements of body weight in 49 anorexia nervosa patients 12 years after inpatient treatment

Estimation	*n*	%	↓	↑
Correct	8	16		
Almost correct, difference ≤1 kg	20	41	17	3
Fairly good, difference >1 and ≤2	11	22	7	4
Poor, difference >2 and ≤4	6	12	5	1
Very poor, difference >4 kg	4	8	3	1

↓, Weight underestimated; ↑, weight overestimated.

searcher represting comparability. We would therefore prefer to examine patients in specific and small homogeneous groups, which requires the use of a more differentiated and multilevel examination instrumentarium in a multilevel approach.

3. Is the course of the disease the same on all levels (physical, psychic, social)?

Such an instrumentarium is not only desirable because of the different types of anorexia nervosa, but the specification of individual categories of symptoms also requires an adequate corresponding instrumentarium: a considerable number of anorexia nervosa symptoms have been included, for example, into the anorexia nervosa definition system of Feighner et al. (1972) and have also been widely covered in the literature (e.g., Vigersky 1977, Vandereycken and Meermann 1984). The symptoms are classified into physical, psychic, and social subgroups and can be examined at different times in the course of the study. This procedure offers the advantage that the most important criteria for anorexia nervosa or the important screening parameters for physical and psychic wellbeing are included in an evaluation system which takes into account the various levels of observation.

4. Is the way the patient copes with anorexia nervosa an important factor?

Finally, there is the necessity of establishing another illness category – one which becomes increasingly more important not only with respect to coping with the illness and attitude towards the illness itself, but also as a financial concern – the attitude towards seeking treatment. Our study showed that this should be an independent category among the symptoms.

The aim of our study was the methodological establishment of a system of criteria that would include the aspects described above (Deter 1981, Deter et al. 1983). This system includes the most important symptoms of the illness known from the literature at the different recording phases at the physical, psychologic, and social level and the treatment-seeking attitude of anorexia nervosa patients. At the same time, this system of criteria reflects the clinical experience gained over many years by a team of therapists treating anorexia nervosa patients.

The system has not yet been tested using a normal random sample or other patient groups with clinical diseases. A more exact psychologic validation (factor analysis) clarifying how the individual parameters are related is still needed. We cannot exclude on principle that further significant categories of symptoms may be missing in this system and that some of the criteria employed may be more important than others. In the anorexia nervosa symptom score (ANSS), each individual criterion has been attributed the same significance for clinical reasons (except for weight).

On the one hand, the inclusion of a large number of criteria renders an examination instrumentarium of this type – developed according to clinical requirements – more differentiated. On the other hand, it also becomes less clear compared to the Morgan-Russell score. A structure was therefore established by subdividing the criteria on four different levels, which offers the possibility to include parameters measurable with real numbers beside the common clinical evaluations. The information obtained on different levels can thus be collected and evaluated in a single evaluation system (the individual levels, which overlap in theory, see also Deter 1991, contain parameters of different validity from a technical point of view). Nevertheless, we think that this novel multilevel evaluation approach justifies the use of such an instrumentarium, which certainly needs to be developed further.

Evaluation criteria should be objectifiable, if possible. As far as psychosomatic problems are concerned, we, like other authors (Rudolf 1981; Rüger 1981), believe that an inclusion of independent evaluation tools allows a qualitatively better observation of the course of the disease. The main objectifiable and specifically physical criteria of anorexia nervosa (Feighner et al. 1972) are: weight reduction to less than 25%,[1] secondary (and sometimes also primary) amenorrhea, and vomiting (including self-induced vomiting after binge eating). The criteria for constipation and abuse of laxatives were evaluated separately for clinical reasons.[2] Bradycardia and hypotension are further criteria; we chose to record the latter since it seemed to be more constant in the course of the examination. We also decided to determine the electrolyte status (especially serum potassium) and hormone levels (follicle stimulating hormone, FSH; luteinizing hormone, LH; plasma cortisol; T3/T3). Increased hepatic values, irritation parameters (blood sedimentation rate, blood count, electrophoresis), substances normally eliminated with the urine and pancreatic/parotic enzymes are not pathognomonic for anorexia nervosa, but indicate nevertheless other functional disorders corresponding to the severity of the basic disorder or to disturbed behavioral patterns (alcohol or laxative abuse, vomiting; Table 2).

Other objectifiable patient data of the disorder are the number of days in the hospital or on sick leave which can be obtained from the health insurance plans under guarantee of data protection.

Other criteria require either the patient's cooperation or anamnesis from another source. It is still possible here to adequately clarify the question of a steady partner, whether the patient has a family of his/her own, whether there is successful integration at the professional or school

[1] Actually a weight reduction measured in relation to the former body weight. We have taken the stricter criteria of age-appropriate normal weight.

[2] But not lanugolike hirsutism and periods of hyperactivity (Feigner's criteria) that appeared to be not variable enough (lanugo hirsutism) for a course control or could not be recorded accurately enough (hyperactivity).

Table 2. Evaluation parameters used in the follow-up study of anorexia nervosa patients

	Definition	Evaluation[a]	
		normal	pathologic
Physical parameters			
Obstipation	Persisting for at least 3 weeks during anorexia nervosa	0	+
Self-induced vomiting	Self-induced vomiting with reported or observed episodes of binge eating	0	+
Laxative abuse	Intake of laxatives (suppositories, tablets, teas) for more than 3 weeks	0	+
Amenorrhea	Absence of the menses for at least 3 months	0	+
Other physical symptoms not related to anorexia nervosa	Moderate to severe physical complaints that show no apparent relationship with the original clinical picture of anorexia nervosa	0	+
Body weight (% of ideal weight)	1. Body weight 75%–90% of ideal weight	0	+
	2. Body weight below 75% of ideal weight		+
Electrolytes	Sodium, potassium, chlorine, calcium, phosphorus in serum (normal values of the clinical laboratory)	0	+
Pancreatic and salivary enzymes	Serum amylase	0	+
Substances eliminated in urine	Serum creatinine and urea	0	+
Transaminases	Serum GOT and GPT	0	+
Inflammatory signs, changes in blood count	Electrophoresis, erythrocyte sedimentation rate, leukocytes, hemoglobin	0 0 0	+ + +
Hormonal changes	FSH, LH, cortisol, T3/T4 in serum	0	+
Blood pressure	Systolic pressure below 105 mmHg	0	+
Psychic parameters[b]			
Severe psychic symptoms	Neurotic anxieties, phobias, compulsive thoughts, compulsive acts hypochondriac fears	0	+
Low self-esteem	Patient convinced of being worth little or nothing		
Suicidal tendencies	Thoughts of suicide at the time of the study or suicidal acts in the history	0	+
Addictive tendencies	Psychic dependence on alcohol, nicotine, tranquilizers, excessive appetite for sweets	0 0	+ +
Degree of psychic suffering	Patient seriously affected by the disease or the physical, mental and social symptoms	0	+
Sexual problems	Lack of sexual desire, anorgasmy or fear of sexual experience or acts	0	+

Table 2 (*continued*)

	Definition	Evaluation[a]	
		normal	pathologic
Disturbed body image	Patient not satisfied with his/her body Despite objectively normal weight, patient feels "too heavy or too plump"	0	+
No opportunity of compensating for psychic symptoms	Patient has no social, individual-creative, familial or professional opportunities to experience success independent of the illness	0	+
Infantility	(Childish) behavior, thinking and feeling inappropriate to the patient's age	0	+
Social parameters[b]			
No social integration	Relationship with friends (of the same age) and colleagues: social activities	0	+
No communication with family	Patient states that s/he does not get on well with either parent	0	+
No steady relationship	No long-term (more than 1 month) friendship or sexual relationship with the opposite sex	0	+
Exaggerated emotional dependence on the family of origin	Age-inappropriate emotional and real dependence on the family of origin	0	+
No founding of a family of his/her own	Patient is not married or has no children	0	+
No professional integration	Patient is not working at the time of the study or has no or only incomplete professional training	0	+
Medical treatment seeking[b]			
More than 10 days per year on sick leave	Number of days on sick leave or absence from school as stated by the patient	0	+
More than 10 days per year in hospital	Number of days in hospital as stated by the patient	0	+

[a] The psychic and social parameters were evaluated with the aid of a five-grade scale (parameter not present or pronounced).
[b] In the last year before examination.
GOT, glutamic-oxaloacetic transaminase; GPT, glutamic-pyruvic transaminase; FSH, follicle-stimulating hormone; LH, luteinizing hormone.

level, or whether social integration is satisfactory (personal contacts, clubs, etc.). These criteria have to some extent been used in follow-up studies of anorexia nervosa patients reported so far (e.g., Thomä 1967; Petzold 1979).

Certain criteria such as the registration of strong psychologic symptoms (especially the fear of eating or gaining weight)[3], the occurrence of physical symptoms independent of those of anorexia nervosa or the estimation of certain capabilities (hobbies or social contacts) are more difficult to assess since the subjective assessment of the patients is associated with a considerable rate of error. For assessing an exaggerated emotional dependence of the patients on their original families or the degree to which the patients can communicate with their original family, the investigator must be more experienced in taking family histories. Nevertheless, these criteria and also the subsequent ones will rather be regarded as "soft", although they may in principle be considered to be harder on the basis of expert training and interrater comparison studies.[4] The following criteria must be assessed by a clinically experienced investigator[4]: degree of suffering motivating desire for recovery, problems concerning sexuality, self-destructive tendencies, addictive tendencies (e.g., increased alcohol consumption, drug abuse etc.), infantility, reduced self-esteem, and body image.[1,5] There are no clear definitions for the classification and rating of the above factors, which are, however, derived from descriptions of typical characteristics of anorexia nervosa patients. We therefore consider these factors to be necessary for a fairly valid evaluation of the disease course in these patients.

This symptom score may be supplemented or refined by various additional test instruments:

1. At the level of independent assessment, it may be useful to establish the precise type of anorexia (abstainer, purger, juvenile anorexia nervosa, bulimia) and to evaluate comorbidity by means of the DSM-III-R or ICD-10 system as well as the severity and type of the familiar disorder.
2. At the level of self-assessment, categories such as eating behavior, disturbed body image, and psychologic symptoms may be analyzed in more detail by appropriate test instruments (see also the chapter by Morgan et al., in this volume).

---

[3] High-quality questionnaires are now available for this purpose e.g., E.D.I. Garner et al. 1983; SIANEX, Fichter 1983).

[4] In order to standardize these criteria, we performed two interrater comparison studies using the SIANEX (Fichter 1983), which contains all criteria of the ANSS for evaluating psychic and social factors (with the exception of the family scales and the infantility criterion). We found a highly significant mean interrater reliability of .81.

[5] This category may be supplemented by psychodiagnostic self-assessment tests on the patients' body image status (e.g., Meermann and Vandereycken 1987).

3. At the level of detailed assessment, psychic defense mechanisms and denial of anorexia nervosa patients can be analyzed more precisely by means of special inverview techniques (psychoanalytic setting).

These additional examinations will not be discussed further.

Examples illustrate the application of the ANSS, a clinical instrument that gives relative priority to objectifiable physical findings in order to obviate the well-known denial associated with anorexia nervosa, an attitude that may even persist in a follow-up study performed 12 years later.

Results

In the following, the results of the ANSS in the follow-up study are correlated with the findings obtained using the Morgan-Russell Score (MRS) and the impairment score (IS) devised by Schepank (1988). The latter, not specific for anorexia nervosa patients, is widely used in the Federal Republic of Germany in epidemiologic investigations, twin studies, and the evaluation of psychosomatic outpatients.

Figure 1 indicates the changes registered in 76 of the 103 still living female anorexia nervosa patients after 12 years with the ANSS and the MRS, whereby the numeric representations of the score are opposite. The pathologic abnormalities recorded by the ANSS show a mean decrease from 20.1 to 8.7 points, while the MRS increases from 2.8 to 10.6 points.

There is thus a high correlation between the two tests of −.73 for the total score and of −.65 for the physical subscore as well as a correlation of −.66 for the psychosocial subscore (Table 3).

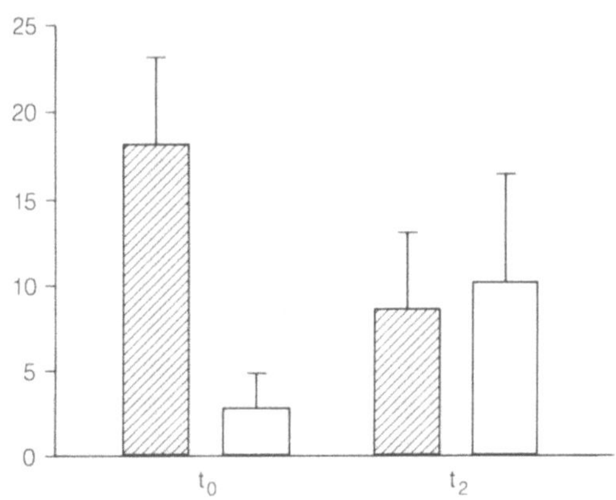

Fig. 1. The Morgan-Russell Score (*open bars*) and the Anorexia Nervosa Symptom Score (*hatched bars*) at two examination times 12 years apart in 76 of 103 anorexia nervosa patients of the Heidelberg-Mannheim Study

Table 3. Selected intercorrelations (according to Spearman) between the Morgan-Russell Score (MRS) and the Anorexia Nervosa Symptom Score (ANSS) as a total and physical and psychosocial findings

	MRS t_2			Body weight
	Total	Physical category A,B	Psychosocial category C,D,E	
ANSS t_2				
Total	−.73*			−.45**
Physical factors		−.65**		−.51**
Psychosocial factors			−.66**	
Body weight	.33*	.81**		

$^{*}p < 0.01$; $^{**}p < 0.001$.

Body weight naturally shows a high correlation (.81) with the two physical scales A and B of the MRS (which only cover body weight and menstrual status), while there is a markedly lower correlation of only −.51 with the ANSS (physical parameters), in which weight is only one of 13 physical factors. The total score of the MRS shows a relatively low correlation with body weight (.33), while that of the ANSS is slightly higher (−.45).

These results indicate that there is a relationship between weight and psychosocial findings in anorexia nervosa, but that it is less pronounced than that between weight and other physical parameters.

A comparison of the anorexia-specific scores (ANSS and MRS) with the general psychosomatic, social and somatic impairment score (IS) reveals the following: there is a highly significant correlation between the psychosocial ANSS scores and IS scores of psychic and sociocommunicative impairment (.81 and .73, respectively), while the correlation with the MRS (−.60 and −.63, respectively) is markedly lower, as is the correlation of the physical scales A and B of the MRS with the physical impairment scale of the IS (−.38). This correlation is higher for the ANSS at .56 (Table 4).

Although the findings are probably not quite independent of each other, since the criteria of the different scores were evaluated by the same examiner, it appears that a general psychic and sociocommunicative impairment, which is also associated with disorders other than anorexia nervosa, is more clearly registered by the ANSS than it is by the MRS. The physical impairment scale of the IS includes not only weight and amenorrhea but the whole range of possible physical disorders; these are not in the foreground in acute phases of anorexia nervosa but become important in a follow-up study, in which the occurrence of syndrome shifts must be taken into consideration.

Table 4. Correlations of the Morgan-Russell Score (MRS) and the Anorexia Nervosa Symptom Score (ANSS) with the Impairment Score (IS) of Schepank

	Impairment score			Total
	Psychic	Sociocommunicative	Physical	
ANSS				
Psychosocial	.81	.73		
Physical			.56	
Total				.82
MRS				
Psychosocial	−.60	−.63		
Physical			.38	
Total				−.73

It is interesting that there is good agreement between the total scores of all three tests (correlation >.70), although they have been designed for fairly different groups of patients.

A look at the different dimensions of the ANSS clearly reveals the wide variation of findings for the individual symptom categories in anorexia nervosa patients on presentation at the hospital (Table 5). The hard symptoms such as underweight and amenorrhea are present in nearly 100% of the patients; most other categories of symptoms are found in 50%–90% of the patients, and only individual physical symptoms are seen in as few as one third to one half of the patients. This result indicates that all categories of symptoms of the ANSS are relevant in the anorexia nervosa patients presenting at a medical department of a university hospital.

A separate factor analysis for the psychosocial (standardized interviews in anorexia nervosa expert assessment, SIANEX (items) of the ANSS isolated the following factors:

1. Psychic symptoms
2. Anorectic behavior
3. Partner- and sexuality-related problems
4. Bulimia/alcohol abuse

A factor analysis for part of the physical items (laboratory parameters and weight) is in preparation.

The follow-up reveals that the symptoms from the psychic categories of the ANSS persist longer than those in the social and somatic categories, although there are again great differences between individual variables. Despite the apparently confusing diversity of criteria, such a detailed test allows a more differentiated evaluation of the course of the disease than a relatively broad category system based mainly on the criterion of weight.

Table 5. The anorexia nervosa symptom score (ANSS): percentage of pathologic findings at referral to the clinic (n = 91 female anorexia patients)

Summary	Mean percentage
Social state	66.5
No sexual partner	73.3
No own family	77.8
No integration into an occupation	51.7
Emotional dependence on family of origin	98.9
No understanding with family of origin	44.9
No integration in social environment	60.7
Medical treatment seeking	
More than 10 hospital days (past year)	51.1
More than 10 lost working days (past year)	72.1
Psychic state	75.7
Psychic symptoms	90.0
Sexual problems	92.0
Suffering pressure	84.4
Self-destructive tendencies	60.7
Addictive tendencies	56.3
Childishness	78.4
Low self-esteem	80.0
Body image disturbance	84.5
No curative compensation	54.0
Somatic state	62.6
Constipation	79.3
Self-induced vomiting	50.0
Laxative abuse	54.7
Amenorrhoea	96.6
Body weight	98.9
Electrolytes	51.8
Amylase	41.0
Urea and creatinine	37.8
Transaminases	39.3
Signs of inflammation, red and white bloodcell counts	54.9
Hormonal disturbances	49.4
Hypotension	69.6
Other symptoms	41.9

Another aspect is the evaluation of two variables that have so far only been investigated in few anorexia nervosa studies (Herzog et al. 1989). We found that anorexia nervosa patients had been hospitalized for a total of 34.8 weeks and, even after 12 years, 25% were hospitalized for a mean of 5.4 days per year, which is markedly longer than in the normal population (for example, 2.8 days/year for a member of the AOK, General Health Insurance Co) but shorter than other patients who have undergone inpatient psychosomatic therapy five years before (9 days/year; Deter et al. 1986). These figures suggest that the coping strategies and the treatment-seeking attitude are important aspects to be considered in the evaluation of patients with chronic anorexia nervosa. Another important variable in this context is the annual number of days on sick leave: 40% of the patients followed up after 12 years had been on sick leave for more than 10 days/year prior to the final examination.

Although we have not yet analyzed all data, the results obtained so far seem to indicate that the way in which patients cope with the illness may be a separate, possibly independent, dimension that plays an important role in anorectic patients.

The detailed analysis of the different anorexia nervosa subgroups has not as yet been concluded, but a preliminary examination has already revealed wide differences between individual courses, a finding which supports the initially stated need for a detailed test for evaluating the course of the disease.

Concluding Remarks

We are aware that the ANSS suggested here and applied in our study uses a selection of symptom categories which may have to be modified and supplemented by additional criteria. We used this evaluation system for the following reasons:

1. To investigate the value of a detailed evaluation tool with 30 symptom categories in the follow-up of anorexia nervosa patients
2. To include more objectifiable parameters in the evaluation of these patients
3. To relate psychic, somatic, and social aspects as well as patient treatment-seeking behavior to the course of anorexia nervosa

In our opinion, this procedure is a valuable extension of the test instruments that have so far been used in the evaluation of anorexia nervosa patients.

References

Browning C, Mittler SJ (1968) Anorexia nervosa: a study in prognosis and management. Am J Psychiatry 124:1128–1132

Deter H-C (1981) Zur Methodik von katamnestischen Untersuchungen bei psychosomatischen Patienten am Beispiel einer Gruppe von 31 Anorexie-Patienten. Psychother Med Psychol 31:48–52

Deter H-C, Sameith W, Maroska U, Ferner H, Reindell A (1986) Katamneseergebnisse von 64 Patienten, 5 Jahre und der stationären psychosomatischen Behandlung. Z Psychosom Med 32:231–248

Deter H-C, Manz R, Becker M, Gladisch R (1991) Soziale Unterstützung in ihrer Beziehung zu köperlichen und seelischen Befunden bei Morbus Crohn Patienten. Z Psychosom Med 37:45–59

Feighner JP, Robins E, Guze SB, Woodruff R, Winnocur G, Wounots R (1972) Diagnostic criteria for use in psychiatric research. Arch Gen Psychiatry 26:57–63
Fichter MM (1980) Klinische Verlaufsuntersuchungen zur Verhaltenstherapie der Pubertätsmagersucht. In: Hautzinger M, Schultz W (eds) Klinische Psychologie und Psychotherapie. DWG, Munich, pp 267–276
Fichter MM, Meermann R (1983) Zur Psychometrie der Anorexia nervosa. In: Meermann R (ed) Anorexia nervosa. Enke Stuttgart
Garner D, Olmsted M, Polivy I (1983) Development and validation of a multidimensional eating disorder inventory or anorexia nervosa and bulimia. Int J Eating Dis 2:14–34
Herzog W, Deter H-C, Manz R, Riecken G, Petzold E (1989) Inanspruchnahme medizinischer Dienste bei Anorexia nervosa. Psychopathology 15:780
Meyer AE (1984) Taxonomic subgroups within psychosomatic disease entities: an alternative strategy to the specificity approach? Psychother Psychosom 42:26–36
Moldofsky H, Garfinkel PE (1977) Problems of treatment of Anorexia nervosa. Can J Psychiatry 19:169–174
Morgan HC, Russell GM (1975) Value of family background and clinical features as predictors of long-term outcome in anorexia nervosa: 4 year follow-up study of 41 patients. Psychol Med 5:355–371
Petzold E (1979) Familienkonfrontationstherapie bei Patienten mit Anorexia nervosa. Vandenhoeck and Ruprecht, Göttingen
Rudolf G (1981) Untersuchung und Befund bei Neurosen und psychosmatischen Erkrankungen. Beltz, Weinheim
Rüger U (1981) Stationär-ambulante Gruppenpsychotherapie. Springer, Berlin Heidelberg New York
Russell G (1977) General management of anorexia nervosa and difficulties in assessing the efficacy of treatment. In: Vigersky R (ed) Anorexia nervosa. Raven, New York
Schepank H (1988) Psychogene Erkrankungen der Stadtbevölkerung. Springer, Berlin Heidelberg New York
Steinhausen HC, Glanville A (1983) Follow-up studies of anorexia nervosa: a review of research findings. Psychol Med 13:239–249
Thomä H (1967) Anorexia nervosa. International University Press, New York
Vandereycken W, Meermann R (1984) Anorexia nervosa. De Gruyter, Berlin
Vigersky R (ed) (1977) Anorexia nervosa. Raven, New York

Follow-up of Males with Eating Disorders

A.E. Andersen

Introduction

Follow-up studies play a vital role in understanding the etiology and natural history of psychiatric disorders and in assessing the effectiveness of treatment interventions. While cross-sectional evaluation is central to the process of diagnosis and treatment planning, only longitudinal study of a disorder can confirm the accuracy of that diagnosis and the response to treatment. Follow-up is especially important in the field of eating disorders where therapies purporting to be effective are still controversial, and no single method has been demonstrated to be obviously superior to others. Also, follow-up helps to evaluate some treatments which are not dramatic in the short run, but prove themselves to be eventually effective. Some recent studies on interventions in coronary artery disease, for example, have begun to reveal results only after the 5- to 10-year observation points (The Multiple Risk Factor Intervention Trial Research Group 1990).

Early exposure to a variety of noxious agents such as lead (Needleman et al. 1990), ionizing radiation, electromagnetic radiation, and psychosocial deprivation have only been appreciated for their full deleterious consequences after a long-term period of observation. While some treatments such as penicillin for streptococcal infection of the pharynx produce prompt and obvious effects, many other, more subtle, but ultimately effective treatments require a much longer period of observation and may be mediated by processes that require considerable time for change to occur. For example, a long-term study of antidepressant medication vs. electroconvulsive therapy (ECT) for depressive illness showed a beneficial effect in favor of ECT (Avery and Winokur 1976).

The goals of this chapter are to review published follow-up studies on males with eating disorders from the scientific literature and to describe the Johns Hopkins experience, as well as to illustrate the natural history and treatment outcome of males with eating disorders by describing selected case histories.

Eating Disorders in Males

Anorexia nervosa has been reported in males since 1689 when Morton described the syndrome in a young man (Morton 1694). This chronology was further continued by Gull (1874), but thereafter, various theoretical biases clouded the concept of eating disorders in males from both public and professional awareness. The condition in males was lost sight of amidst a flurry of newly developed psychodynamic interests which required the obviously gender-biased motif of fear of oral impregnation for diagnosis (Scott 1986). Also, males could not meet criteria that required amenorrhea for the diagnosis of anorexia nervosa.

A reemergence of interest in males with eating disorders occurred in the 1970s, exemplified by studies by Beumont (1970) and Crisp and Toms (1972) and others who described the psychological and endocrine features of male patients. Bulimia nervosa has only more recently been recognized formally, although Stunkard (1959) described a "night eating syndrome" and a "binge eating syndrome" and Hilde Bruch (1973) also mentions this abnormal pattern of binge eating in her seminal work, *Eating Disorders*. Russell (1979), however, definitively described bulimia nervosa in a series of patients with binge and purge symptoms, and clearly related it to the syndrome of anorexia nervosa by emphasizing the shared psychopathological motif of a morbid fear of fatness underlying both of these disorders.

Vandereycken and Van den Broucke (1984) have critically reviewed literature reports of 107 males and found, despite the presence of only a moderate number (37/107) of well-diagnosed cases, that the group shared substantial features with 148 females, a finding confirmed by Fichter et al. (1985) as well as Margo (1987).

Some preliminary questions regarding males with eating disorders might be the following:

1. *What is the prevalence of males with eating disorders?* The first caution here is to reaffirm the rule that the type and sex ratio of cases referred to a university hospital clinic or other specialty service may not be synonymous with the occurrence in the general community. In our Johns Hopkins experience, consistent over 14 years, approximately 1 in 10 patients with an eating disorder has been a male, both in the consultation clinic and in the inpatient program. Crisp et al. (1986) disagree with this number and suggest a lower ratio of 1%–5% of cases being male. Recent community-based studies using careful epidemiological techniques, such as the reports by Rastram et al. (1989) and Nielsen (1990), have confirmed that our clinic-based ratio of 1 male to 10 or 11 females is also an accurate estimate of the occurrence in the general population. Halmi et al. (1981), using older DSM-III criteria, found in a survey of 355 college students that 5% of the males experienced all the major symptoms of bulimia, and they comprised 13% of all bulimic cases.

The absolute prevalence of eating disorders remains controversial and largely depends on the rigor of the criteria used. A reasonable range of estimates would be that, in the United States, approximately 100000 to 500000 males suffer from some form of an eating disorder, using as a consensus figure a 1% incidence of eating disorders overall for the age range covering older school children through young adults, with 1 out of 10 of these cases being male (Fairburn and Beglin 1990).

2. *Why are there so few males with eating disorders*? Definitive answers to this question are not available at the present time, but some clinical and research observations may offer preliminary responses. First, there appears to be a close "dose-response" relationship between the amount of socio-cultural reinforcement promoting behaviors leading to thinness and the probability of developing an eating disorders (Hsu 1989; Andersen and DiDomenico 1991). When subgroups of males are exposed to situations requiring weight loss, such as occurs with wrestlers, swimmers, runners, and jockeys (Yates et al. 1983; King and Mezey 1987; Steen et al. 1988), then a substantial increase in the behaviors of self-starvation and/or bulimic symptomatology follows, suggesting that behavioral reinforcement, not gender, is the crucial element.

Secondly, Sours (1980) has noted that the process of transition through puberty in Westernized society is harder on girls than boys in a number of ways. As more ample nutritional resources have become available to the population in general, young girls have steadily decreased in the age at which they experience menarche. Younger and younger girls are therefore becoming sexually mature while being, conversely, psychologically more immature because of the lowered age of puberty, all the while being exposed to a society that is both highly sexualized and narcissistic in its view of the body. Boys, in contrast, mature sexually about 2 years later, and this critical difference in age may allow them to undergo the process of puberty with less distress. In addition, young boys tend to be less conflicted about secondary sexual characteristics and the onset of sexuality, even though they experience their own share of identity issues.

Wherever the causes of the gender-related difference in prevalence of cases, the fact of this substantial difference in ratio of cases according to sex conveys important information about the nature of the illness. Either substantial genetic factors are involved, or alternatively, there are present culturally related norms that lead to particular behaviors which are differentially distributed between the sexes. Even more interesting is the possibility of complex genotype-environment interactions, a subject explored in antisocial behavior disorders (Cadoret 1982) and posttraumatic stress disorders (Goldberg et al. 1990) but only tentatively in eating disorders (Strober and Humphrey 1987).

3. *Why do certain males in particular develop eating disorders*? The issue of symptom choice is a vexed and convoluted one in dynamic psychiatry, and completely satisfactory explanations have not been found. The males

that we have studied have, in general, borne some shared and some differing gender-related risk factors, compared to females. These factors can be understood as predisposing, initiating, or perpetuating to the illness, especially in regard to different reasons for dieting. Males who were obese and teased for their obesity as children are in our experience, more vulnerable to dieting (Andersen 1988). More than 60% of the males in our series of eating disordered patients experienced childhood or adolescent obesity. In general, women who develop eating disorders *felt* fat before they began dieting, but were only 10%–12% above average weight before dieting, while the majority of males *were* medically overweight.

Secondly, more males than females dieted in relationship to athletic achievement. Thirdly, more males dieted in order to achieve a body definition which they felt would be pleasing to a homosexual partner. Finally, more males than females dieted to ward off medical illnesses that they feared might fall their way because of excess weight. They took to heart, with preoccupying seriousness, warnings they had heard addressed to their parents about coronary disease, cancer, and diabetes, concerns that were inappropriate to young boys and teens in good health. In contrast, we have found relatively few young women who dieted to avoid medical illness. It may be possible to derive an empirical mathematical equation that predicts risk, as described by Andersen (1990a).

Characteristics of Good Follow-up Studies

First and foremost, good follow-up studies must contain sufficient subjects and be conducted long enough in order to be worthwhile. This requires, as Dr. Sten Theander has said (personal communication), that an investigator begin follow-up studies while young and energetic, and then be sure to live to a ripe old age to obtain results. In addition, adequate follow-up studies need to have comprehensive measures of outcome assessment rather than single or narrow definitions. In certain illnesses, a single outcome might be quite appropriate, but with eating disorders, which touch on virtually every aspect of existence from the molecular to the psychosocial, broad-based assessment of biological, psychological, and social modalities are required. Thirdly, interval measurements are necessary to accurately sketch out the longitudinal course of these eating disorders. Knowledge of the final death rate after 30 years would by itself be less useful than the knowledge of which particular cumulative 5-year periods in which the deaths take place. The same questions can be asked of documented clinical improvement – does it happen relatively soon after diagnosis and treatment intervention, or later on, or at different rates for different symptoms, and perhaps independent of, or in spite of, treatment.

Some examples of comprehensive follow-up studies of females include Theander (1985), Hall et al. (1984), Tolstrup et al. (1985), Casper et al. (1988), Toner et al. (1986), Engel (1988), and Morgan and Russell (1975). More subtle predictors of outcome have been investigated, including personality characteristics (Edwin et al. 1988) and ego functioning (Sohlberg and Norring 1989). Outcome of female patients after treatment for bulimia nervosa are just beginning to appear (Mitchell et al. 1989; Hsu and Sobkiewicz 1989).

Review of Follow-up Studies of Males

Burns and Crisp (1984, 1985) conducted a thorough follow-up study on 27 consecutive males diagnosed with anorexia nervosa, evaluating subjects from 2 to 20 years after initial presentation. This study meets the criteria of an adequate number of patients who were followed for a long enough period of time. In addition, they were treated in a consistent manner by the same clinicians, by a well-defined and published treatment. Finally, they were tested on a variety of outcome measures.

Overall, the outcome of this group of male anorectics was similar to that reported for female anorectics: 44% were rated good on outcome, having a stable weight and normal sexual functioning; 26% were intermediate in outcome, weighing within 15% of a matched population weight but with diminished sexual activity or having a lower than normal weight without regular sexual activity. The poor outcome group, with both low weight and absent or substantially diminished sexual activity, was found to be characteristic of 30% of the patients.

Poor outcome patients experienced a longer duration of illness, a greater number of previous treatments and substantially more weight loss during their illness. Two consistent factors associated with poor outcome were a disturbed relationship with parents during childhood and the absence of normal sexual activity during adolescence prior to the onset of illness (Burns and Crisp 1985). Interestingly, vomiting was associated with good outcome in males, but poor outcome for females (Crisp et al. 1986).

Hall et al. (1985) conducted one of the few other comprehensive studies of outcome in males with anorexia nervosa that meet the previously mentioned criteria. Their 9 male patients, aged 14 to 24 years of age, represented only 5% of the eating disorders patients presenting for diagnosis or treatment. They manifested a characteristic loss of libido during illness; at follow-up, the patients' strength of libido was related to testosterone levels. Their male patients were similar to females in age at onset and clinical features. None were homosexual, in marked contrast with studies by Herzog et al. (1984). Features common to the males in the study by Hall and colleagues were an intense focus on bodily appearance before the onset of the illness, and a problematic sense of control or personal identity.

Mitchell and Goff (1984) describe the outcome of 12 male bulimic patients ages 18 to 34. Only one subject (9%) was married; a high rate of chemical dependency was found in the study subjects as well as their first degree relatives. Good outcome occurred in two thirds of their patients, while one third failed to comply with treatment referrals. They noted that one quarter of their patients had initiated bulimic symptoms in the setting of athletic training, a finding supported by Andersen (1988).

The Johns Hopkins Follow-up of Males with Eating Disorders

We have confirmed only one death in 86 male patients over a total of approximately 14 years of treatment experience. The initial long-term outcome on ten patients aged 18–47 was reported in 1983 (Andersen and Mickalide). This follow-up of up to 6 years after treatment found that patients did at least as well as females (Andersen et al. 1983). The average weight of the males in the series, 92% of a matched population weight, was well within the normal range. Seventy-five percent of the group were within 90%–110% of a matched population. Twenty-five percent experienced persisting bulimic behavior, but only one had a markedly poor social adjustment. Although social and psychological function generally improved, there was persisting lowering of testosterone or infertility in two males, despite return to normal weight. Whether this represents a pre-illness vulnerability remains to be determined.

Summary of Follow-up Studies of Males with Eating Disorders

The three largest follow-up studies of males all suggest a global outcome of general improvement similar to that for females. Most studies have found for both males and females something close to the "rule of thirds": one third with good outcome, one third with intermediate outcome, and one third with poor outcome. These reports are from specialized clinics where a substantial proportion of patients have chronic illness. We believe early diagnosis of milder cases will have an even better outcome, but this impression needs to be confirmed by additional studies. Certainly there is no evidence from the limited studies completed to date that anticipation of outcome in males with eating disorders should have a pessimistic tone. In general, there were few surprises on follow-up: the best outcome was associated with individuals having a shorter duration of illness, less severe symptomatology, fewer psychiatric comorbidities, better family functioning, and more adequate sexual adjustment prior to illness.

Illustrative Cases

Case History – Greg

Outcome: Excellent outcome at 4-year follow-up (anorexia nervosa, restricting subtype)

Greg was a 14-year-old boy of Greek background referred to Johns Hopkins Hospital for severe weight loss. His family included father, age 42, with whom he had a distant relationship, mother, 40, a librarian, and two sisters, one 17 who was diagnosed as having a mild case of anorexia nervosa and another, 12, without psychiatric symptoms. The family atmosphere was characterized by constant dieting by all members of the family. Father dieted to improve his high blood pressure and lost considerably more weight than his physician had requested. He passed on to his son an anxious concern for avoiding the medical consequences of overweight. Mother was always "crash dieting" and warned her children not to eat too much, especially snack foods, or they would become fat. Greg summarized the home atmosphere by saying "there are constantly fights over food. Everyone is always on a crash diet."

In personal history, Greg was always thin as a child, and, in fact, took special status from his thinness. Despite intermittent asthma, he was generally well. He underwent puberty beginning at age $11\frac{1}{2}$ with a sense of guilt and perplexity. He occasionally thought of dating girls but was anxious about asking them out and was in many ways asexual. In personality, he was perfectionistic, sensitive and somewhat withdrawn.

His illness began in the eighth grade when he lost his position as the thinnest person in the school. Despite being only 110 pounds at 5 feet 2 inches tall, he restricted his food intake and additionally began a strenuous program of 2 h of exercise a day. Despite concern by others, he was not worried about his lowered weight and reduced to $83\frac{1}{2}$ pounds over the next year, meanwhile growing to 5 feet 4 inches tall. His food intake was calculated to be approximately 600 calories on a typical day. He experienced severe distortion of body image despite pursuit of lower and lower weights, saying, "I thought everything I put into my mouth was going to make me fat. Once I started cutting down, I couldn't stop." He suffered from medical symptoms of feeling cold in the summer as well as decreased mental concentration. He intermittently thought of suicide but made no plans or attempts.

On mental status examination, Greg was a very thin, anxious young man, dressed in a sweat suit. His mood was depressed. He was self-critical but without delusions or hallucinations. He had intermittent suicidal thoughts. He compulsively exercised for several hours a day and also felt he had to look at the license plates on trucks in order to memorize them as they passed. He had pervasive fear of fatness and distortion of body image, saying, "If I gain weight, I may be left out of the crowd."

Diagnosis. Anorexia nervosa, restricting subtype; major depressive illness; vulnerable personalty traits, including perfectionistic, sensitive and avoidant features.

Course and Follow-up. Greg did well in comprehensive treatment, responding to nutritional rehabilitation, psychotherapy, antidepressant medication, and behavioral retraining (Andersen 1990b). He began to develop awareness that an identity based on extreme thinness was unhealthy and could see, at least intellectually, the need for a more appropriate basis for a secure identity. He began to differentiate himself from his family, not taking on the worries that his father had about being overweight or responding to his mother's weight preoccupation. He recognized his perfectionistic and compulsive features and began to socialize a bit more.

At Follow-up. Greg, 4 years later, had increased his weight to 125 pounds, staying at 5 feet 4 inches tall. He pursued a supervised program of moderate exercise in order to increase his muscle definition and strength. He progressed in his social life, developing close friendships with other students and occasionally dating in small groups. He developed a sense of identity based on healthy factors in place of the former extreme thinness, such as his friendships and shared interests. He made plans for going to college and developed a career plan. On several return visits to the clinic over 4 years, he demonstrated that he had successfully put behind him his fear of fatness, pursuit of thinness, and distortion of body image. In addition, he discovered a sense of humor and a self-effacing but confident manner, with a touch of bravado.

Comment. This case illustrates the development of anorexia nervosa as a solution to a crisis in maturation, especially the need to develop an identity more appropriate to his current age. Greg was not overweight as a child but did grow up in a family that was relentless in its dieting. It also illustrates the effect of pervasive family dieting on young children. He "caught" his father's concern about the dangers of overweight despite its complete inapplicability to him. He illustrates the frequent finding of the two most common comorbidities: an Axis I mood disorder, and vulnerabilities of personality on Axis II. (No formal diagnosis of personality disorder was given patients under 18.) Greg shows how a young male can abuse exercise in the process of weight loss, but also, conversely, can constructively use exercise in the process of recovery. His case also illustrates the increased probability of a good outcome with prompt, thorough treatment. He had to deal with the twin conflicting desires of becoming heavier through adding muscle bulk to achieve a stereotyped ideal or becoming very thin to regain his former special status, but finally settled for a normal weight, with moderate muscular development.

Case History – Sam

Outcome: Moderate outcome at 10-year follow-up (anorexia nervosa, bulimic subtype)

Sam was a 26-year-old, white male referred to Johns Hopkins Hospital for treatment of low weight and binge-purge behavior after 10 previous hospitalizations. The father died when the patient was 2½ years old. The patient never knew why, only that father had disappeared. Father was frequently drunk and belligerent. Mother, age 63, was debilitated with coronary artery disease. She was compulsive, persevering, very close to the patient, and controlling of his behavior and feelings. He frequently felt guilty believing that he should have prevented his mother's illness. A half brother had episodes of depressive illness and a half sister was alive and well. There was a strong history of both alcoholism and depressive illness in the family.

His personal history included normal milestones and a distinguished high school career, receiving academic and athletic awards. Except for a brief problem with emptying of his stomach at age 3, of unknown etiology, he had no other gastrointestinal symptoms. He held two jobs for a brief time and did not initially abuse alcohol or drugs. He was heterosexual but with limited sexual interaction. In personality, he described himself as being both a loner but also feeling lonely. His family described him as very health conscious, indecisive, fearful of criticism, and perfectionistic.

His eating disorder began after high school when he went off to college. He dropped out after 4 days, feeling homesick and lonely, then enlisted in the Marine Corps, weighing 130 pounds at 5 feet 6 inches tall. Despite being in good physical condition, he felt the need to become even more fit and followed a program of extensive physical training even more strenuous than the one prescribed for new recruits. In addition, he restricted his food intake and quickly lost 15 pounds. After his initial weight loss, he experienced the onset of binge eating, which provoked his fear of loss of control as well as guilt about getting fat. He then began to induce vomiting. A pattern of regular binge eating and purging four to six times a day was initiated. He attempted suicide with an overdose of aspirin. For the next 8 years he was treated in a series of 10 hospitalizations at other institutions with virtually all the available methods including behavioral modification, hyperalimentation, and psychotherapy. His weight declined to 80 pounds. His eating pattern became ritualized, having to eat particular foods at certain times while wearing specific clothing.

On mental status examination, Sam was emaciated but alert, anxious and sad. He described his mood as very low and had persistent thoughts, but not delusions, about his worthlessness. He felt guilty about his own illness as well as his mother's. He experienced thoughts of suicide, but no specific plans. Sam had pronounced fear of fatness and distortion of body image. He described, as well, recurrent obsessional fears of leaving home, combined with a desire to leave.

Diagnosis. Sam met criteria for anorexia nervosa, bulimic subtype. In addition, he suffered from a major depressive illness without bipolar features. Finally, he met criteria for personality disorder with mixed features from cluster B (the antisocial, narcissistic, borderline categories) and cluster C (the obsessional, anxious, perfectionistic and sensitive groups).

Course and Follow-up. Sam required one more additional hospitalization after leaving Johns Hopkins but kept in touch with the hospital staff. He went though a 3-year phase of extensive alcohol abuse and abuse of prescription drugs. Six years after leaving Johns Hopkins, he "hit bottom" and made a decision to implement what he had been taught in the past, saying "I've always known what I have to do." He inhibited vomiting and with this response-prevention method of behavioral change found that his bingeing decreased. He calculated he could have paid for his college education with the cost of food that he had eaten in binge episodes, and this further motivated him. He was not able to stop his bingeing immediately but behaviorally shaped it with the help of an outpatient therapist until it was reduced to one small binge, late in the day, and then finally, none. He recovered a sense of purpose for life, enrolling in college. He used physical exercise to increase his self-esteem as well as reacquire moderate muscle bulk and sharper muscle definition. He began to appropriately separate from his mother and to develop an individual identify. When he returned to Johns Hopkins for follow-up 10 years after treatment, he weighed 147 pounds at 5 feet 6 inches tall. His body fat was 6%. He was enrolled in college and began to hold a part-time job. His scores on the Eating Attitudes Test and Eating Disorders Inventory were within the normal range on follow-up.

Comment. Sam illustrates how an eating disorder may require a long and difficult treatment process, but also that every aspect of it has the potential for eventual improvement. Specific male issues involved in this case were Sam's use of exercise both to initiate his weight loss and to help restore his physical health and self-esteem. At his current stable weight, he is accepting of an above average weight for his age and height because of satisfaction with his body shape, a typical male feature of being less worried about weight than shape. Although the combination of an absent or abusive father and an overly close, controlling mother has been described as the cause of a variety of different psychiatric disorders, this adverse parental experience certainly contributed to difficulty in developing a separate, confident, male identity, and, we hypothesize, had a major role in the onset and maintenance of his illness. In addition, he inherited family genetic vulnerabilities that predisposed him to depressive illness, alcoholism and an abnormal personality, factors we have found to be associated with a more severe course of illness.

Case History – Eric

Outcome: Fair outcome at 2-year follow-up (bulimia nervosa)

Eric was a 26-year-old, Hispanic male referred for treatment of binge eating, depression, and drug abuse.

He grew up in a family where the parents' marriage was characterized by constant bickering and verbal abuse. Father, a salesman aged 56, frequently belittled his son and criticized him for being overweight. Mother, 54, felt helpless at intervening and experienced intermittent depressive episodes. A brother, an electrical engineer, was considered the role model and hero of the family.

Eric's early development was characterized by frequent fights in school and reading difficulties. Overweight as a child he was teased mercilessly by family members and schoolmates. Beginning in high school he experimented with marijuana and cocaine. At one time charges were brought against him for sexual abuse of a younger female cousin, but these were later dropped. He held a number of short-term jobs, most of which were terminated when he became violently angry at any criticism of his performance.

In personality, he was emotionally volatile, with frequent rage reactions. He developed intense short-term friendships that were broken off suddenly when he became angry. He would often not keep promises to friends and occasionally stole small items impulsively. Despite these problematic features, he was able to show some warmth and occasional insight and professed that he wanted to improve his life.

He was put on a series of unsuccessful diets beginning at 9 years of age. At 18, he began a program of severely restricted calorie intake as well as moderate exercise and reduced his weight from 210 pounds at 6 feet tall to 152 pounds. For a brief time, weight went down to 135 pounds, but was restored to 150. As his body weight was gradually reduced, he began to experience binge eating followed by purging and occasional laxative abuse. He noticed that smoking marijuana increased his appetite, but that cocaine diminished it. After 4 days of intermittent cocaine ingestion, when he ran out of money, he would develop a ravenous appetite and gain 5 to 10 pounds quickly. He broke his collarbone in an argument with a drug dealer over his share of the profits. He was referred to Johns Hopkins Hospital after local outpatient treatment for several years was unsuccessful.

On mental status examination, Eric was a tall, extraverted, white male who would sometimes tease the examiner, but at other times become angry if he was not treated with what he considered to be sufficient respect or attention. He described his mood as being low about half of the time, but with intermittent episodes of 4–7 days in length of elevated mood not related to drug use. He experienced a lower mood in the winter, but tended to have brief episodes of elevated mood, as well as rapid thinking, in the spring and summer. He denied compulsive exercising or obsessional thoughts. There were no hallucinations or delusions. He had pronounced fear of fatness, declaring that he "never, ever," wanted to be teased again or criticized for his weight.

Diagnosis. Bulimia nervosa; bipolar II mood disorder with seasonal component; personality disorder with antisocial and borderline features (cluster B).

Course and Follow-up. Eric did only moderately well in hospital, testing the staffs' patience many times, and exceeding limits placed on his behavior, but showing some ability to interrupt his abnormal patterns of emotions and behavior through consistent limits and support by staff. Over a 2-month treatment period, his mood improved and stabilized on a combination of a tricyclic antidepressant and lithium. Psychotherapy was reality oriented and directed toward recognizing how his mood swings produced a precondition for binge eating, as well as recognizing his "all or none" tendency in thinking and behavior. He was confronted about his irresponsible behavior.

On follow-up at 2 years, he described how he had briefly relapsed into drug abuse after discharge, but then voluntarily entered a drug treatment program. Since that time, he has finally moved out of his parents' home, found a small apartment, and began supporting himself for the first time in his life on his own salary. He attends both outpatient treatment and a support group and is currently drug and alcohol free.

Comment. Eric illustrates the frequent finding of actual, rather than perceived, pre-illness obesity in males who later develop bulimia nervosa. His personality vulnerabilities were limited to cluster B, rather than a mixture of B and C, as with Sam. Although problematic in his behaviors and causing distress to others, he had some capacity for warmth and responsible behavior. He made a fair adjustment on follow-up and also began to take responsibility for his life for the first time. Finally, he illustrates how males have a higher probability of involvement with drug and alcohol abuse, as well as antisocial behavior, and how the drug abuse can affect eating behavior, depending on the pharmacological class.

Summary and Conclusions

The twin spotlights of empirical, scientific studies, and broad clinical experience can be brought to focus on either the similarities or the dissimilarities between males and females with eating disorders. When individuals of either sex are very ill, suffering from emaciation or abnormal electrolytes and other medical complications, they appear very similar and require similar treatment. As patients become medically more healthy and the symptoms of their eating disorders are "deconstructed," then the individual life story behind each patient unfolds, and then, male vs. female differences in predisposition, onset, and course of illness become more apparent. Males are relatively less exposed to the general sociocultural pressure promoting thinness, but are exposed in their own specific subgroups where they may experience just as intense a behavioral reinforcement for slimming as do females more widely. In general, there is an apparent "dose-response" relationship between the amount of sociocultural reinforcement for thinness and the probability of developing an eating disorder, with the intermediate variable being dieting behavior. The process of transition from a normal behavior such as dieting, which is not by itself abnormal, into a fixed illness meeting DSM-III-R criteria for an eating disorder, has not yet been well defined.

Predisposing features that are more common in males than females include a history of actual pre-illness obesity rather than perceived obesity, participation in sports requiring weight loss, slimming to become more attractive to a homosexual lover, and reducing in weight to prevent the appearance of medical illnesses seen in older generations in the family.

Males, during the natural course of their eating disorder, especially if it involves substantial weight loss, do not experience the "on-off" phenomenon of loss of menstrual periods, such as females do, which may alert the family, the physician or the individual to an eating disorder problem. Rather, testosterone is gradually reduced, with a resulting stepwise decline in libido and sexual performance. Also, the predominant male concern during the development of the eating disorder, except in individuals who were substantially overweight prior to illness, is usually the development of sharper body definition rather than achievement of excessive weight loss, although that may occur. Many males who are normal or slightly heavier than average in weight prior to illness would elect to keep their body weight unchanged if in exchange they could specify an extraordinary degree of muscle definition. There is seldom in males the trauma comparable to that of young girls who are psychologically threatened by increasing from "double digit" pounds (below 100) to triple digits. There is a comparable phenomenon even in countries using kilograms or "stones," with a vague but widespread emotional investment by women especially in specific body weights and sizes of clothing. Men seem not to be as psychologically distressed by a mere change in the size of clothing as women experience, but are very sensitive to the perception of a "big gut" or abdominal obesity.

As the male with an eating disorder becomes physically well, as his Axis I affective disorders, anxiety states, and drug and alcohol abuse are treated and the Axis II personality disorders are ameliorated, only then can attention can be placed on understanding the individual male's experience with the culture or subculture he lives in, and its demand, or perceived demands on him regarding body shape, body weight, "macho" behavior, and emotional expression. It would not be inaccurate to say that the male stereotype of physical ideal is different from the female, but is equally impossible to achieve. It appears that males in our society especially those with an "external locus of control," perceive they are being asked to develop a fit, mesomorphic athletic appearance, indifference to danger, an egalitarian attitude toward women, along with a sense of humor, and the sensitivity of a poet.

Our follow-up studies at Johns Hopkins have shown, so far, only one death in males known to us. Also, in our completed formal study of 10 male patients, there was no significant difference in regard to body weight compared to a matched population weight, or in overall sexual and psychological outcome between males and females. Larger studies of follow-up on all of our male subjects are in progress. Based on our clinical experience, as well as other published studies, there is currently no reason for undue pessimism regarding the treatment or prognosis of males with eating disorders. They generally do not do well, however, or avoid treatment, if they are considered by staff to be indistinguishable from the more typical young female patients, if they feel they will be treated for a "female" disorder, or if they are considered in advance to have a homosexual orientation prior to evaluation.

The experience of males vs. females is clearly different from the time of conception throughout the rest of life. When dieting behavior is begun during critical phases of a male's development, especially during adolescence, and especially in individuals rendered more vulnerable from the presence of pre-illness obesity, other psychiatric disorders, or distressed family situations, then the probability of development of an eating disorder is increased, especially when he belongs to some societal subgroup emphasizing weight loss.

Additional long-term follow-ups are needed in males with eating disorders to affirm confidently our observation that they have as good a prognosis as do females, and to study how they differ from females in their pre-illness vulnerabilities, their onset of illness, the natural course of their illness, and their transition from normal dieting to a diagnosable disorder. We hypothesize that males who develop eating disorders may have more extremes of personality traits than females, especially obsessional or antisocial features, and will not have had as much sexual abuse as females, but will have experienced other issues of sexuality, especially identity, involved in the onset and maintenance of their eating disorders.

References

Andersen AE (1988) Anorexia nervosa and bulimia nervosa in males. In: Garfinkel PA, Garner DM (eds) Diagnostic issues in Anorexia nervosa and bulimia nervosa. Brunner/Mazel, New York, pp 166–207

Andersen AE (1990a) A proposed mechanism underlying eating disorders and other disorders of motivated behavior. In: Andersen AE (ed) Males with eating disorders. Brunner/Mazel, New York, pp 221–254

Andersen AE (1990b) Diagnosis and treatment of males with eating disorders. In: Andersen AE (ed) Males with eating disorders. Brunner/Mazel, New York, pp 133–162

Andersen AE, DiDomenico L (1991) Diet vs shape content of popular male and female magazines: a dose-response relationship to the incidence of eating disorders? Int J Eating Dis (in press)

Andersen AE, Mickalide AD (1983) Anorexia nervosa in the male: An underdiagnosed disorder. Psychosomatics 24(12):1066–1076

Andersen AE, Hedbloom JE, Hubbard FA (1983) A multidisciplinary team treatment for patients with anorexia nervosa and their families. Int J Eating Dis 4(4):181–192

Avery D, Winokur G (1976) Mortality in depressed patients treated with electro-convulsive therapy and antidepressants. Arch Gen Psychiatry 33:1029–1037

Beumont PJV (1970) Anorexia nervosa in male subjects. Psychother Psychosom 18:365–371

Bruch H (1973) Eating disorders. Obesity, anorexia nervosa and the person within. Basic Books, New York

Burns TP, Crisp AH (1984) Outcome of anorexia nervosa in males. Br J Psychiatry 145:319–325

Burns TP, Crisp AH (1985) Factors affecting prognosis in male anorexics. J Psychiatr Res 19(2/3):323–328

Casper RC, Pandey GN, Jaspan JB, Rubenstein AH (1988) Eating attitudes and glucose tolerance in anorexia nervosa patient at 8-year follow-up compared to control subjects. Psychiatry Res 25(3):283–299

Cadoret RJ (1982) Genotype-environment interaction in antisocial behaviour. Psychol Med 12:235–239

Crisp AH, Toms DA (1972) Primary anorexia nervosa or weight phobia in the male: report on 13 cases. Br Med J 1:334–338

Crisp AH, Burns T, Bhat AV (1986) Primary anorexia nervosa in the male and female: a comparison of clinical features and prognosis. Br J Med Psychol 59:123–132

Edwin D, Andersen AE, Rosell F (1988) Outcome prediction by MMPI in subtypes of anorexia nervosa. Psychosomatics 29(3):273–282

Engel K (1988) Prognostic factors in anorexia nervosa. Psychother Psychosom 49(3–4): 137–144

Fairburn CG, Beglin SJ (1990) Studies of the epidemiology of bulimia nervosa. Am J Psychiatry 147(4):401–408

Fichter MM, Daser C, Postpischil F (1985) Anorexic syndromes in the male. J Psychiatr Res 19(2/3):305–313

Goldberg J, True WR, Eisen SA, Henderson WG (1990) A twin study of the effects of the Vietnam war on posttraumatic stress disorder. JAMA 263(9):1227–1232

Gull WW (1874) Anorexia nervosa. Trans Clin Soc (Lond) 7:22–28

Hall A, Slim E, Hawker F, Salmond C (1984) Anorexia nervosa: long-term outcome in 50 female patients. Br J Psychiatry 145:407–411

Hall A, Delahunt JW, Ellis PM (1985) Anorexia nervosa in the male: clinical features and follow-up of nine patients. J Psychiatr Res 19(2/3):315–321

Halmi KA, Falk JR, Schwartz E (1981) Binge-eating and vomiting: a survey of a college population. Psychol Med 11:697–706

Herzog DB, Norman DK, Gordon C, Pepose M (1984) Sexual conflict and eating disorders in 27 males. Am J Psychiatry 141(8):989–990

Hsu LKG (1989) The gender gap in eating disorders: why are the eating disorders more common among women? Clin Psychol Rev 9:393–407

Hsu LKG, Sobkiewicz TA (1989) Bulimia nervosa: a four- to six-year follow-up study. Psychol Med 19:1035–1038

King MB, Mezey G (1987) Eating behaviour of male racing jockeys. Psychol Med 17:249–253

Margo JL (1987) Anorexia nervosa in males. A comparison with female patients. Br J Psychiatry 151:80–83

Mitchell JE, Goff G (1984) Bulimia in male patients. Psychosomatics 25(12):909–913

Mitchell JE, Pyle RL, Hatsukami DK, Goff G (1989) A 2–5 year follow-up study of patients treated for bulimia. Int J Eating Dis 8(2):157–165

Morgan HG, Russell GFM (1975) Value of family background and clinical features as predictors of long-term outcome in anorexia nervosa: four-year follow-up study of 41 patients. Psychol Med 5:355–371

Morton R (1694) Phthisiologica: or a treatise of consumptions. Smith and Walford, London

Needleman HL, Schell MA, Bellinger D, Leviton A, Allred EN (1990) The long-term effects of exposure to low doses of lead in childhood An 11-year follow-up report. N Engl J Med 322(2):83–88

Nielsen S (1990) The epidemiology of anorexia nervosa in Denmark from 1973 to 1987: a nationwide register study of psychiatry admission. Acta Psychiatr Scand 81:507–514

Rastam M, Gillberg C, Garton M (1989) Anorexia nervosa in a Swedish Urban Region. A population-based study. Br J Psychiatry 155:642–646

Russell G (1979) Bulimia nervosa: an ominous variant of anorexia nervosa. Psychol Med 9:429–448

Scott DW (1986) Anorexia nervosa in the male: a review of clinical, epidemiological and biological findings. Inter J Eating Dis 5(5):799–819

Sohlberg S, Norring C (1989) Ego functioning predicts first-year status in adults with anorexia nervosa and bulimia nervosa. Acta Psychiatr Scand 80:325–333

Sours JA (1980) Adolescence and the anorexia nervosa syndrome. Jason Aronson, New York, pp 336–338

Steen SN, Oppliger RA, Brownell KD (1988) Metabolic effects of repeated weight loss and regain in adolescent wrestlers. JAMA 260(1):47–50

Strober M, Humphrey LL (1987) Familial contributions to the etiology and course of anorexia nervosa and bulimia. J Consult Clin Psychol 55(5):654–659

Stunkard AJ (1959) Eating patterns and obesity. Psychiatr Q 33:284–292

The Multiple Risk Factor Intervention Trial Research Group (1990) Mortality rates after 10.5 years for participants in the multiple risk factor intervention trial. JAMA 263(13):1795–1801

Theander S (1985) Outcome and prognosis in anorexia nervosa and bulimia: some result of previous investigations, compared with those of a Swedish long-term study. J Psychiatr Res 19(2/3):493–508

Tolstrup K, Brinch M, Isager T, Nielsen S, Mystrup J, Severin B, Olesen MS (1985) Long-term outcome of 151 cases of anorexia nervosa: the Copenhagen anorexia nervosa follow-up study. Acta Psychiat Scand 71:4:380–387

Toner BB, Garfinkel PE, Garner DM (1986) Long-term follow-up of anorexia nervosa. Psychosom Med 48(7):520–529

Vandereycken W, Van den Broucke S (1984) Anorexia nervosa in males. A comparative study of 107 cases reported in the literature (1970 to 1980). Acta Psychiatr Scand 70:447–454

Yates A, Leehey K, Shisslak CM (1983) Running – an analogue of anorexia? N Engl J Med 308(5):251–255

Part II
Follow-up Studies of Anorexia and Bulimia Nervosa

The Heidelberg-Mannheim Study: Long-Term Follow-up of Anorexia Nervosa Patients at the University Medical Center – Background and Preliminary Results

H.-C. Deter, W. Herzog, and E. Petzold

Background

More than two decades ago, one of us (E.P.) at the Heidelberg University Medical Center began to develop an interest in the clinical picture of anorexia nervosa. Soon after, in addition to methods of individual and group therapy, family assessment and family therapy were applied and developed (Petzold 1979). The specific organizational structure of the Psychosomatic Department at the Heidelberg University Hospital in the early 1970s made it possible to develop an examination and treatment concept which would meet the manifold clinical needs of a hospital geared to regional care. As such, this approach differed greatly in its structure from the many established psychosomatic facilities as well as from the "clinics for eating disorders" whose number has been steadily increasing in Germany in recent years.

The patients were referred to the Psychosomatic Outpatient Clinic by physicians in private practice and other hospitals as well as from other internal wards of the university hospital through the psychosomatic consultation-liaison service. The therapeutic procedure differed according to the severity of the condition and motivation for treatment of each patient (Fig. 1).

Patients who were less seriously ill (weight, severity of psychic disturbance) or who did not wish to undergo special psychosomatic inpatient therapy received individual and/or family outpatient therapy which, if necessary, was also performed within the framework of the psychosomatic consulting service.

For patients whose illness was serious enough to render inpatient treatment absolutely necessary, the following differential indications were established:

- If the somatic disorder was predominant (weight, physical complications), the patient was treated in an internal medicine unit, including behavioral therapy and elements of a strict medical regimen, also tube feeding (Frahm 1965). In addition, individual and family therapy sessions were conducted by a well-coordinated team under the supervision of an experienced liaison-psychosomatic specialist.

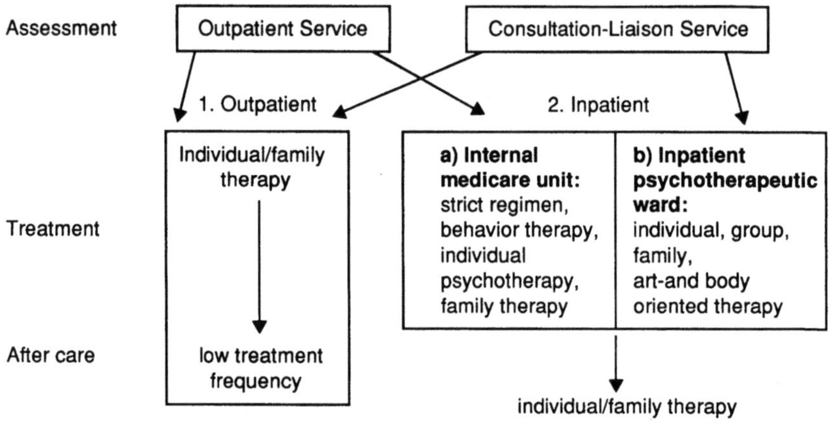

Fig. 1. Psychosomatic care at the Heidelberg University Medical Center 1970–1980

- Subsequently (or initially in cases where the psychic disturbance predominated and the somatic impairment was less serious), a special inpatient psychotherapy was carried out within the framework of a psychotherapeutic community for 2 or 3 months as a rule (sometimes up to 6 months); it included individual, group, family, art and a body-oriented therapy.

At the end of inpatient treatment, patients were offered the opportunity to receive further outpatient aftercare with individual and family therapy.

After this concept of psychosomatic care had been instituted, a parallel program of clinical research was initiated to examine in particular the therapeutic effects in the short term (Deter and Petzold 1978; Deter 1981; Deter et al. 1989) and the inclusion of a family treatment approach (Petzold 1979). At the same time, the documentation of this program formed the point of departure for our follow-up study.

The treatment procedure was systematically assessed in 1980/1981. The results were satisfactory, and we therefore continued the procedure initiated for treating anorexia nervosa patients. The follow-up time, however, was too short for some patients (1–2 years). We found, for example, a highly significant correlation between the improvement rate and the follow-up duration.

In a third phase, we decided to review all the material documented up to that point and to perform a third investigation. The aim here was to examine both clinical and methodological aspects of the clinical course of anorexia nervosa patients (Vandereycken and Meermann 1984). The prerequisite was a minimal follow-up of 9 years for each patient and an average of 12 years for the total sample. It was not a prospectively planned study in the true sense of the word, and the lack of a control group rendered it impossible to check the results pertaining to the treatment setting. Nevertheless, we hoped to gain important information about the "natural course of the illness" in a regionally mixed, successive population seeking treatment at a medical center with an active psychosomatic department. We were aware that we could only cover one sector from the pool of anorexia nervosa patients of a region, namely, the group with the most somatic problems and with a relatively high age. Other patients, who did not come to us, were admitted to the psychiatric hospital, the clinic for child and adolescent psychiatry, or a special psychotherapeutic clinic in Heidelberg.

Our investigation has addressed four questions:

1. How does the clinical picture of anorexia nervosa evolve in a period of at least 9 years following psychosomatic treatment on a psychic, a social and, particularly, on a somatic level?
2. What factors are prognostically important at the beginning of treatment for long-term outcome?
3. What influence is exerted on the outcome by the social situation, particularly by the social service system and social support?
4. Is it possible on the basis of relapses, as indicated by renewed hospitalization and psychotherapy, to identify those persons who are particularly predisposed for a clinical course of chronic recurrence?

Since the investigation has not yet been completed, we would like to present some of the results on the first two questions. Elsewhere in this book, Herzog et al. will elaborate on the somatic aspects of questions 1 and 2, and Manz et al. will deal with aspects of the third question.

The Study

Design

All 103 patients (91 female, 12 male) with anorexia nervosa according to the DSM-III-R diagnosis were included in the study. They underwent consecutive treatment between January 1, 1971, and October 31, 1980, at the Heidelberg University Medical Center either as inpatients or as outpatients at the Psychosomatic Outpatient Clinic of this department. Diagnostic evaluation according to DSM-III-R was done retrospectively. In 1980–1981, Deter and coworkers (1983) had performed a preliminary follow-up after an average of 3.6 years, applying the diagnostic criteria of anorexia nervosa by Feighner et al. (1972). Four female patients in this sample did not fulfill the DSM-III-R criteria and had to be excluded from the study. They were replaced by four female patients with anorexia nervosa (according to DSM-III-R) who underwent treatment immediately after the end of the originally planned assessment period at the Medical

Clinic. It may therefore be viewed as a representative random sample of the treatment-seeking clientele of a university medical department. Selection effects were present with regard to the severity of the disease, since patients who did not respond to previous therapy were usually referred to our center, and the sample also included an altogether large number of patients with acute or severe somatic disturbances. Eighty-nine of the patients received inpatient treatment and 14 underwent outpatient therapy (see Fig. 1). Table 1 gives the characteristics of the random sample.

At the first follow-up examination (1980–1981), by which time 12 patients had already died, we were able to assess 51 patients personally, 9 by questionnaire, and 3 by information obtained from third parties. Thus we had a follow-up rate of 69.2%.

Method

The former patients were contacted in writing by their previous therapists (H.C.D. and E.P.) and invited to the clinic for a follow-up examination. If the patient could not be reached by telephone, a visit at home was

Table 1. Basic data of the random sample of anorexia nervosa patients according to DSM-III-R critera

Patients (n = 103)	
Sex distribution	91 female, 12 male
Age at onset of illness	18.0 ± 5.0 years (12–38 years)
Age at first admission	20.8 ± 6.1 years (12–48 years)
Weight at first admission	65.7 ± 9.9% ABW (45.5%–89%)
Additional severe somatic disorders	9 (8.7%)
Number of deaths during first admission to the university clinic	5 (4.9%)
Family status	Single 90 (87.4%)
	Married 11 (10.7%)
	Separated or divorced 2 (1.9%)
Social status	Lower class 9.1%
	Upper lower class 34.3%
	Lower middle class 24.2%
	Middle middle class 12.1%
	Upper middle class 19.2%
	Upper class 1.0%
Educational status	10th grade not completed 5.1%
	10th grade completed 41.8%
	Secondary school completed 24.5%
	Diploma qualifying for university entrance 28.6%
Average duration of follow-up	12.7 ± 3.7 (9–19 years)

ABW, average body weight.

Table 2. Rate of participation in long-term follow-up study

Type of follow-up	No. of patients	(%)
Clinical examination	68	(78.2)
Extensive interview at home	9	(10.4)
Short interview at home	3	(3.4)
Direct interview (hospital and at home)	80	(92.0)
Telephone interview	5	(5.8)
Extensive interview with the mother	1	(1.1)
No information (failure-to-trace)	1	(1.1)
Total	87	(100.0)

suggested in order to arrange an appointment for an examination at the clinic.

The following assessments were performed:

- Medical examination (case history, physical examination, blood sample)
- Structured interview (SIANEX; Fichter and Meermann 1983)
- Interview with former therapists
- Questionnaires: Eating Attitudes Test (Garfinkel and Garner 1982), Eating Disorder Inventory (Garner et al. 1983), Psychosomatic State Questionnaire (Hehl 1981), Family Climate Scales (Moos 1974), Network Questionnaire (Manz and Schepank 1989)
- Bone-density examination

A direct-interview rate of 92% and a failure-to-trace rate of 1.1% was obtained in 87 of the patients still alive (78 women, 9 men). The participation rate is summarized in Table 2.

Results

In the following, several important aspects of the results that are illustrated particularly well by our central test instrument, the Anorexia Nervosa Symptom Score (ANSS; see Deter, this volume), are highlighted.

Clinical Course (Question 1)

Based on the three description levels of the ANSS and the three examination times, a variance analysis was carried out at three measuring times according to a 3 × 3 design. This revealed a general, main effect through the passage of time and secondary effects through differences between the ANSS values at the somatic and psychic levels as well as at the social level.

This can be ascribed to the effect that the values of the ANSS showed a mean continuous and significant improvement in the course of time. In

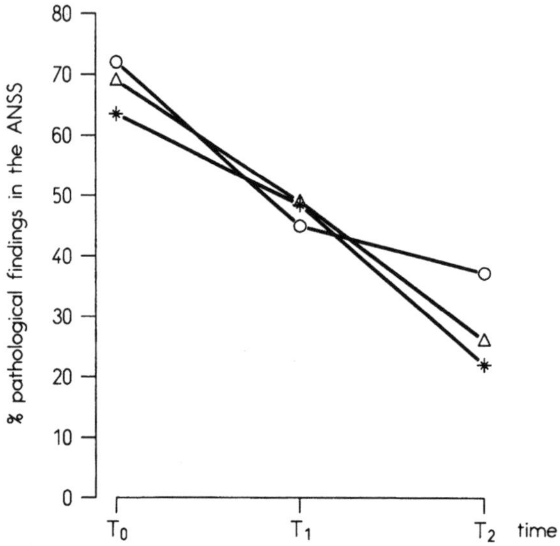

Fig. 2. Course of pathological findings in the anorexia nervosa symptom score (ANSS, analysis of variance). Main effect: T_0 to T_1 to T_2. Interaction: (T_0) psychic (*circles*) to somatic (*asteriks*); (T_2) psychic to somatic; (T_2) psychic to social (*triangles*)

part, however, there were also significant differences between the psychic and somatic factors:

The psychic factors are more strongly pronounced at the time of presentation (T_0) and still persist to a more marked degree than the physical factors 12 years later (T_2). It now becomes evident that the social disturbances also normalize to a significantly greater degree than the psychic ones (pathological psychic status) in over a third of the values on the average). With the ANSS it is also possible to obtain a much more differentiated description of the alterations of the parameters individually recorded at three different examinations (Table 3).

In the social field, the family items "no family of their own" and "above average emotional dependence on the family of origin" remained pathological in almost half of the patients; one third of all patients had no sexual relations and were not socially integrated. In contrast, professional activity and contact to the family of origin – in terms of mutual understanding – had normalized at a much higher percentage.

The factor medical treatment seeking, which is calculated in hospital days, is rated only in 2.5% of the patients above the relatively high limit of 10 hospital days per year. In contrast, almost half of the former patients were absent for more than 10 days from work due to illness in the previous year.

The psychic symptoms still persisted 12 years later in more than 60% of the patients (the low values at T_1 might be due to the fact that only

Table 3. Changes in the psychic state according to the Anorexia Nervosa Symptom Score (ANSS) – percentage of pathological findings ($n = 87$)

	T_0 (at referral to the clinic; %)	T_1 (3.6 years later; %)	T_2 (12 years later; %)
Psychic symptoms	90.0	53.5	64.0
Sexual problems	92.0	54.8	45.8
Suffering pressure	84.4	58.1	55.3
Self-destructive tendencies	60.7	34.9	13.3
Addictive tendencies	56.3	56.1	47.3
Childishness	78.4	51.2	49.3
Low self-esteem	80.0	42.9	21.3
Body image disturbance	84.5	45.2	38.0
No curative compensation	54.0	34.1	32.9
Psychic state (total)	75.7	47.8	41.2
Social state (total)	66.5	51.5	29.7
Somatic state (total)	62.6	46.7	23.5

51 patients who may have been in a better state of health were included in the analysis because of a lower follow-up rate). Also, the psychic level of suffering and an extraordinary intellectual and emotional infantility still persisted in more than half of the patients, whereas self-destructive tendencies as well as a lack of self-esteem had clearly decreased. Finally, in up to two thirds of the former anorexia nervosa patients psychic disorders could still be detected 12 years later with the ANSS. This percentage was markedly lower for physical symptoms (see also Deter, this volume).

The preliminary analysis of 50 former patients from self-assessment questionnaires such as the Eating Attitudes Test (EAT), Eating Disorder Inventory (EDI) and the standardized interviews in connection with the anorexia nervosa expert assessment (SIANEX) confirm similar tendencies, although they do not measure changes in this case but differentiate the values of the follow-up group (T_2) and compare them with other populations. In the EAT as well as in the EDI, the three outcome groups differentiated by Morgan and Russell's (1975) criteria exhibited completely different curve profiles. On the subscales diet behavior, bulimia, thoughts about food, and oral control of the EAT, the patients with a poor treatment result are shown to be in the highest range, the patients with a good treatment result in the lowest one. The intermediate group is in the middle of the scale (Fig. 3). The EDI shows similar findings (Fig. 4).

The factor analysis of the SIANEX we performed yielded four scales: anorectic symptoms, psychic symptoms, partnership/sexuality, bulimia. The differentiation in the various outcome groups yielded different values for the individual factors in this case, too (Fig. 5).

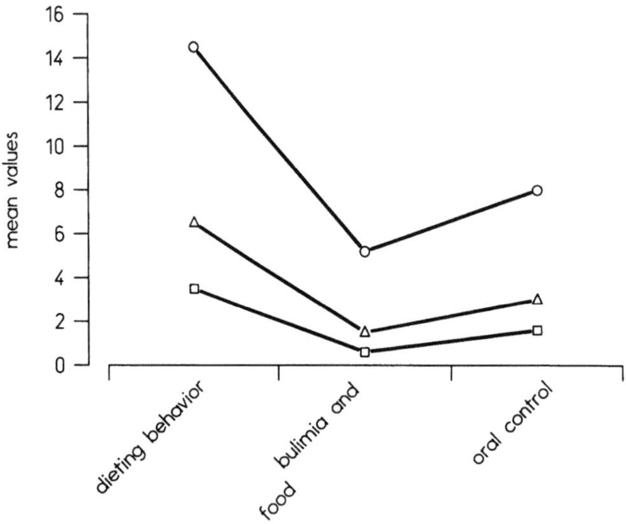

Fig. 3. Eating Attitudes Test (EAT) scales. Comparison of outcome groups (mean values of scales). *Circles*, poor outcome; *triangles*, intermediate outcome; *squares*, good outcome

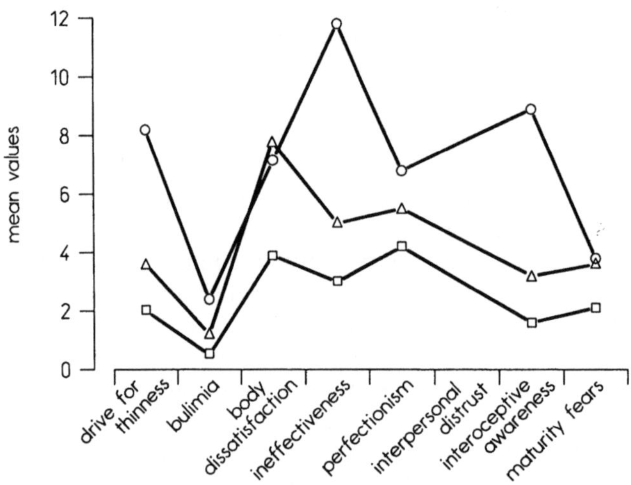

Fig. 4. Eating Disorder Inventory (EDI) scales. Comparison of outcome groups (mean values of scales). *Circles*, poor outcome; *triangles*, intermediate outcome; *squares*, good outcome

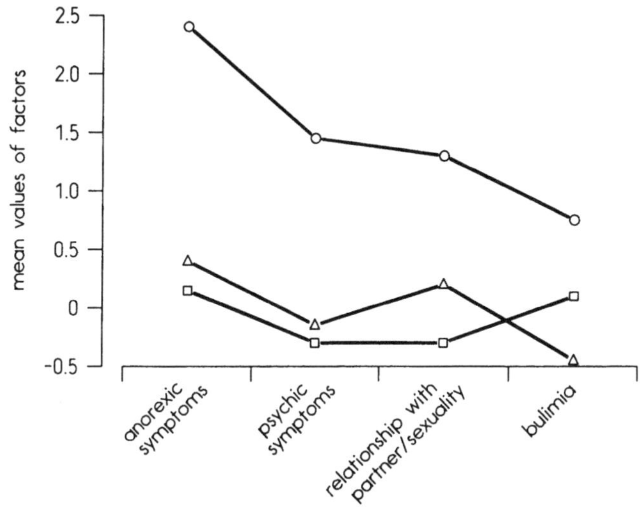

Fig. 5. Standardized interviews in connection with the anorexia nervosa expert assessment (SIANEX) values of factors. Comparison of outcome groups (mean values of factors). *Circles*, poor outcome; *triangles*, intermediate outcome; *squares*, good outcome

Psychiatric comorbidity, 26 patients showed psychopathological disorders that required treatment. Eight percent of the total group suffered from major depression, 12% from phobic disorder, 14% from substance abuse, and 22% from personality disorder. Three patients suffered from acute (two paranoid hallucinatory, one manic), one from chronic psychosis.

Table 4 shows the diagnoses for the total group and those of the Morgan-Russell general outcome categories. Here, outcome is likewise only assessed according to weight and menstrual status, but within the good outcome group, those patients who are considered cured (with regard to their eating symptoms) are differentiated as well as those who, although they may have normal menstruation and weight between 85% and 115% of the average weight standardized according to age and height, continue to have eating disorders (e.g., vomiting, abuse of laxatives, bulimia). The psychopathological diagnoses were established by the investigator according to the standardized and the free interview. These diagnoses are regarded as comorbidity: the immediate eating symptoms were not taken into consideration in this case.

Psychopathological disorders requiring treatment are found in 11.5% of the patients in the recovered group, 47.8% in the intermediate outcome group, and 100% in the poor outcome group. Particularly striking in the total group were substance abuse with 14.1% and personality disorders

Table 4. Psychiatric comorbidity

Outcome		Psychopathology					Psychiatric diagnoses (%)
		Major depression (%)	Phobic disorder (%)	Substance abuse (%)	Personality disorder (%)	Psychosis (%)	
Recovered	(n = 52)	0	3.8	5.8	1.9	1.9	11.5
Intermediate	(n = 23)	0	13	13	26.1	13	47.8
Poor	(n = 10)	60	50	50	90	0	100
Total	(n = 85)	8.2	11.8	14.1	22.4	4.7	31.8

with 22.4%. There were three cases of acute psychoses at the time of the investigation. These female patients were unable to work, and one of them lived in a nursing home. Another three female patients had suffered psychotic episodes during the follow-up time. In the meantime two of these patients were working in academic professions; the third chronic ill patient could only do unskilled work.

In summary of the total clinical evaluation after a mean evaluation time of 12 years, 16 patients had died, 10 continue to suffer from chronic anorexia nervosa, 23 have improved but are still partially anorectic, 13 are free from the physical symptoms of anorexia nervosa but continue to have some eating-related disturbances, and 39 are regarded as being cured from eating disorders. Fourteen percent of the patients in the total group suffer from bulimia nervosa and 5% have become obese.

Predictors of Outcome (Question 2)

The answer to this question will be confined mainly to psychosocial predictors (for further results, see Herzog, this volume). The predictive value will be examined for the severity of pathological findings in the ANSS at the present follow-up assessment (T_2). In a factor analysis of variables, which were regarded as important predictors on the basis of the literature, and in the subsequent multiple regression analysis of the factors, it was shown that a factor representing the duration of illness at the clinical presentation had the highest predictive value.

Discussion

On the basis of the assessment criteria of the ANSS, we were able to perform a detailed description of the course of disease of 103 anorexia nervosa patients who underwent treatment in our hospital from 1971–1980. Separate statements about the changes in the physical, psychic, and social areas were possible on the basis of a multilevel test battery. The sample showed significant improvements in the course of time, but on the whole, pathological disturbances were observed for a markedly longer period time in the psychic area than in others. The evaluation of psychiatric comorbidity underscores these results. The classification of patients (according to Morgan and Russell 1975) yielded diagnoses of psychopathological conditions in 31.8% of the cases, whereby all patients with a poor treatment outcome had been diagnosed as having a psychic disturbance, as had 47.8% with an "intermediate" treatment outcome and 11.5% with a "good" assessment. Even in the group who had been classified "good" (according to the Eckert criteria), 2.6% of the patients still had diagnoses of psychiatric disorders. Addicts and patients with personality disorders were found most frequently in our sample.

Finally we were able to demonstrate great differences among the individual patients of the follow-up group with regard to physical factors, eating attitudes (EAT), tendency toward bulimia as well as various personality factors characteristic for eating disorders (EDI, SIANEX). This seems to support former experiences that the physical classification criteria correspond largely with the factors of the EAT (Russell 1977). However, the self-evaluation methods such as EDI and EAT that are centered on eating disorders do not appear to be sufficient to permit specific statements about the extent of the psychopathological condition. Only expert ratings assessing the extent of psychopathological factors will yield valid findings. The improvement rates in our patients are comparable to those seen in other long-term studies (Theander 1985; Remschmidt et al. 1988; Eckert 1990; Russell, this volume) using the Russell-Morgan scale for evaluation. Individual findings are discussed in other chapters of this volume (see Herzog et al., this volume); however, the ANSS clearly shows that individual parameters from one area may be quite different in the course of follow-up. Although we have not yet concluded our investigations, it can be said that the necessarily comprehensive clinical assessment should be supplemented by a differentiated examination of the long-term course of all parameters investigated.

The question of prognostic factors is discussed in detail by Russell in this volume. In our study, which allows only a preliminary statement because it has not yet been concluded, only the duration of the disease turned out to be a strong predictor (additional somatic predictors are discussed in the chapter by Herzog et al., this volume).

These results correspond only partially with the predictors found in the London studies, but they are in accordance with the results obtained by Strober (1990), who found that only the duration of anorexia was an important prognostic indicator. These differences might be explained by the fact that the patients studied by the authors differed with respect to their attitude towards seeking treatment, which in turn requires a different weighing of the various factors. Further studies are therefore required to elucidate these differences in the findings.

So far we can only present a preliminary analysis of our findings. Although it is still too early to draw final conclusions, it is clear that long-term course studies of anorexia nervosa patients require extreme precision in the analysis of study parameters, including a description of the study population and especially its selection effect as determined by the type of attitude towards seeking treatment, the extent of psychopathological and somatic comorbidity, and the differentiation of the course of disease parameters at different levels. Our investigation is aimed at offering a basis for discussing and giving partial answers to these questions. Irrespective of the relevance of the results presented here, it can already be said that the results of our long-term investigation have influenced considerably the clinical treatment of anorexia nervosa patients in our clinic and have led to decisive changes in the self-concept of our staff and the therapeutic aims and expectations in the treatment of anorexia nervosa patients newly admitted to the clinic.

References

Deter HC (1981) Zur Methodik von katamnestischen Untersuchungen bei psychosomatischen Patienten am Beispiel einer Gruppe von 31 Anorexie-Patienten. Psychother Med Psychol 31:48–52

Deter HC, Petzold E (1979) Klinische Befunde und Verlaufsuntersuchungen bei Patienten mit Anorexia nervosa. Therapiewoche 29:6320–6331

Deter HC, Petzold E, Hengst R, Breiden O, Lanzinger G (1983) Katamnestische Ergebnisse einer klinisch-psychosomatischen Behandlung von 103 Patienten mit Anorexia nervosa. Innere Med 10:3–12

Deter HC, Petzold E, Hehl FJ (1989) Differenzierung der Langzeitwirkungen einer stationären psychosomatischen Therapie von Anorexia nervosa-Patienten. Z Psychosom Med Psychoanal 35:68–91

Eckert E (1990) Long-term follow-up study. 4th International Conference on Eating Disorders. 28 April 1990, New York

Feighner JP, Robins E, Guze SB, Woodruff R, Winokur G, Munoz R (1972) Diagnostic criteria for use in psychiatric research. Arch Gen Psychiat 26:57–63

Fichter MM, Meermann R (1983) Zur Psychometrie der Anorexia nervosa. In: Meermann R (ed) Anorexia nervosa. Enke, Stuttgart

Frahm H (1965) Ergebnisse einer systematisch durchgeführten somatisch orientierten Behandlungsform bei Kranken mit Anorexia nervosa. In: Meyer JE, Feldmann H (eds) Anorexia nervosa. Thieme, Stuttgart

Garner DM, Garfinkel PE (1982) Anorexia nervosa. A multidimensional perspective. Brumer/Mazel, New York

Garner D, Olmsted M, Polivy I (1983) Development and validation of a multidimensional eating disorder inventory for anorexia nervosa and bulimia. Int J Eating Dis 2:14–34

Hehl F, Wirsching M (1981) Der psychosomatische Einstellings-Fragebogen. Hogrefe, Göttingen

Manz R, Schepank H (1989) Soziale Unterstützung, belastende Lebensereignisse und psychogene Erkrankung in einer epidemiologischen Stichprobe. In: Angermeyer MC, Klusmann D (eds) Soziales Netzwerk. Ein neues Konzept für die Psychiatrie? Springer, Berlin Heidelberg New York

Moos R (1974) The Social Climate Scales. Consulting Psychologists, Palo Alto

Morgan HC, Russell GM (1975) Value of family background and clinical features as predictors of long-term outcome in anorexia nervosa: 4 year follow-up study of 41 patients. Psychol Med 5:355–371

Petzold E (1979) Familienkonfrontationstherapie bei Patienten mit Anorexia nervosa. Vandenhoeck and Ruprecht, Göttingen

Remschmidt H, Wienand F, Wewecker C (1988) Der Langzeitverlauf der Anorexia nervosa. Monatsschr Kinderheilkd 136:726–791

Russell G (1977) General management of anorexia nervosa and difficulties in assessing the efficacy of treatment. In: Vigersky F (ed) Anorexia nervosa. Raven, New York, pp 270–289

Strober M (1990) Long term follow up of eating disorders. 4th International Conference on Eating Disorders. 28 April 1990, New York

Theander S (1985) Outcome and prognosis in anorexia and bulimia: some results of previous investigations compared with those of a Swedish long term study. I Psychiat Res 19:493–508

Vandereycken W, Meermann R (1984) Anorexia nervosa. de Gruyter, New York

Somatic Findings at 12-Year Follow-up of 103 Anorexia Nervosa Patients: Results of the Heidelberg-Mannheim Follow-up*

W. Herzog, H.C. Deter, D. Schellberg, S. Seilkopf, E. Sarembe, F. Kröger, H. Minne, H. Mayer, and E. Petzold

Introduction

In the past 2 decades results from a large number of long-term follow-up studies have been published (Steinhausen et al. 1990; Herzog et al., this volume). The significance of somatic investigation is exemplified by mortality rates of 18% (Theander 1985), a third of which is due to suicide and about two thirds directly and indirectly related to somatic complications of anorexia (Andersen et al. 1990). Besides psychiatric comorbidity, somatic complications such as susceptibility to infection, renal failure, and osteoporosis also determine the quality of life of these patients during the course of the illness. Whereas somatic considerations have found their way into prospective short-term follow-ups, this is not true for long-term assessment. There are a great number of individual case studies (regarding osteoporosis see, for example, Brotman and Stern 1985; Rigotti et al. 1984; Olmos et al. 1990; Beumont, this volume); however, systematic consideration of somatic aspects in the investigation of large samples and long course of anorexia is a seldom occurrence. The same applies to outcome studies from departments of internal medicine (for example, Köhle and Mall 1983; Bassoe and Eskeland 1982). Fichter and Pirke (1982) and Meyer et al. (1986a,b) presented the results of laboratory investigations from follow-up of parts of larger patient cohorts. An integral part of the Heidelberg-Mannheim Follow-up consisted in describing the course of anorexia continually at numerous levels. Besides the psychologic and social findings, the data from somatic examination were collected directly and as completely as possible.

A complete somatic examination is relevant for the study as a whole for a number of reasons. First it provides criteria and control variables for the assessment of outcome. Given the known tendency of patients with eating disorders to deny their condition, weighing them is more reliable than receiving this information over the telephone; normal menstruation is doubtful if levels of follicle stimulating hormone (FSH) or

*This study was supported in part by grants from the German Ministery of Technology and Research.

luteinizing hormone (LH) are extremely low (if no ovulation inhibitors are being taken); and serum levels of potassium provide a control variable for information regarding vomiting and use of laxatives. Hence, data regarding symptoms are more reliable if they are derived through physical examination or laboratory tests.

Second, the somatic findings from a large coherent sample can be systematically investigated and reported.

Third, from a clinical point of view the question of somatic comorbidity, that is, of diagnosis, which is raised at follow-up is interesting: To what extent do other physical illnesses occur? How great an effect do these illnesses have?

Fourth, it stands to be investigated whether somatic variables can serve as predictors of long-term outcome.

In this chapter we present a picture of the somatic aspects of anorexia in the long term, describing the condition from various perspectives – for example, individual findings, somatic global scores, non-psychiatric diagnoses, mortality rates and causes of death, and somatic predictors of course.

Subjects and Methods

In the preceding chapter Deter et al. outlined the sample and participation rates of the Heidelberg-Mannheim Follow-up Study in detail: Accordingly, 103 patients (12 men) with anorexia nervosa (DSM-III-R criteria) were

Table 1. Participation in Heidelberg-Mannheim Follow-up (%)

Kind of participation	Source of somatic data		
	Own findings	Hospital records[a]	Findings from GP[a]
Clinic examination (n = 68)	97	1.5	–
Visit with detailed interview (n = 9)	22.2	33.3	11.1
Visit with short inverview (n = 3)	–		67
Telephone interview (n = 5)	–	–	20
Interview with relatives (n = 1)	–	–	–
No information (n = 1)	–	–	–
Total sample[b] (n = 87)	78.1	4.6	4.6

GP, general practitioner.
[a] Data from the previous 6 months before catamnesis.
[b] Sixteen former patients had died.

examined at on average 12.7 years after key admission, the sample likely being a negative selection of patients with a higher rate of physical illnesses.

In Table 1 the results of the follow-up are presented according to kind of participation: whether physical examination and laboratory tests were carried out personally (column 1); if it was not possible to carry out these tests personally, the results of such tests were taken from examinations which had been carried out during hospitalization (column 2) or by the general practitioner (column 3) if such tests had been carried out within 6 months of catamnesis. In total, 78.1% of the former patients who were alive were examined by us and for 9.2% of patients test results from hospital records or the general practitioner were available. For 12.6% of the subjects still living no examination could be carried out nor could recent test results from the general practitioner or hospital be requested.

In the physical examination were included:

Table 2. Laboratory tests

Electrolytes
Sodium, potassium, chloride, calcium, iron, phosphate
Renal retention
Urea, creatinine, uric acid
Serum lipids
Cholesterol, triglycerides
Blood glucose
Pancreatic enzymes
Lipase, amylase, pancreatic amylase
Hepatobiliary function
LDH, GOT, GPT, AP, γ-GT, bilirubin
Immunoglobulins
IgA, IgM, IgG
Serum protein and infection parameters
Total protein, albumin, blood sedimentation rate, C-reactive protein, transferrin, ferritin
Blood count and differential blood count
Leukocytes, erythrocytes, hemoglobin, HK, differential hemogram, thrombocytes, reticulocytes
Hormones
T3, T4, TSH, FSH, LH, basal cortisol, STH, estrogen, prolactin, parathyroid hormone, testosterone, LH-RH test[a], TRH test[a]
Osteologic parameters
Osteocalcin[a], vitamin D

[a] These tests were only carried out in part of the sample.
LDH, lactate dehydrogenase; GOT, glutamine-oxaloacetic transaminase; GPT, glutamic-pyruvic transaminase; AP, alkaline phosphatase; γ-GT, glutamyl traspeptidase; HK, hematocrit; FSH, follicle-stimulating hormone; LH, luteinizing hormone; LH-RH, luteinizing hormone releasing hormone; TRH, thyrotropin-releasing hormone.

1. Taking a thorough patient history (including eating symptoms and course of weight and menstruation status as well as all information regarding hospital admissions since first admission and psychotherapy)
2. Physical examination
3. Laboratory tests (see Table 2)
4. Standardized orthostatic test
5. Dental examination
6. Bone density measurement
7. Global ratings (Anorexia Nervosa Symptom Score, ANSS; Morgan-Russell Outcome Score)

Results

Mortality

In the description of the sample it was already pointed out that during the first hospital stay five (4.9%) patients died. By second follow-up 16 patients had already died (15.5%). The causes of death are outlined in Table 3.

Whereas in our sample the mortality for anorexia only was 11.7%, mortality in the subgroup of patients ($n = 9$) with anorexia and severe somatic illness independent of the anorexia amounted to 55.5%.: one male and one female with Crohn's disease, one diabetic patients, one patient with acute intermittent porphyria and another with anorexia subsequent to tuberculosis had died in the meantime. Necropsy was performed in 11 of the 16 patients.

The patients from the sample who died were older than those still alive, and the duration of illness was longer and the weight lower at

Table 3. Causes of death ($n = 16$)

Causes of death in anorexia nervosa	Number
Suicide	3
Ileus	2
Schock in extreme cachexia	2
Pneumonia	2
Sepsis	1
Terminal renal insufficiency	1
Causes of death in patients with severe somatic illness and concomitant anorectic reaction	
Lung embolism	2
Suicide	1
Sepsis	1
Fat embolism	1

Necropsy was performed in 11 patients.

first admission. The total ANSS at first examination (T0) was higher in the patients who died than in those still alive. Interestingly, there was no significant difference in the subscores from the somatic ANSS (ANSSs), the psychic ANSS (ANSSp) and the social ANSS (ANSSz) at first admission.

Weight, Menstruation, and Eating Symptoms

Morgan and Russell developed a system of classifying outcome based on the physiologic parameters of weight and menstruation, which can be adequately correlated with more complex psychologic and social outcome categories (Morgan and Hayward 1988). The outcome results of the Heidelberg-Mannheim Follow-up in terms of the Morgan-Russell General Outcome Categories are illustrated in Fig. 1 (see also Morgan, this volume).

The outcome results indicated a relatively high mortality rate compared to other long-term follow-ups. This can be explained, however, by the high number of patients who died during first hospitalization after having been admitted in critical physical condition. The interrelationship of good, intermediate, and poor outcome in the Heidelberg-Mannheim Follow-up, in which the data were derived from the clinic of internal medicine, is not essentially different from that from adult psychiatric clinics.

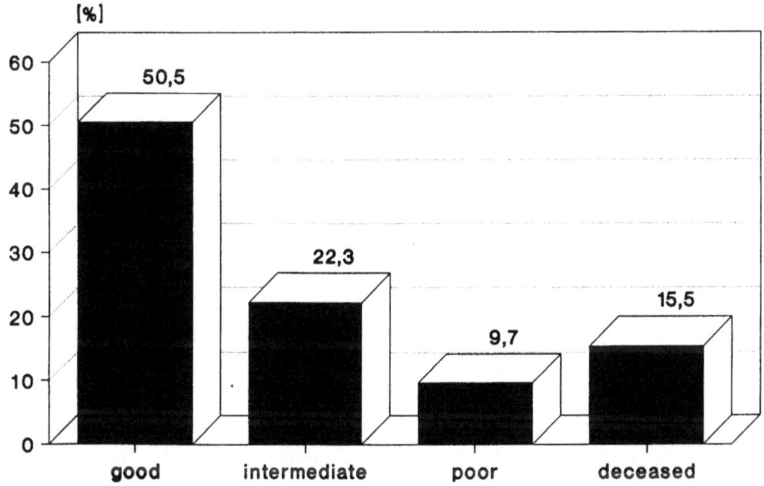

Fig. 1. Morgan-Russell Outcome Scores in the Heidelberg-Mannheim Follow-up

Vomiting, laxative abuse, and bulimia were found at first admission and at second follow-up with the following frequency:

- At first admission about half of the patients were vomiting, and at follow-up about one fourth. At second follow-up six subjects indicated they vomited daily, three twice a week, two once a week, and nine once a month.
- At follow-up about one tenth of the examined patients said they used laxatives: two subjects twice a week, three subjects once a week, two subjects once a month, and two subjects daily.
- After careful retrospective analysis of the patient records there was likely a 40% rate of bulimia in the patients at first admission. This percentage had decreased at second follow-up (T2) to about 15%. The proportion of bulimic patients at first contact was significantly higher in the group of patients who died than in those still alive.

Laboratory Findings

In the literature reference is often made to the pathologic laboratory findings in anorexia nervosa (see, for example, Fichter and Pirke 1982; Brotman et al. 1985; Palla and Litt 1988; Andersen et al. 1990). The results of systematic laboratory testing in the Heidelberg-Mannheim Follow-up are compiled in Table 4 for electrolytes, serum creatinine, hepatobiliary function, amylase, blood count, and serum albumin. The percentage of pathologic laboratory findings for the particular category is also indicated.

For serum electrolytes more than 50% of the findings were pathologic at first examination, whereas the potassium levels were decreased in 35.9%. At second examination the proportion of pathologic electrolyte disturbances had decreased, potassium and chloride levels were still decreased in more than 10%.

Levels of serum creatinine were increased in 18.8% of the sample at first examination. The clinical significance of renal functional disturbances as a predictor of serious complications in anorexia nervosa is not sufficiently reflected by these numbers and must be explained further: There were two cases of terminal renal insufficiency in our sample. One of these patients has died, and the other is on regular dialysis. After having had glomerulonephritis another patient received a kidney transplantation. The transplanted kidney functioned perfectly. In two other cases the creatinine levels were chronically slightly high, both also showed a chronic hypokalemia.

Hepatobiliary function was disturbed in 40% of patients at the initial examination. At second follow-up this rate was markedly reduced. Transaminase changes were seen in about 10% of patients at second follow-up.

Serum amylase levels were increased at first examination in almost half of the patients. The percentage had decreased to less than 25% at second follow-up. These two percentages about correspond to the proportion of bulimic patients at both examination times.

The blood test yielded the following results at admission: 19.8% of patients were anemic and 10% had leukocytosis. In conjunction with the increased blood sedimentation rate (after Westergren) this may be taken to indicate an increased susceptibility to infection. The 12.1% of patients with leukopenia at first admission had decreased at second follow-up.

Table 4. Pathologic laboratory findings

	Pathological laboratory findings (%)
	Admission
K^+ (mmol/l)	
<3.5	35.9
>5.0	6.5
Na^+ (mmol/l)	
<135	18.0
>145	10.1
Cl^- (mmol/l)	
<97	24.2
>110	0
Serum creatine (mg/dl)	
>1.3	18.8
Serum GOT (U/l)	
>20	21.1
Serum GPT (U/l)	
>30	27.4
Serum bilirubin	
>1	15.7
Serum amylase (U/l)	
>110	47.8
Hemoglobin (g/dl)	
<12	19.8
Leukocytes (per mm^3)	
< 4.000	12.1
>10.000	9.9
Sedimentation rate (Westergren/1.hour; mm)	
>10	22.8
Serum albumin (g/l)	
<35	13.4

See Table 2 for abbreviations.

Decreased levels of albumin were found in 13.4% at first admission.

Pulse, Blood Pressure, and Orthostatic Measurement

Low blood pressure and bradycardia are among the typical physical findings in the acute phase of anorexia nervosa. The means at the various examinations can be seen in Table 5. Systolic blood pressure increased such that at second follow-up only 16.7% of patients had systolic blood pressures of less than 105 mmHg. Bradycardia was only registered in 21.1% at first admission (minimum 32/min) and at recent examination amounted to 13.5%. For further analysis of blood pressure and heart rate regulation schellong's test, a standardized orthostatic test, was performed using a tilting table with automatic blood pressure and pulse measurement.

The subjects laid down on the tilting table. The cuff was put on their right upper arm to measure blood pressure, and the ECG was taken using Nehb's method. Blood pressure and pulse were measured every minute for 6 min with the patient lying down and then for 10 min in standing. Blood pressure was determined according to Riva-Rocci criteria, whereby the cuff was selected according to the circumference of the patient's upper arm. So that the external conditions were always the same, the tests were always carried out by the same assistant, the room temperature was constant, and the measurements were taken between 9 and 11 a.m. to avoid deviations in daily rhythm.

For 41 patients of the sample there are results from standardized orthostatic examination both at admission and at second follow-up.

Table 5. Blood pressure and heart rate

	Admission	First follow-up	Second follow-up
Systolic blood pressure (mmHg)			
Mean	99	108	116
Range	60–170	85–140	85–200
SD	17.0	13.3	17.5
<105	69.8%	52%	16.7%
Diastolic blood pressure (mmHg)			
Mean	67	69	77
Range	0–120	40–105	60–110
SD	14.7	11.3	8.2
Heart rate (bpm)			
Mean	71	67	67
Range	32–112	48–84	47–104
SD	17.5	9.4	12.7
<60	21.1%	16.0%	13.5%

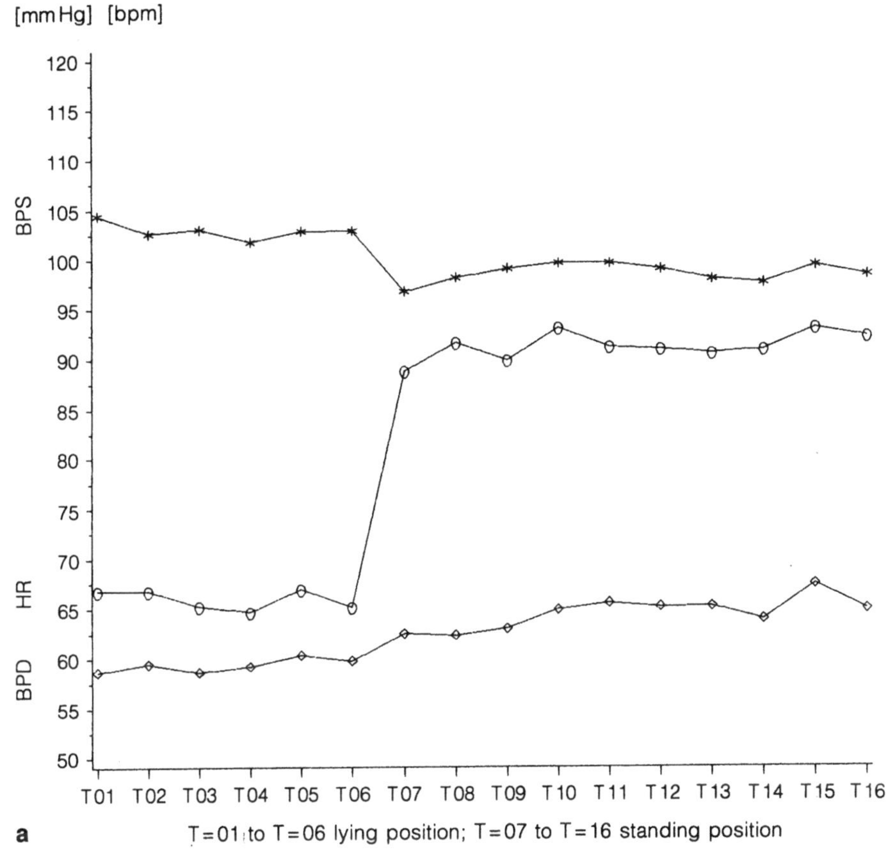

Fig. 2a,b. Standardized orthostatic test of anorexia nervosa patients (n = 41) at **a** admission and **b** follow-up. *systolic blood pressure (mm HG, *BPS*); <> diastolic blood pressure (mm HG, *BPD*); O, heart rate (beats per minute, *HR*); *T*, time (in minutes). **b** see p. 94

Although the orthostatic examination was carried out in all patients, we are limiting our analysis here to the 41 patients tested both at admission and at second follow-up. The body mass index (BMI) for the group of patients included in the analysis was 14.6 (kg/m^2) at first admission.

The results of the orthostatic examination were compared with those from an age-matched control group (n = 41) for which the results of two standardized orthostatic tests were available. In the control group BMI was 21.5 (kg/m^2). The subjects from the control group had been treated on a psychosomatic ward.

At second follow-up the BMI for the anorectic group was on average 19.9 (kg/m^2); one patient was under 16 (kg/m^2).

The results of blood pressure and pulse monitoring are shown in Table 6 and Fig. 2.

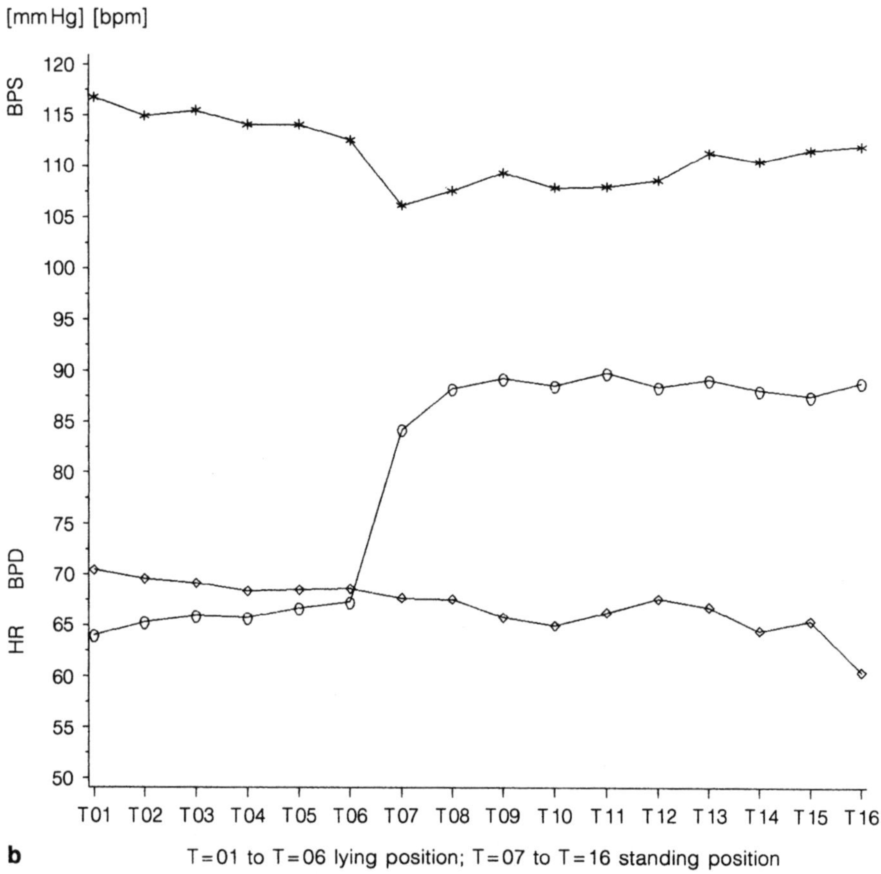

Fig. 2b. Legend see p. 93

At first examination systolic blood pressure was significantly lower in the anorectic group than in the controls in both the lying and standing positions. At second examination in the follow-up there were no longer any significant differences between the former anorexia patients and the controls. The increase in systolic blood pressure from measurement time 0 to measurement time 2 in the anorectic group is significant.

Significant differences in diastolic blood pressure between the anorexia patient group and controls were only seen in the lying position at follow-up. In standing position there was no significant difference ascertained either at first examination or at follow-up. There is a significant difference between anorexia patients and the control group in pulse at both times of measurement in lying and in standing.

Significant differences were found in blood pressure amplitude between the anorectic group and controls in both lying and standing at first examination. At follow-up only amplitude in the lying position of

Table 6. Results of blood pressure and heart rate testing

Time	Position	Group	Mean	SD	Analysis of variance (p)
Systolic blood pressure (mmHg)					
T0	L	AN	104.22	5.11	0.0001
T0	L	C	117.38	4.38	
T0	S	AN	98.14	5.23	0.0001
T0	S	C	110.87	5.94	
T2	L	AN	114.63	3.44	0.2240
T2	L	C	117.91	4.85	
T2	S	AN	109.61	6.23	0.6395
T2	S	C	110.86	6.49	
Diastolic blood pressure (mmHg)					
T0	L	AN	59.47	5.25	0.4626
T0	L	C	57.71	5.92	
T0	S	AN	64.06	4.96	0.8861
T0	S	C	65.00	6.98	
T2	L	AN	69.00	3.06	0.0016
T2	L	C	61.36	5.74	
T2	S	AN	65.66	5.52	0.9242
T2	S	C	65.46	6.19	
Heart rate (beats per minute)					
T0	L	AN	67.11	5.49	0.0001
T0	L	C	80.79	5.16	
T0	S	AN	92.68	6.97	0.0032
T0	S	C	104.05	6.76	
T2	L	AN	65.73	2.23	0.0001
T2	L	C	79.25	4.97	
T2	S	AN	89.09	4.69	0.0186
T2	S	C	97.97	6.17	
Blood pressure amplitude systolic-diastolic (mmHg)					
T0	L	AN	44.75	9.03	0.0001
T0	L	C	59.66	15.38	
T0	S	AN	34.07	8.44	0.0001
T0	S	C	45.87	12.45	
T2	L	AN	45.63	6.82	0.0002
T2	L	C	56.36	15.29	
T2	S	AN	45.95	5.52	0.5132
T2	S	C	45.41	11.54	

T0, key admission; T2, follow-up; L, lying position; S, standing position; AN, anorexia nervosa group; C, control group.

Table 7. Spearman's correlation: orthostatic indexes SL and K vs age and BMI

	Correlation coefficient			
	AN group		Control group	
	T0	T2	T0	T2
SL vs age	−.05	−.07	.26*	−.07
SL vs BMI	−.06	−.31**	−.14	.16
K vs age	−.38**	.07	.22	.26*
K vs BMI	.05	.17	.38	.19

T0, key admission; T2, follow-up; AN, anorexia nervosa group; SL, orthostatic index (Stoll and Legeler); K, orthostatic index (Kirchhoff); BMI, body mass index.
$^{*}p < 0.10$; $^{**}p < 0.05$.

the former anorexia patients was significantly lower than that of the control group. Comparing the anorectic group at the various times of measurement in standing position there was a significant increase from 34.07 mmHg to 43.95 mmHg.

In determining orthostatic lability there was a significant difference between the control and anorectic groups in terms of the orthostatic index SL[1] of Stoll and Legeler (1971) at both times of measurement as well as between the means within the anorectic group at first examination and follow-up. In terms of Kirchhoff's and Beckmann's (1965) orthostatic index the results at second measurement were significantly lower than at first measurement within the anorectic group; in comparison to the control group a lower orthostatic lability was found in the anorectic group at second examination using Kirchhoff's criteria.

Spearman's rank correlation coefficient was used to answer the question of age and weight dependency of the orthostatic indexes of Kirchhoff and Stoll and Legeler (Table 7). The Kirchhoff indexes were correlated with age in the anorexia group at first examination and yielded significant results. At second measurement no significant correlation was found. Only at follow-up measurement was the Stoll and Legeler index significantly correlated with weight ($r = 0.31$). In the control groups there was a trend towards age-dependent correlation of the orthostatic indexes at both examinations but it was not significant.

The results of the standardized orthostatic tests at two measurement times (admission and second follow-up) in comparison to the control groups can be summarized as follows:

[1] $SL = 3.6 - 0.02a - 0.007b + 0.01c$, where a = systolic blood pressure at the end of the standing phase; b = percent change in blood pressure amplitude at the end of the standing phase; and c = pulse at the end of standing minus pulse in lying position.

1. Systolic blood pressure differed at first measurement in lying and standing significantly from the control group. At follow-up there is no longer a significant difference between former anorexia patients and controls. The changes within the anorectic group from time of first and second measurements become significant.
2. Pulse in the anorectic group is significantly lower than that of the control group at both measurements in both positions. Within the anorectic group there was no significant change in pulse.
3. Blood pressure amplitude differs between anorectic patients and controls at first measurement in both standing and lying.
4. The orthostatic index of Stoll and Legeler as measure of lability in orthostatic load is significantly higher at first and second examination in the anorectic group than in the controls.
5. At follow-up orthostatic lability in terms of the Stoll and Legeler index was significantly correlated with weight within the anorectic group. This does not apply to first admission, but it must be taken into account that at that time there was very little variation in weight.

While orthostatic lability, hypotension, and bradycardia decrease at follow-up in the anorectic group, there are still deviations from the control group. Blood pressure and pulse regulation prove to be areas in which the consequences of anorexia are mirrored.

Bone Density

Osteopenia is a common finding in acute cases of anorexia (Ayers et al. 1984; Brotmann 1985; Davies et al. 1990; Newman and Halmi 1989; Rigotti 1984; Rüegsegger 1988; Szmuker 1985; Treasure 1987). Clinically it has far-reaching implications since painful and debilitating pathologic fractures may result from the condition (Brotman 1985; Rigotti et al. 1984; Olmos et al. 1990; Beumont, this volume).

Though osteopenia is an almost certain finding in anorexia and mineral loss is estimated to occur at a rate of about 3% per year of disease (Rueggsegger et al. 1988; Andersen et al. 1990), the question of which mechanisms give rise to osteoporosis in anorexia nervosa and to what extent osteoporosis is reversible in this condition is still open. The etiologic factors in osteoporosis which have been suggested are long-standing estrogen deficiency (Biller et al. 1989; Treasure and Russell 1987; Beumont, this volume), competing nutritional factors such as uptake of calcium, fat, vitamin D (Beumont, this volume; Davies et al. 1990), and hormonal factors such as hypercortisolism (Biller et al. 1989; Beumont, this volume) and hyper-prolactemia (Klibanski and Greenspan 1986). Physical exercise seems to have a protective effect (Rigotti et al. 1984; Drinkwater et al. 1984). As to the reversibility of osteopenia in anorexia nervosa two studies with conflicting findings have been published (Treasure and Russell 1987;

Rigotti et al. 1988): Rigotti rejects the reversibility of loss of forearm bone density; Treasure and Russell found indications that osteopenia is reversible after anorexia has been healed. In a prospective study (Scharla et al. 1990) it was shown that lumbar bone density loss due to estrogen deficiency over a limited, 6-month period of time (treatment of endometriosis and uterine leiomyoma with long-acting gonadotropin-releasing hormone agonists) in premenopausal women was completely reversible after 6–8 months.

It was the aim of the Heidelberg-Mannheim Follow-up to procure bone density data for the entire follow-up group if possible. Since our patients were treated for the first time between 1971 and 1980 there are no bone density data from the time of first admission and so an intraindividual assessment of course cannot be made directly. However, our patients belong to the group of more severely affected individuals and the 3% loss of bone density per year of anorexia (see above) can likely be applied to first admission data. The bone density data in the good, intermediate, and poor outcome groups (according to the Morgan-Russell General Outcome Categories) which have been measured and reported with standard deviation from age-matched controls (z-score) allow, however, indirect conclusions as to reversibility of the assumed bone density loss to be made.

The following questions need to be answered:

1. What information on bone density does dual photon absorptiometry of the lumbar spine provide in a defined clinical sample of anorexia nervosa patients at least 9 years after first admission in the good, intermediate, and poor outcome groups (according to Morgan-Russell General Outcome Score)?
2. Are there any clinical parameters from the admission examination which predict bone density at second follow-up?
3. Are there any course of disease parameters which predict bone density at second follow-up?
4. Can the resumption of menstruation (and possibly also weight gain) in the last years before follow-up examination (at least 9 years after admission) be expected to have a positive effect on bone density at second follow-up?

On preliminary analysis of the data on bone density of the lumbar spine the following trends appear:

1. Bone density in the poor outcome group is significantly different than that of the good and intermediate outcome groups.
2. The ANSSs at first admission seems to be a predictor of bone density at follow-up.
3. Of the variables of outcome the sum measurements of both weight and menstruation status seem to be predictors.
4. The resumption of menses, along with the relative time of exposure to estrogen (ratio of actual and maximum possible estrogen exposure time

in months), in the years preceding second follow-up examination seem to contribute significantly to the prediction of bone density. This, and indirectly the findings in point 1., can serve as an indication of the possibility of the reversibility of loss of bone density in anorexia nervosa if the condition improves.

Anorexia Nervosa Symptom Score – Somatic Section

Deter et al. (1983, 1989) have compiled in the ANSS 31 somatic, psychodiagnostic, and social criteria which can be rated as either pathologic (1) or nonpathologic (0). The individual ratings are added up to make a total ANSS score. The scores are then given in percentages, where 0% corresponds to no disturbance and 100% corresponds to disturbances in all rated areas.

According to the definition of the somatic section (ANSSs) of the ANSS the results of the somatic global score are reported for all three examinations in the Heidelberg-Mannheim Follow-up. The ANSSs global score of course of anorexia nervosa is comprised of the following items; the rating consists of thorough history, physical examination, and laboratory tests:

1. Weight
2. Amenorrhea
3. Vomiting
4. Laxative abuse
5. Obstipation
6. Electrolytic disturbance
7. Amylase
8. Kidney function
9. Transaminase
10. Changes in blood test results and signs of infection
11. Hormonal status
12. Hypotension
13. Physical symptoms not related to anorexia nervosa

The assessment of the individual items as pathologic (1) or nonpathologic (0) proceeds as follows:

1. If weight is less than 85% of the normal weight, the rating of "1" is given. If weight is less than 75% of the normal weight, the rating of "2" is given. For assessment purposes the directly measured weight is decisive.
2. Secondary amenorrhea is assumed if for 3 continuous months there is no menses. If ovulation inhibitors are being taken and use has not been discontinued in the past 12 months so as to check menstruation status, when in doubt amenorrhea is assumed.
3. If the patient indicates that in the past 2 months she has induced vomiting, once a month (and more), the item is rated as pathologic.

4. If laxatives have been taken once (or more) per month, the item is rated as pathologic.
5. If indicated that in the past 2 months bowel movements occur every 3–4 days (or more seldom), this item is rated as pathologic.
6. If the results of laboratory tests indicated that one or more of the following items were not within the normal range, the item is rated as pathologic: potassium, sodium, chloride, calcium, phosphate.
7. If amylase levels are not within the normal range, the rating "1" is given.
8. If serum creatinine or serum urea are not within the normal range, the rating "1" is given.
9. If glutamine-oxaloacetic transaminase (GOT) or glutamic-pyruvic transaminase (GPT) or γ-glutamyl transpeptidase (γ-GT) are not within the normal range, the rating of "1" is given.
10. If one of the following laboratory findings are pathologic, the rating of "1" is given: hemoglobin, leukocyte count, blood sedimentation rate, total protein, serum albumin.
11. If one of the following hormone tests yields pathologic results, the rating of "1" is given: T3, T4, plasma cortisol (morning levels), basal LH, FSH.
12. If systolic blood pressure in lying is under 105 mmHg, the rating "1" is given.
13. If there are other pathologic findings in amnesis or at physical examination which are not related to anorexia nervorsa or if other somatic conditions are diagnosed, the item receives the rating "1".

The ANSSs as a somatic global score is given in percentage of ratings which are pathologic with respect to the items rated. The score is only meaningful if there is a valid rating for at least nine items.

The results of the ANSSs, i.e., the global somatic scores at first admission and first and second follow-up, are illustrated in Fig. 3.

Although in terms of the ANSSs physical condition on the whole improves significantly from first admission to second follow-up, this factor

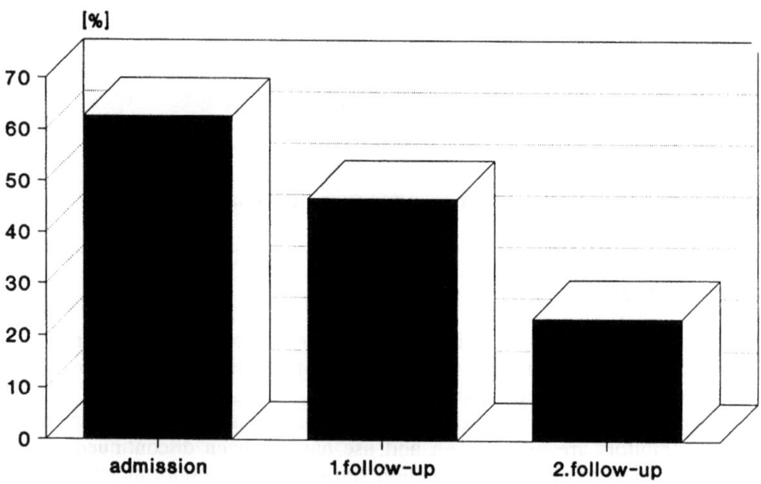

Fig. 3. Anorexia Nervosa Symptom Score-Somatic Section in the Heidelberg-Mannheim Follow-up

cannot be neglected. In our sample an ANSSs rating of 33% distinguishes patients with good outcome from those with poor outcome according to the Morgan-Russell General Outcome Cateogories: all former anorexia patients with a good course have an ANSSs rating of under 33% at follow-up; those with a poor course have an ANSSs rating of above 33%. The ANSSs ratings correlate – as is to be expected – with weight: the correlation coefficient is $r = -.45$ ($p = 0.0001$) both at first admission and second follow-up.

If, in addition to weight, menstruation status is taken into account, the correlation with ANSSs is not any higher at first admission.[2] At second follow-up, however, there is better correlation ($r = -.81$) between the ANSSs and the combined criteria of weight and menstruation according to Morgan-Russell Outcome Scores at this time, too. This finding can be interpreted to mean that at follow-up (also considering that three fourths of the patients in the sample have recovered or improved) there is a high concordance between ANSSs and the combined weight-menstruation criteria. In the acute phase, in contrast, the ANSSs leads to a differentiated assessment, which cannot be produced in the same manner by weight and menstruation alone.

Somatic Comorbidity

In the Heidelberg-Mannheim Follow-up we were able to compile most of the current examination findings for 87.4% of subjects who are still alive ourselves. The types of conspicuous findings and diagnoses which were determined in these examinations are listed in Table 8. A total of 26 subjects (34.7% of the 75 examined, who had no severe disease other than anorexia nervosa at key admission) were affected. The serious conditions of the kidneys and skeletal system which have particular impact on prognosis should be noted:

- Terminal, dialysis-dependent renal failure in two patients, one of whom died
- Multiple pathologic fractures due to extreme osteoporosis which led to permanent disability in two patients
  Furthermore, a tendency to be susceptible to infection can be read from the diagnoses.

In the meantime one of the patients with type 1 diabetes mellitus had died of sepsis and juvenile diabetes mellitus had been newly diagnosed in another individual. The insulin levels of both of the remaining

[2] By definition amenorrhea is a diagnostic factor in anorexia.

Table 8. Somatic comorbidity

Type of diagnosis	Number
Neurological	4
Dermatological	3
Endocrinological	5
Gastrointestinal	3
Respiratory	1
Renourogenitary	7
Osteological	3
Vascular	3
Other	1

diabetic patients were stable and neither patient was anorectic or bulimic at follow-up.

Course of Anorexia Nervosa and Predictors of Outcome

The Heidelberg-Mannheim Follow-up is a cross-sectional study with three examination times: at admission to the Medical Clinic of the University of Heidelberg (T0), first follow-up after an average of 3.6 years (T1), and second follow-up after 12.7 years (T2). We did succeed, though, in longitudinally determining some base data on weight and menstruation: Besides the three direct examinations mentioned, all patients were questioned about hospitalizations since key admission. If patients consented, hospital records and data from health insurance companies were requested for these hospitalizations. About 500 records (70% of registered hospital admissions) were available.

Every year after first admission minimum weight and poorest menstruation status were determined for each patient. If no objective findings from catamnesis or physician's records were available, the subjective data from the patient was relied upon. Reference data from first admission and data from second catamnesis (T2) were available from direct examination. Based on minimal weight and poorest menstruation status each patient was categorized according to Morgan-Russell General Outcome Categories every year. The proportion of former anorexia patients with good outcome increases up to the 12th year of catamnesis up to about 50%, whereby an asymptote of 55% is presumed.[3]

Whereas the proportion of good outcome patients increases with increasing time of catamnesis – this applying also to the patients who

[3] In individual cases improvement may also be possible after a longer period of time, as was the case, for example, in one patient in the 17th year of catamnesis.

died – this increases at the cost of the intermediate and poor outcome groups. The time dynamics in the course of anorexia are in agreement with studies which used similar methods and in which patients were examined on the basis of well-defined diagnostic criteria, whereby a high direct interview rate and a low failure to trace rate was achieved (see Herzog et al., this volume). Our results agree with other studies, which report that in terms of Morgan-Russell General Outcome Categories no real recovery can be expected in the first 2 years of catamnesis (Theander 1985; Vandereycken and Pierloot 1983; Tolstrup et al. 1985).

As a measure of outcome which also reflects the course of the illness in the follow-up period, each patient received a Morgan-Russell sum score, that is, the sum of the Morgan-Russell General Outcome Categories received each year: The scores for good outcome "0," for intermediate outcome "1," for poor outcome and in the event of death "2." This provided us with a measure of outcome which was based on simple physiologic parameters (weight and menstruation) for every year and which also reflected the aspect of course. A speedy recovery led to a lower sum score than did a late recovery and an early death led to a higher sum score than a later death. Such a Morgan-Russell sum score is a continually distributed measure of outcome (range 2–40) on which multiple regressions can be calculated.

A criticism which can be applied to our construct of outcome variables is that these are only very simple physiologic parameters (weight and mensturation) and that fatal and chronic course of anorexia are weighted very heavily. As counterargument to the first criticism Morgan and Hayward (1988) were able to show that the Morgan-Russell General Outcome Categories correlate highly with the social and psychologic partial scores of the Morgan-Russell Outcome Assessment Schedule. It is indeed so that death and chronic course are rated high in the Morgan-Russell sum score. But the effect of the high rating of deaths and chronic course is desired: In determining predictors those variables are rated higher which especially contribute to differentiating between good and improved outcome and poor and fatal outcome.

A stepwise multiple regression of the Morgan-Russell sum score as a dependent variable indicates that, for example, serum albumin at T0 and length of illness at key admission are good predictors. With respect to these variables social class, weight at key admission, serum phosphate and hemoglobin levels at key admission, and age at onset of illness are not significant.

While the analysis mentioned above included all subjects, it was shown that by limiting it to the surviving patients the eating symptom variables, especially bulimia, for prediction of outcome in the Morgan-Russell sum score are more important.

Summary

In the Heidelberg-Mannheim Follow-up 103 patients with anorexia nervosa were examined in a medical clinic on average 12.7 years after first contact. At a low failure to trace rate of 1% and a high direct interview rate of 92% somatic findings were compiled from own examinations in 78.1% and from the examination results from other clinics or the general practitioner in 9.2% of the sample. On the basis of these data it is possible to systematically describe the somatic status of a larger sample of anorexia nervosa patients from various perspectives:

1. The mortality rate in our sample was 15.5%. This high rate is within the range of other long-term studies of course (see Russell, Theander, and Engel et al., this volume). The cause of death was suicide in one fourth of the cases, in another fourth infection, and furthermore ileus and extreme cachexia. In our sample the 55% mortality in the subgroup of patients with severe physical illnesses who also presented with an anorectic reaction was particularly high.
2. In about 50% of the patients followed up weight and menstruation normalized. In about 40% the eating symptoms had disappeared in addition to normalization of weight and menstruation. The DSM-III-R criteria for bulimia were fulfilled in less than 20% of the sample, and occasional vomiting was indicated by about one fourth of the follow-up sample. Chronic anorectic course with underweight and amenorrhea was present in about 10% of patients. Hence the condition in about one fourth of the patients in the sample had taken an extremely unsatisfactory course, i.e., chronic anorexia or death.
3. Serum electrolyte disturbance was seen in more than 10%. Serum amylase levels were increased in less than 25%. The increased amylase levels were associated with the increase in salivary amylase as, for example, described by Kronwall (1988) for bulimia. Transaminase levels were increased in about 10% of patients, the rate for anemia and leukocytosis being about the same.
4. Hypotension and orthostatic lability decreased in the course of disease. The extent of orthostatic lability at second follow-up correlates significantly with weight.
5. Bone density was lower in the follow-up group than in age-matched controls. Patients with poor outcome in terms of the Morgan-Russell General Outcome Categories had significantly lower bone density than patients in the intermediate and good outcome groups. In regression analyses the relative estrogen deficiency throughout the entire period of illness and nutritional factors contributed to prediction of bone density.
6. The ANSSs, the somatic section of the anorexia nervosa symptom score, decreased from examination at first admission to first follow-

up to second follow-up. Some 35% of follow-up patients have non-psychiatric diseases according to ICD criteria. Besides the less serious diagnoses, there were also cases of terminal renal insufficiency and disabling osteoporosis which carry a serious prognosis. Although such severe conditions were a seldom occurrence, a certain tendency of susceptibility to acute and chronic infection was noticed. One case of chronic pancreatitis was diagnosed at another university clinic. On the whole, pancreatic conditions seemed to occur less often then was previously assumed.

7. In the sample as a whole, serum albumin levels at first admission and length of illness before first admission seem to be predictors for the course of anorexia nervosa using the Morgan-Russell sum score. If the regression is limited to subjects who are alive, then the symptom variables bulima and potassium levels at hospital discharge have proven to predictors.

The present study has attempted to directly compile the somatic findings of a sufficiently large cohort of former anorexia patients after a catamnesis interval of over 10 years. Besides individual findings (weight, menstruation, laboratory test results, blood pressure and pulse/Schellong's test, bone density), the outcome was described in terms of the ANSSs somatic global score, current ICD diagnosis, mortality rate (and causes of death), and predictor analysis. Whereas psychiatric comorbidity was described in the previous chapter, in this one the far-reaching somatic consequences, which not seldom lead to death or permanent disability, have been elucidated. It was also shown, though, that it is possible to recover (in about 40% of cases) and that then improvement can be expected in areas in which improvement was previously considered doubtful, such as osteopenia.

Although the patient sample used in the Heidelberg-Mannheim Follow-up was derived from a medical – psychosomatic clinic, the proportions of good, intermediate, and poor outcome in terms of the Morgan-Russell General Outcome Categories are comparable to those of other long-term course studies (see Herzog et al., this volume). Hence the results of further somatic investigations can be generalized for the long-term course of anorexia.

References

Andersen A, Copeland P, Rosell F (1990) The medical complications of eating disorders 4th International Conference on Eating Disorders, New York, 27–29 April 1990

Ayers JWT, Gidwani GP, Schmidt IMV, Gross M (1984) Osteopenia in hypoestrogenic young women with anorexia nervosa. Fertil Steril 41(2):224–228

Bassoe HH, Eskeland I (1982) A prospective study of 133 patients with anorexia nervosa. Acta Psychiatr Scand 65:127–133

Biller BMK, Saxe V, Herzog DB, Rosenthal DI, Holzman S, Klibanski A (1989) Mechanisms of osteoporosis in adult and adolescent women with anorexia nervosa. J Clin Endocrinol Metab 68(3):548–554

Brotman AW, Stern TA (1985) Osteoporosis and pathologic fractures in anorexia nervosa. Am J Psychiatry 142(4):495–496

Brotman AW, Rigotti N, Herzog DB (1985) Medical complications of eating disorders: outpatient evaluation and management. Compr Psychiatry 26(3):258–272

Christian P (1989) Anthropologische Medizin. Chapter 2.1: Anorexia nervosa – der Körper als Phänomen. Springer, Berlin Heidelberg, New York, 40–46

Davies KM, Pearson PH, Huseman CA, Greger NG, Kimmel DK, Recker RR (1990) Reduced bone mineral in patients with eating disorders. Bone 11:143–147

Deter HC, Petzold E (1979) Klinische Befunde und Verlaufsuntersubhungen bei Patienten mit Anorexia nervosa. Therapiewoche 29:6320–6331

Deter HC, Petzold E, Hengst-Theis R, Breiden U, Lanzinger-Rossnagel G (1983) Katamnestische Ergebnisse einer klinisch-psychosomatischen Behandlung von 103 Patienten mit Anorexia nervosa aus internistischer Sicht unter besonderer Berücksichtigung der Mortalität. Innere Med 1:3–12

Deter HC, Petzold E, Hehl FJ (1989) Differenzierung der Langzeitwirkungen einer stationären psychosomatischen Therapie von Anorexia-nervosa-Patienten. Z Psychosom Med Psychoanal 35:68–91

Drinkwater BL, Nilson K, Chesnut CH, Bremner WJ, Shainholtz S, Southworth MB (1984) Bone mineral content of amenorrheic and eumenorrheic athlets. N Engl J Med 311:277–281

Fichter MM, Pirke KM (1982) Somatische Befunde bei Anorexia nervosa and ihre differentialdiagnotische Wertigkeit. Nervenarzt 53:635–643

Kirchhoff HW, Beckmann D (1965) Regulationsstörungen des Herzens und des Kreislaufs. Ambrosius Barth, Munich, pp 11–81

Klibanski A, Greenspan SL (1986) Increase in bone mass after treatment of hyperprolactinemic amenorrhea. N Engl J Med 315:542–546

Köhle K, Mall H (1983) Follow-up study of 36 anorexia nervosa patients treated on an integrated internistic-psychosomatic ward. Int J Eat Disorders 2(4):215–219

Kronwall P (1988) Elevation of serum salivary amylase in bulimia nervosa. Third International Conference on Eating Disorders, New York, 22–24 April (Abstract) 1988

Meyer AE, von Holtzapfel B, Deffner G, Engel K, Klick M (1986a) Amenorrhea and predictors for remenorrhea in anorexia nervosa: a psychoendocrinological study in inpatients. Psychother Psychosom 45:149–160

Meyer AE, von Holtzapfel B, Deffner G, Engel K, Klick M (1986b) Psychoendocrinology of remenorrhea in the late outcome of anorexia nervosa. Psychother Psychosom 45:174–185

Morgan HG, Hayward AE (1988) Clinical assessment of Anorexia nervosa. The Morgan-Russell Outcome Assessment Schedule. Br J Psychiatry 152:367–371

Newman MM, Halmi KA (1988) The endocrinology of anorexia nervosa and bulimia nervosa. Endocrinol Metab Clin North Am 17(1):195–212

Newman MM, Halmi KA (1989) Relationship of bone density to estradiol and cortisol in anorexia nervosa and bulimia. Psychiatry Res 29:105–112

Noordenbos G (1989) Improving the process of recovery of patients with anorexia nervosa. Br Rev Bulimia Anorexia Nervosa 4(1):17–31

Olmos JM, Pesquera C, Amado JA, Riancho JA, Gonzales-Macias J (1990) Pathologic fracture of the sternum in a patient with anorexia nervosa. Rev Clin Esp 186(1): 23–25

Palla B, Litt IF (1988) Medical complication of eating disorders in adolescents. Pediatrics 81(5):613–623
Rigotti NA, Nussbaum SR, Herzog DB, Neer RM (1984) Osteoporosis in women with anorexia nervosa. N Engl J Med 311(25):1601–1605
Rigotti NA, Neer RM, Ridgway L, Herzog DB, Nussbuam SR (1988) Osteopenia does not reverse with recovery from anorexia nervosa: a short study. Third International Conference on Eating Disorders, New York, 22–24 April (Abstract) 1988
Rüegsegger P, Müller A, Dambacher MA, Ittner J, Willi J, Kopp HG (1988) Knochenabbau bei Patientinnen mit Anorexia nervosa. Schweiz Med Wochenschr 118(7):233–238
Scharla SH, Minne HW, Waibel-Treber S, Schaible A, Lempert UG, Wüster CH, Leyendecker G, Ziegler R (1990) Bone mass reduction after estrogen deprivation by long-acting gonadotropin-releasing hormone agonists and its relation to pretreatment serum concentrations of 1,25-dihydroxyvitamin D3. J Clin Endocrinol Metab 70(4):1055–1061
Steinhausen HC, Rauss-Mason C, Seidel R (1991) Follow-up studies of anorexia nervosa: a review of four decades of outcome research. Psychol Med (in press)
Stoll KD, Legeler HJ (1971) Der Stehversuch nach Schellong in klinischer und statistischer Sicht zur objektiven Bewertung des Therapieerfolgs von 9-alpha-Fluorhydrocortison bei orthostatischen Regulationsstörungen. Arzneimittelforschung 21:1444–1449
Szmukler GI, Brown SW (1985) Premature loss of bone in chronic anorexia nervosa. Br Med J (Clin Res) 5:26–27
Theander S (1985) Outcome and prognosis in anorexia nervosa and bulimia: some results of previous investigations, compared with those of a Swedish long-term study. J Psychiatr Res 19(2/3):493–508
Tolstrup K, Brinch M, Isager T, Nielsen S, Nystrup J, Severin B, Olesen NS (1985) Long-term outcome of 151 cases of anorexia nervosa. The Copenhagen anorexia nervosa follow-up study. Acta Psychiatr Scand 71:380–387
Treasure JL, Russell GFM (1987) Reversible bone loss in anorexia nervosa. Br Med J 295:474–475
Umeki S (1988) Biochemical abnormalities of the serum in anorexia nervosa. J Nerv Ment Dis 176(8):503–506
Vandereycken W, Pierloot R (1983) Long-term outcome research in anorexia nervosa. The problem of patient selection and follow-up duration. Int J Eat Disorders 2(4):237–242
Waibel-Treber S, Minne HW, Scharla SH, Bremen Th, Ziegler R, Leyendecker G (1989) Reversible bone loss in women treated with GnRH-agonists for endometriosis and uterine leiomyoma. Hum Reprod 4(4):384–388
Wellmann W, Pries K, Freyberger H (1981) Die Kombination von Morbus Crohn und Anorexia-nervosa-Symptomatik. Dtsch Med Wochenschr 106:1499–1502

A Prospective Follow-up Study in Early-Onset Eating Disorders

H.-C. Steinhausen and R. Seidel

Introduction

Our follow-up project, the Berlin Study, is restricted to patients with early-onset conditions, i.e., adolescent patients suffering from eating disorders. The study originated from two sources: a retrospective outcome study of former adolescent anorectic patients that served as a pilot study and allowed the feasibility of the instruments and the methods to be tested (Steinhausen and Glanville 1983b,c) and a comprehensive analysis of 45 follow-up studies published between 1953 and 1981 (Steinhausen and Glanville 1983a). In addition, the Berlin Study considered a number of demands directed at design and methodology that had arisen in other critical reviews concerning the issue of outcome studies (Hsu 1980; Schwartz and Thompson 1981; Swift 1982).

The criteria for research on outcome in anorexia nervosa included the following requirements regarding follow-up methodology: (1) direct method of assessment, (2) corroborating information from multiple sources, (3) objective ratings and standardized scales, (4) duration of follow-up no less than 4 years, and (5) a low failure-to-trace rate or, respectively, controls for drop-out effects. In addition, outcome parameters were to include (6) anorexia-related symptoms, i.e., eating behavior, body weight, and menstrual functioning, (7) further psychiatric symptoms, and (8) psychosocial functioning.

At the time the Berlin Study was started, there were few studies dealing specifically with *early-onset* anorexia nervosa and knowledge concerning its outcome was limited. In the majority of studies, the age range of the patients assessed differed enormously (see Steinhausen and Glanville 1983a), which might well limit the interpretation of data concerning outcome. In addition, our analysis of follow-up studies indicated that contradictory findings were reported regarding the prognostic significance of *age of onset*.

The purpose of the Berlin Study, therefore, was manifold. Firstly, based on a prospective study, we were interested in increasing the available data regarding long-term outcome in early-onset eating disorders. Secondly, we aimed to evaluate therapeutic intervention through repeated measurements that served to monitor the continuous changes in specific aspects of the clinical profile. And finally, we hoped to amend some of the major deficits in the design and methodology of previous follow-up studies that confounded an evaluation of data with regard to outcome.

Methods

Subjects

The entire consecutive series of patients with eating disorders admitted to two adolescent psychiatric wards at the Free University of Berlin between 1978 and 1988 was included in the study. During the past decade repeated assessments have been made for a total of 60 patients. The last data entry will occur in 1992. Thus, at the present time, no final statement with regard to outcome can be made about this series of patients. Final analysis, including the significance of prognostic factors, has to be reserved for the future. Accordingly, this chapter provides an interim report.

Subject characteristics of the series are given in Table 1. Almost three fourths of the sample is of middle class origin and more than two thirds have a high educational status. Patients were originally diagnosed according to DSM-III and DSM-III-R criteria. Since the recently proposed ICD-10 criteria were felt to be more discriminating for the classification of eating disorders, all files were reanalyzed according to ICD-10. However, neither DSM nor ICD criteria are satisfactory with regard to the classification of patients suffering from both anorexia and bulimia concurrently. Whereas DSM-III-R advocates using both diagnoses in such an instance, ICD-10 does not provide an adequate classification for such patients and,

Table 1. Characteristics of patients in sample ($n = 60$)

Characteristic	No. of patients	(%)
Socioeconomic status		
Lower class	16	(26.7)
Middle class	44	(73.3)
Educational status		
Low	5	(8.4)
Middle	14	(23.3)
High	41	(68.3)
Diagnostic classification		
Anorexia nervosa	48	(80.0)
Anorexia nervosa with bulimic features	6	(10.0)
Bulimia with normal weight	5	(8.3)
Anorectic reaction	1	(1.7)

therefore, for the purpose of the present study the term "anorexia nervosa with bulimic features" was introduced.

Accordingly, the series is comprised of 60 former patients who had suffered from either anorexia nervosa, anorexia nervosa with bulimic features, bulimia with normal weight, or anorectic reaction. Furthermore, all patients were hospitalized for the first time. Mean age at onset of the disease was 14.0 (SD = 1.6) years and the mean interval between onset and admission to inpatient treatment was 1.7 years. The mortality rate so far is 5% for the total sample: two deaths by suicide and one as a direct result of the disease. Among 34 patients followed up for at least 4 years, three refused to cooperate, resulting in a 9% drop-out rate. The mortality rate for this long-term follow-up sample is 3%, i.e., one patient. Because only preliminary data are presented and the drop-out rate is small, separate sample characteristics are not reported for these cases. Only after completion of the data collection will it be possible to analyze whether or not the attrition rate jeopardizes the representativity of the sample.

As shown in Table 1, a minority of patients suffered from bulimia with normal weight or from anorexia nervosa coupled with bulimic features. Only one patient suffered from an anorectic reaction – a *forme fruste* of the full-blown clinical picture of anorexia nervosa. The mean percentage of ideal body weight (i.e., height in centimeters minus 100, minus 15%) at the time of admission was 67.8 (SD = 7.9) for 55 patients with a diagnosis of anorexia nervosa, including one patient suffering from anorectic reaction, and 98.2 (SD = 7.9) for five patients suffering from bulimia with normal weight.

Treatment

As stated above, all patients were admitted to one of two wards for the treatment of child and adolescent psychiatric disorders. The wards accommodated patients of both sexes, and the types of psychiatric disorders treated there were diverse. There were usually only one or two anorectic patients in treatment at any one time and, accordingly, neither the milieu nor the organization scheme of the treatment setting were comparable to that of a treatment center for eating disorders.

Treatment was multidimensional and individually tailored to the specific needs of the patients and their families. It included a high caloric diet, but no tube feeding. Patients received supportive psychotherapy; family therapy and parent counseling were also provided. Weight restoral was supported by such behavioral measures as contingent reinforcement, but without strong reprimands or severe restriction of privileges. Furthermore, the therapeutic milieu of the adolescent psychiatric inpatient treatment settings included body-oriented therapy and music therapy, additionally enriching the treatment package for the individual patients.

Mean duration of inpatient treatment was 14 (SD = 9.8) weeks. There was no general aftercare program. Depending upon individual needs, some patients were seen again as outpatients during the first months after discharge, while others were referred to psychotherapists in private practice.

Follow-up

As stated above, data collection for the entire sample will be completed only in 1992. Currently, two subsamples have been analyzed: a group of patients with long-term follow-up of 4 years (n = 30) and a second group of patients with short-term follow-up ranging from 6 to 24 months (n = 20). Short-term follow-up assessments were not introduced at the beginning of the study. All former patients were reassessed either at home or at the clinic by the same interviewer (R.S.) who had personally supervised their inpatient treatment and who personally knew most of the patients from this period.

Follow-up assessments were conducted on the basis of an interview schedule and standardized questionnaires. In this chapter we report on data coming from an unpublished semantic differential that was originally designed by a US-American collaborative study group for the assessment of the perception of body and personality. The German version of this self-evaluation instrument was analyzed in a recent paper (Steinhausen 1991). In addition, two further questionnaires were used, namely, the Eating Attitudes Test (Garner and Garfinkel 1979) and the Eating Disorder Inventory (Garner et al. 1983). These questionnaires were not available at the beginning of the study, therefore, sample sizes for these follow-up assessments are lower. The semi-structured interview resulted in ratings of twelve areas of functioning, each rated on a four-point scale. The latter procedure was modified after Sturzenberger et al. (1977). The twelve areas were assessed and the rating scales will be introduced with the findings.

Results

Findings with regard to follow-up status and diagnostic classification will be presented first, and thereafter, the results of the former patients' self-evaluations, based on the questionnaires. Normalization of weight, i.e., within the 85%–115% range of ideal body weight, occurred in 80% of the subsample of 30 patients that had been followed up for at least 4 years. Further outcome parameters are summarized in Table 2. Normalization of the different parameters varies considerably. Clear symptoms indicative of persistent eating disorders occur only in a minority of cases. However, a

Table 2. Interview findings at long-term follow-up of 30 patients (in percent)

Finding	None	Slightly	Moderate	Severe
Abnormal reduction of food intake	56.7	23.3	16.7	3.3
Disturbed eating attitude	60.0	3.3	26.7	10.0
Vomiting	90.0	–	3.3	6.7
Bulimia	86.7	6.7	3.3	3.3
Laxative abuse	90.0	–	3.3	6.7
Abnormal menstruation	73.3	–	10.0	16.7
Avoidance of sexual matters	50.0	23.3	16.7	10.0
Avoidance of active sexual behavior	36.7	16.7	13.3	33.3
Unsatisfactory family relationships	56.7	26.7	13.3	3.3
Unsatisfactory establishment of independence from the family	76.7	13.3	10.0	–
Unsatisfactory social contacts	70.0	10.0	20.0	–
Impairment of career	90.0	–	–	10.0

Table 3. Diagnosis at admission and at long-term follow-up (n = 30)

	Diagnosis at admission	
Diagnosis at follow-up	Anorexia nervosa (n = 27)	Anorexia nervosa with bulimic features (n = 3)
Anorexia nervosa	2	0
Anorexia nervosa with bulimic features	1	0
Partly anorectic	4	1
Partly bulimic	0	1
Healthy	20	1

certain degree of abnormal reduction of food intake was still persistent in almost half of the subjects, and up to 40% showed varying degrees of disturbed attitude toward eating. On the other hand, vomiting, bulimia, and laxative abuse occurred relatively rarely. There was a marked avoidance of sexual matters and sexual activity reported for a large number of the subjects. Similarly, a considerable proportion of the sample had unsatisfactory family relationships. The disease had impaired the patient's career in only 10% of the sample.

Table 3 shows changes in the clinical picture of the same patients that occurred between the time of admission and long-term follow-up. Since the group of patients suffering from anorexia nervosa with bulimic features is very small, conclusions must be limited at this time. It is noteworthy, however, that among the subjects seen at follow-up, two of the three who had intermittently suffered from bulimia were healthy at follow-up while the third was still partly anorectic at that time. One further subject showed

Table 4. Classification at the time of short-term and long-term follow-up

	Short-term ($n = 20$)		Long-term ($n = 30$)	
	n	%	*n*	%
Anorexia nervosa	7	35	2	6.6
Anorexia nervosa with bulimic features	1	5	1	3.3
Bulimia with normal weight	1	5	0	0
Partly anorectic	3	15	5	16.6
Partly bulimic	1	5	1	3.3
Healthy	7	35	21	70.0

bulimic features at follow-up that did not exist at the time of admission. Among three patients who presented with bulimic features at the time of admission, none presented with a full-blown clinical picture at the time of long-term follow-up.

A comparison of findings coming from short-term and long-term follow-up results is given in Table 4. Taking into consideration that the two subgroups are only partly identical, there is some indication that long-term results are more favorable than short-term results. After 4 years, 70% of the former patients were assessed as clinically healthy. The second largest group consists of subjects who are of normal weight but continue to have certain anorectic symptoms such as disturbed eating behavior, irregular menstruation, and laxative abuse. One subject retained some bulimic features and was classified as partly bulimic because the ICD-10 criteria for a diagnosis of bulimia nervosa were not strictly fulfilled. Only 10% of patients is still anorectic at 4-years follow-up. Table 4 shows that the percentage of healthy subjects at the time of the short-term follow-up is only half that of the long-term follow-up.

While the findings pertaining to diagnostic classification were based on clinicians' ratings, the findings presented in the following are based on the self-evaluations of the former patients. Again, findings for the short-term follow-up and the long-term follow-up subgroups are presented. Findings yielded from data obtained at discharge from inpatient treatment are included. Figure 1 shows the total scores for the perception of body and personality as assessed by the semantic differential. High scores indicate negative self-evaluation. The findings show a significant improvement of body perception during the course of inpatient treatment that remains constant within the short-term follow-up interval. There is a similar trend for the long-term group. Self-evaluation of personality, however, does not change over time in either of the subsamples.

There is a similar trend for eating attitudes, as shown in Fig. 2. Again, a highly significant improvement takes place from admission to discharge that remains stable for both short-term and long-term follow-up periods.

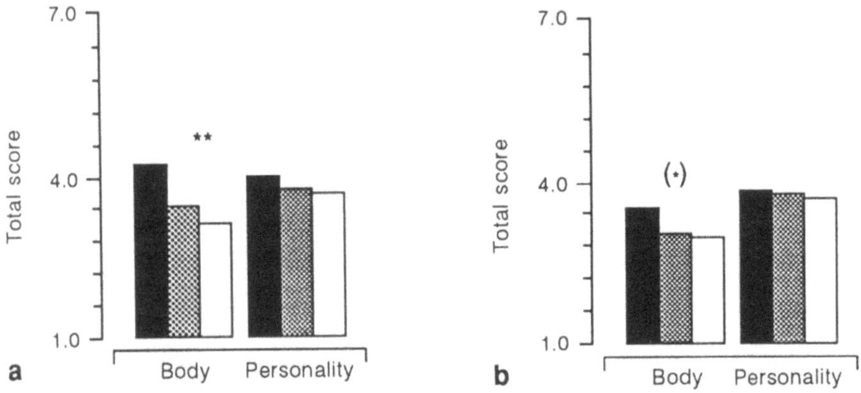

Fig. 1a,b. Total scores for perception of body and personality in two patient subgroups (**a** short-term, $n = 22$; **b** long-term, $n = 21$), based on semantic differentials at admission (*black bars*), discharge (*stippled bars*), and short-term and long-term follow-up (*open bars* in **a** and **b**). $^{**}p = 0.01$; (*) $p = 0.08$

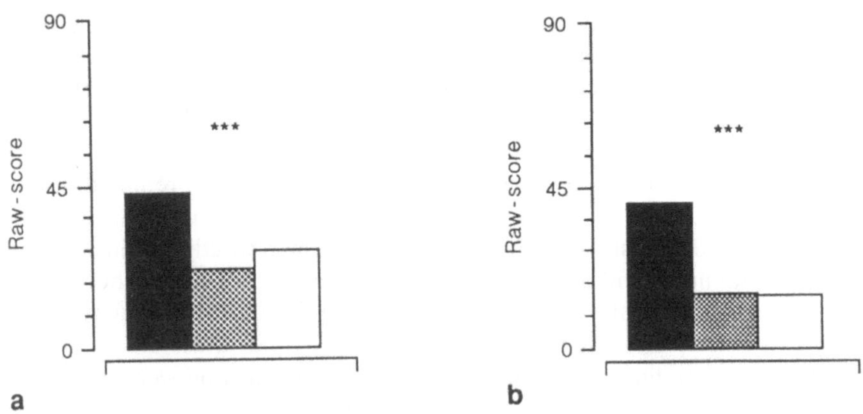

Fig. 2a,b. *EAT*, total score at admission (*black bars*), discharge (*stippled bars*), and short-term (**a**) and long-term (**b**) follow-up (*open bars* in **a** and **b**) in two patient subsamples (**a**, $n = 20$; **b**, $n = 18$). $^{***}p = 0.001$

Finally, a similar picture emerges from the multidimensional Eating Disorders Inventory. As shown in Fig. 3 changes that occur take place during inpatient treatment and are maintained up through follow-up. In both the short-term and the long-term subsamples, there is improvement with regard to the subscales drive for thinness, bulimia, and interpersonal distrust. In addition, the sense of ineffectiveness diminishes between admission and discharge for patients in the short-term follow-up subgroup.

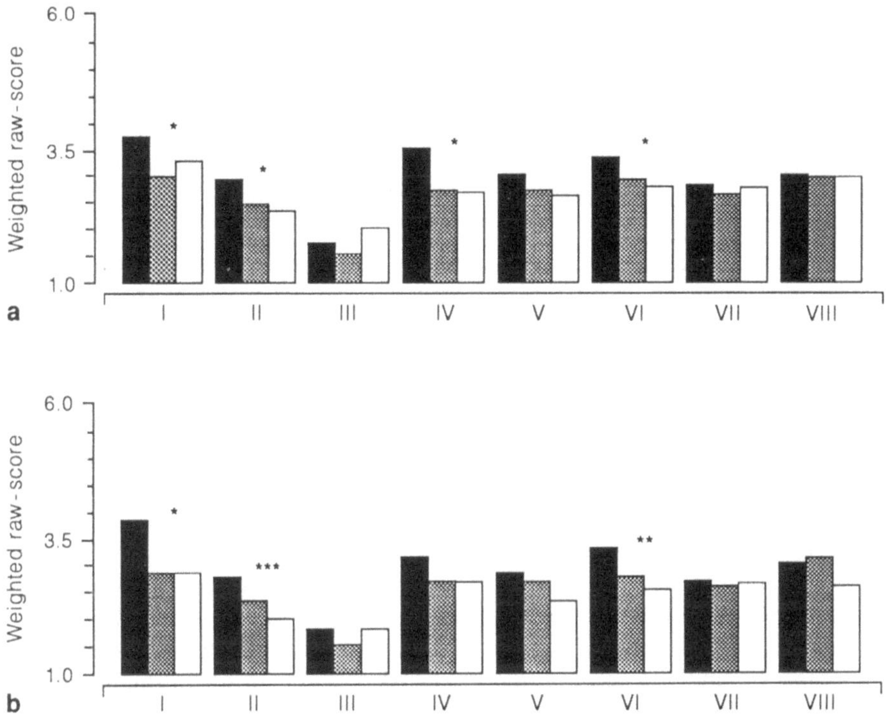

Fig. 3a,b. *EDI* scores at admission (*black bars*), discharge (*stippled bars*), and short-term (**a**) and long-term (**b**) follow-up (*open bars* in **a** and **b**) in two patient subsamples (**a**, $n = 21$; **b**, $n = 15$). *I*, drive for thinness; *II*, bulimia; *III*, body dissatisfaction; *IV*, ineffectiveness; *V*, perfectionism; *VI*, interpersonal distrust; *VII*, interoceptive awareness; *VIII*, maturity fears. $^{***}p = 0.001$; $^{**}p = 0.01$; $^{*}p = 0.05$

Discussion

As stated above, this chapter provides only an interim report on the outcome of eating disorders originating during adolescence. However, the design of the Berlin Study offers some advantages over previous studies. Repeated assessments, a multidimensional approach, and a rather narrow age range have provided new insights into the course of early-onset eating disorders.

In the subsample that has been followed up for at least 4 years, a variety of outcome parameters have been assessed. Some of the findings can be compared to the results of the studies which were summarized in our two reviews (Steinhausen and Glanville 1983a; Steinhausen et al. 1991). Given a rather large variation of results in other studies, there is some indication that our own findings indicate more favorable outcome than the rates reported for the middle range of the majority of previous studies. For instance, the mean rate of remission amounts to only 50% in the literature, while it is 70% in our own subsample. Correspondingly, our own rate of normalization of weight (80%) and menstruation (73%) is in the upper range of rates reported in the literature. Similarly, the 10% chronicity rate in our own subsample is only half of that in the middle range reported in the literature. The 3%–5% mortality rate in this sample, however, is lower than the mean of 10% reported in the majority of the studies.

Furthermore, our interview findings at long-term follow-up show that a more discerning look on the outcome data reveals certain areas where some of the former patients continue to encounter problems. These are mainly related to abnormal eating behavior and abnormal attitudes toward eating. Psychosocial functioning, especially in the areas of sexual behavior and attitudes toward sex, is another realm in which difficulties persist. In accord with our previous retrospective study (Steinhausen and Glanville 1983b), we again find a considerable number of patients who continue to be impaired with regard to psychosexual functioning – an area that has not received sufficient attention in research on eating disorders.

Conclusions with regard to the comparison of diagnoses at admission and at long-term follow-up are jeopardized by the rather small subsample. Currently, there is no indication that patients with eating disorders originating in adolescence are more at risk for acquiring or maintaining bulimic features during the course of the disease. However, definite conclusions will have to await our final analyses of the total cohort of former patients.

Whereas the vast majority of previous follow-up studies restricted themselves to only one follow-up assessment based on samples with considerable variation regarding the duration of follow-up, (see Steinhausen and Glanville 1983; Steinhausen et al. 1991), data presented in this report allow a comparison of the course of short-term and long-term outcome. There is some indication in our data supporting our general conclusion that long-term outcome is more auspicious than short-term status.

In addition to these findings, which were obtained on the basis of clinical interviews, questionnaires afforded some further information. A comparison of the various indices of eating disorders, i.e., perception of body and personality, attitudes toward eating, etc., that were obtained by self-reports at the time of admission and discharge from the hospital and intermittently during follow-up, indicates that ensuing changes occurred only during inpatient treatment. No major change occurred between discharge from the hospital and later follow-up with regard to body perception, attitude toward eating, or to the subscales of the Eating Disorder Inventory.

Again, it would be premature to draw any definite conclusion from this remarkable finding until data for the entire sample are available. The final data analysis will specifically look for the correlation of the self-report measures with clinical ratings coming from the interviews and the diagnostic status. It may be mentioned in passing that in accordance with earlier reports on this cohort (Steinhausen 1983, 1984, 1985) the Eating Attitudes Test and the Eating Disorder Inventory scores in this sample are considerably lower than those of the Canadian sample of Garner et al. (1982, 1983).

In conclusion, the present report, based on preliminary data, indicates that it is worth continuing this multidimensional approach to the study of outcome in adolescent eating disorders. Some evidence coming from this study supports the view that eating disorders originating in adolescence have a more favorable outcome than those starting in later life.

References

Garner DM, Garfinkel PE (1979) The Eating Attitudes Test: an index of the symptoms of anorexia nervosa. Psychol Med 9:273–279

Garner DM, Olmstead MP, Bohr Y, Garfinkel PE (1982) The Eating Attitudes Test. Psychometric features and clinical correlates. Psychol Med 12:871–878

Garner DM, Olmstead MP, Polivy J (1983) Development and validation of a multidimensional eating disorder inventory for anorexia nervosa and bulimia. Int J Eating Dis 2:15–34

Hsu LKG (1980) Outcome of anorexia nervosa – a review of the literature (1954 to 1978). Arch Gen Psychiatry 37:1041–1046

Schwartz DM, Thompson MG (1981) Do anorexics get well? Current research and future needs. Am J Psychiatry 138:319–323

Steinhausen H-C (1983–1985) Anorexia nervosa. Transcultural comparisons. Pediatrician 12:157–162

Steinhausen H-C (1984) Transcultural comparison of eating attitudes in young females and anorectic patients. Eur Arch Psychiatry Neurol Sci 234:198–201

Steinhausen HC (1985) Eating attitudes in adolescent anorectic patients. Int J Eating Dis 4:489–498

Steinhausen H-C (1991) The assessment of body-image distortion using a semantic differential: a methodological study. Child and youth psychiatry: European perspectives. In: Remschmidt H, Schmidt M (eds) Anorexia nervosa 1. Huber & Hogrefe, Toronto, pp 95–102

Steinhausen H-C, Glanville K (1983a) Follow-up studies of anorexia nervosa – a review of research findings. Psychol Med 3:239–249

Steinhausen H-C, Glanville K (1983b) A long-term follow-up of adolescent anorexia nervosa. Acta Psychiatr Scand 68:1–10

Steinhausen H-C, Glanville K (1983c) Retrospective and prospective follow-up studies in anorexia nervosa. Int J Eating Dis 2:221–235

Steinhausen HC, Rauss-Mason C, Seidel R (1991) Follow-up studies of anorexia nervosa: a review of four decades of outcome research. Psychol Med 21:447–454

Sturzenberger S, Cantwell PD, Burroughs J, Salkin B, Green JK (1977) A follow-up study of adolescent psychiatric inpatients with anorexia nervosa. J Am Acad Child Psychiatry 16:703–715

Swift WJ (1982) The long-term outcome of early onset anorexia nervosa – a critical review. J Am Acad Child Psychiatry 21:38–46

Long-Term Outcome in Anorexia Nervosa Inpatients

K. Engel, A.E. Meyer, M. Hentze, and M. Wittern

As anorexia nervosa is a chronic affliction with pronounced psychosomatic aspects, a long-term follow-up with comprehensive assessment of somatic, mental, and social features seem indispensable.

Such a comprehensive evaluation has been advocated by many researchers (e.g., Dally and Sargant 1966; Pierloot et al. 1975; Steinhausen and Glanville 1983).

Subjects and Treatment

The total of 218 female inpatients diagnosed and/or treated between 1963 and 1979 at the University Medical Hospital in Hamburg constituted our present sample. Male patients (n = 20) were not included. Diagnostic criteria corresponded to those described by Feighner et al. (1972), except that we did (and still do) not agree with the proposed age limit. In retrospect we found only two patients in whom the diagnosis was doubtful. Our standard treatment contained both medical and some psychotherapeutic elements. The former are: liquid nourishment through a remaining nasoduodenal tube (between 1800 and 2800 kcal/day = 7500 to 11 700 kJ/day) with administration via the tube of phenothiazines (150–300 mg/day, depending on weight and motor restlessness). Absolute bed rest serves both medical (reduction of energy expenditure) and psychologic purposes (suppression of vomiting, hiding food and/or abusing laxatives, social and perceptual isolation). Normal clinic meals are routinely offered in addition to tube feeding. In the introductory interview the psychosocial origin and the family's contribution are stressed in order to explain the prohibition of visitors, mail or phone contacts during the whole therapy. Then a "strict regime," in which the whole staff cooperates, denounces the patient's anorectic values and life style, and positively reinforces eating and normal weight (for details see Frahm 1966; Engel and Meyer 1982). This standard treatment is called *type I* and 164 patients (72.5%) received it (see Table 1).

Type II was employed for patients who showed a clear weight gain already during the diagnostic phase. Here, tube feeding and phenothiazine

Table 1. Data of inpatient therapy (n = 218)

	Type I (n = 164)		Type II (n = 20)		Diagnosis only (n = 34)	
	Mean	SD	Mean	SD	Mean	SD
Duration of illness (years)	3.4[b]	4.2	4.2	4.5	4.8[b]	3.5
Age of admission (years)	23.0[b]	7.1	23.5	7.2	27.5[b]	8.6
Total inpatient period (weeks)	7.6[a,b]	3.3	4.4[a]	2.3[a]	1.4[b]	0.7
Weight at admission (kg)	37.1[a]	5.6	41.9[a,c]	6.2	39.8[c]	7.4
Difference from "ideal weight" (%)[d]	−36.9	8.9	−28.0	7.2	34.2	11.5
Weight change during therapy (kg)	11.3[a,b]	5.2	5.4[a,c]	5.5	1.8[b,c]	3.1
Difference from admission (%)	32.5	18.0	14.5	15.6	4.8	8.5

[a] Statistically significant difference in type I vs type II.
[b] Statistically significant difference in type I vs diagnosis only.
[c] Statistically significant difference in type II vs diagnosis only.
[d] See text for definition.

medication were omitted, but all the other modalities applied. Only 20 patients within our sample (9.2%) qualified for this alleviation.

A third group of 34 patients (15.6%) stayed for diagnostic evaluation only and did not receive either of the inpatient treatments described above. This sample is called "diagnosis only" and is a heterogeneous group. Some of its members more or less plausibly argued that they would seek therapy elsewhere or later (e.g., after having organized caretaking for their children) or that they would solve their problem on their own. Others rejected our therapy or had to be transferred to a closed psychiatric ward because of suspected suicidal risk. Table 1 lists some important characteristics for these three subsamples. There is no consensus on how to operationalize either "normal" or "ideal" weight. For this study we defined ideal weight as BROCA-10% = .9 × BROCA; where BROCA = height (in cm) − 100 = weight (in kg).

Our three samples differ already before therapy (i.e., when admitted to our clinic), and hence might not be comparable. This is obviously true but ethically unavoidable. To impose tubefeeding with phenothiazines for research purposes on patients who did not need this (= Type II) would be highly unethical. In this (and similar) clinical situations we only have the alternative of not evaluating therapy at all or trying as best we can to assess differences in outcome and weigh these in respect to pretherapy differences. These differences indicate that the anorexia patients requiring type I treatment are more cachectic in a more acute course, and in spite of the more radical treatment (tube feeding plus phenothiazines added) they take longer to obtain the same weight. Or formulated reciprocally: Type II patients are more compliant.

The patients in the diagnosis-only subsample are older than those in the other two (27 vs 23 years), are ill longer than type I subjects, have a weight deficit inbetween the others (−34.2% vs −36.9% and −28.0% respectively), evidently stay in the hospital for a shorter length of time (1.4 weeks), gain very little weight (1.8 kg) and are discharged with a still substantial weight deficit (40.1 kg = −30.3%).

Follow-up Methods

As stated, one of our aims was to evaluate a maximum of former patients. Thus, if our first letter came back "addressee unknown," we contacted registration offices, relatives, the doctors who originally referred the patients, and/or visited former neightbors to obtain the present address (or the fact and circumstances of death) of our former patients. If a second letter remained unanswered, we phoned the former patients and tried to motivate them for full follow-up. If they refused, we then tried to persuade them to fill in a standardized interview for anorexia nervosa questionnaire for self-assessment (SIAN-S; see below). If they disagreed to the latter, we asked them questions pertaining to core variables. This detective work – for which we thank our dedicated, ingenious, and painstaking technical staff – enabled us to obtain information about 202 (92.7%) of the 218 former patients in spite of the long period since treatment or observation (mean = 9.8 years, range 1 to 17 years).

The full follow-up assessment comprised a personal interview, a standardized interview (SIAN-Ex, a rating form for experts, i.e., trained interviewers, in our case trained and supervised undergraduates in psychology or medicine), SIAN-S, and an endocrine checkup. Both SIAN-Ex and SIAN-S were developed by M.M. Fichter, whom we thank for making them available to us. For methodological details see Fichter et al. (1989).

In the patients for whom we obtained both SIAN-Ex and SIAN-S (n = 90) we checked for significant differences between the two methods (Wilcoxon matched pairs) and found none. Numerically no difference between paired arithmetic means surpassed 0.3. This equally applies to our composite scores (see below), e.g., global rating of symptoms attained a mean score of 1.0 in the SIAN-Ex and a mean score of .8 in the SIAN-S (Hentze 1983).

Naturally, the willingness for and the corresponding degree of cooperation varied across our sample of former patients. For instance, only a subsample (n = 38) consented to the time-consuming endocrine assays. As these results are published separately (Meyer et al. 1986), this subsample is not described here.

Table 2 describes the degree of cooperation obtained. Whereas over 50% consented to a personal interview (plus questionnaires) which usually took place in our clinic (exceptionally in the former patient's home), only

Table 2. Degree of follow-up cooperation (n = 218)

	No. of patients	Percentage
Full information (except endocrinology)	114	52.3
Questionnaire only (SIAN-S)	22	10.1
Minimum information (by telephone)	36	16.5
Deceased	30	13.8
Not retrievable	16	7.3
Total	218	100.0

SIAN-S, standardized interviews for anorexia nervosa-self-assessment.

10% consented to fill in the SIAN-S (plus some additional questions), and another 16% could only be persuaded to give information by phone. Here we focussed on topics concening somatic (weight, menstruation) and psychosocial variables (personal situation and working environment), which allowed a crude rating of their general state of health.

In a subsample (n = 105) with longer follow-up (mean interval = 13.5 years) Hentze (1983) checked (with Kruskal-Wallis one-way ANOVA) whether the degree of cooperation covaried with our two global categories (see Tables 3 and 5) and with our "rating of overall deficit" which combined the former two into one assessment. We found significant differences between the arithmetic means for our rating of overall deficit with the highest (i.e., more pathological) mean of 2.0 for the "phone only" responders and the lowest mean of 1.1 for "questionnaire only." The "full information" subgroup was in between (mean = 1.3). This was above all due to the somatic category. The means for the psychologic category went in the same direction without reaching significance. We think that this inverted U curve depicts reality and is not a product of dissimulation or aggravation. If this is true (and if it can be generalized), this has consequences for follow-up studies in anorexia nervosa. The cooperating patients who cooperate best are those who are better but not quite well, whereas those who are almost well only make a small contribution. Patients who are still very ill or have experienced a relapse tend to be reluctant to be followed up.

Our former patients were asked to answer the SIAN-Ex and SIAN-S questions twice: once for the present (= follow-up) and a second time for the period when their illness was worst. Usually choosing the moment of admission to our inpatient treatment. We are fully aware that this retrospective information constitutes very "soft" data. From the SIAN-Ex or SIAN-S (the latter for patients who only consented to fill in questionnaires, see Table 2) we chose the items reproduced in Table 3.

Our choice of items was guided by the experience and recommendations given by other authors (Deter 1981; Pierloot et al. 1975; Schütze 1980; Selvini Palazzoli 1974).

Table 3. Topics explored in follow-up for the present and retrospectively for the beginning of therapy

- Somatic symptoms[a]
  - Weight difference to ideal according to BROCA
  - Menstrual disorders
  - Obstipation
  - Vomiting
  - Laxative abuse
  - Bulimia
- Psychological symptoms
  - Psychological symptomatology
    - Striving for slimness
      - Obsessional thinking concerning eating, food, calories
      - Ideal of slimness
      - Deliberate fasting
    - Distorted body image
      - Distortions of body image
      - Lack of insight into being ill/denial of illness
    - Distorted perception or cognitive interpretation of proprioceptive stimuli
      - Distorted perception of proprioceptive stimuli
      - Hyperactivity
      - Importance of achievement
  - Sexual disorders
    - Decrease of libido
    - Sexual anxieties
  - Lack of self-esteem/depressed state of mind
    - Sense of personal ineffectiveness, weakness, helpnessness
    - Depressive mood
    - Self-esteem
    - Self-confidence
- Social variables
  - Lack of strong and satisfactory partnership
    - Relationship with partner[b]
    - Character of relationship to the actual partner[b]
    - Common activities/internal integration
    - Contentment with marriage/partnership
  - Lack of occupational integration
    - Occupational/school performance[b]
  - Emotional dependence on family of origin
    - Living situation
    - Closeness of mother relation
    - Closeness of father relation
    - Problem solving
  - Lack of integration into social environment[b]
    - Social integration
    - Amount of social contacts
    - Social withdrawal

[a] In forming global ratings, we calculated, the arithmetic means of the sum of single ratings; in those cases where single ratings were particularly high these values were used for the whole field concerned.

Rating of weight:

(0)	= 0%–9%	difference from "ideal" weight
(1)	= 10%–19%	
(2)	= 20%–29%	
(3)	= 30%–39%	
(4)	= 40%	

[b] Of the psychosocial items only these are employed in the rating of overall deficit (see Table 5).

To form the three global and overall ratings (see Table 5) the following procedure was employed. For a rating of somatic disturbances A 1 (weight), A 4 (vomiting) and A 5 (laxative abuse, see Table 3) received priority: if one (or several) of these three ratings were higher (i.e., more pathologic) than the remaining five, then the highest value was taken as representative for somatic disturbance. If there was no such maximum, the arithmetic mean of the six somatic items was employed. For the psychologic and social variables (in Table 3) the same procedure of maximal score was employed, but only within subsets, namely B 1.1, B 1.2, B 1.3, B 2, B 3, C 1, C 2, C 3, and C 4. Then the arithmetic means for B became the *gobal psychosocial impairment*. The arithmetic mean of A plus B became *global rating of symptoms*. For the *rating of overall deficit*, i.e., our global outcome measure, only the arithmetic mean of C 1.1, C 1.2, C 2.1 and of the rating of C 4 was used. This partially sociologic mean was added to the value for the global rating of (somatic plus psychological) symptoms and both then averaged.

Results

In the following sections we describe the outcome for the 172 patients from whom we obtained at least information by telephone (see Table 2). Hence, this sample is only a subset of the one portrayed in Table 1.

Somatic Variables

In Table 4 the course of four somatic variables is listed for the three subsamples. Concerning *weight* before therapy (i.e., at admission to our hospital), type I patients had a significantly lower weight (37.2 vs 42.3 and 40.3 kg) and type II patients a significantly higher one (see Table 4). At follow-up we find a partial reversal: here the diagnosis only group is lowest, whereas type I and II patients are on par (this would be significant if type II had a large N, see Table 4). Clinically, however, the follow-up weight differences are not impressive as they do not surpass 3 kg. Probably more relevant is the fact that for all three groups average follow-up weight is still rather low (50.7, 50.6, and 48.5 kg), indicating that some of these women are still clearly underweight (see Table 4). In regard to intrasample differences over time, type I and type II patients significantly gain weight during their inpatient phase: type I patients (in addition) and the diagnosis only group (exclusively) during follow-up.

Menstrual disorders before therapy are somewhat, but not significantly less frequent in diagnosis only patients (82% vs 95%). This relation still persists at follow-up on a lower plane: 23% vs 43% in type I patients ($p <$ .10). Incidentally, 23% is the quota of menstrual disorders we found in a

Table 4. Comparison of somatic core variables (n = 172)

	Type I (n = 128)						Type II (n = 19)						Diagnosis only (n = 25)					
	Before therapy		After therapy		At follow-up		Before therapy		After therapy		At follow-up		Before therapy		After therapy		At follow-up	
	Mean	SD	Mean	SD	Mean	SD	Mean	SD	Mean	SD	Mean	SD	Mean	SD	Mean	SD	Mean	SD
Weight (kg)	37.2[a,b,d]	5.4	48.1[d,f]	5.2	50.7[b,f]	7.4	42.3[a,c,d]	6.1	47.4	5.6	50.6	7.5	40.3[b,c]	7.0	41.8[e]	7.3	48.5[b,e]	9.9
Difference from ideal weight (%)	−36.8	8.8	−18.4	7.8	−13.8	12.4	−27.8	7.4	−19.6	6.4	−13.3	12.7	−32.7	10.9	−28.8	11.8	−18.3	18.3
Menstrual disorder (%)	96.3[f]				43.6[f]		94.1[f]		–		42.9[f]		82.4[f]		–		23.5[f]	
Vomiting (%)	42.9[f]				19.2[f]		58.8[f]		–		21.1[f]		40.0[f]		–		23.5[f]	
Obstipation (%)	75.0[f]				30.4[f]		64.7[f]		–		27.8[f]		56.3		–		35.3	

[a] Statistically significant difference in type I vs type II.
[b] Statistically significant difference in type I vs diagnosis only.
[c] Statistically significant difference in type II vs diagnosis only.
[d] Statistically significant difference from before to after therapy.
[e] Statistically significant difference from after therapy to follow-up.
[f] Statistically significant difference from before to after follow-up.

normal control group. As can be seen in Table 4, all the within-sample reductions are significant ($p = < .001$).

Vomiting before beginning with therapy is higher in type II than in the other two groups (59% vs 43% and 40%; not significant) and drops to roughly 20% at follow-up (see Table 4). The reductions in vomiting from before therapy to follow-up are all significant ($p < .001$, .01, and .01).

Obstipation is significantly reduced in type I (75%–30%) and type II (64%–28%) patients. The diagnosis only group has a lower (not significant) pretherapy quota of obstipation than the other two samples and its reduction at follow-up (56%–35%) is not statistically significant.

Global Rating of Symptoms, Psychosocial Impairment, and Overall Deficit

We used frequency (of patients) to assess the outcome of treatment for anorexia nervosa, assigning the rates to categories of increasing impairment. As death is the absolute extreme of negative outcome, it seemed logical and necessary to include mortality as a sixth category. For the statistical comparison of our three samples we dichotomized the six categories into none plus slight versus the remaining (more disturbed) categories, then totaled their frequencies and computed χ^2 (with df = 1). As pointed out above, sample differences are already apparent before therapy. Thus, with regard to global psychosocial impairment the diagnosis only sample contains a significantly higher quota of less disturbed patients than the other two samples (52.6% vs 28.9% and 18.7%). The similar advantage for the diagnosis only group in rating overall deficit before therapy (24% vs 4.8% and 15.8%, which only attains significance versus type I patients) is spurious. This is a consequence of the described difference in global psychosocial impairment of which four items are included to form rating of overall deficit. Concerning significant differences within our three samples from before therapy to follow-up, both type I and type II show significant improvements in the global ratings of symptoms and in global psychosocial impairment, and consequently also in their combination, i.e., the rating of overall deficit. For the diagnosis only sample we only find a significant gain in global rating of symptoms (from 0% to 36.7%), but no gain in psychosocial impairment (52.6%–59.4%). Their combination into the rating of overall deficit (although there is a clear gain from 24% before to 46.8% at follow-up) does not attain significance.

Comparing our three samples in respect of their follow-up results, we find that the differences in mortality (type I = 16%, type II = 0%, and diagnosis only = 23.3%) are just short of being statistically significant. In the global rating of symptoms at follow-up, the diagnosis only group (with 36.7% of its patients in the none plus slight categories) lies significantly behind type I (= 62.3%). The type II sample (with 50%) lies inbetween and differs from neither. The follow-up quotas for minor social impairment

Table 5. Patient frequencies in the outcome rating categories (total $n = 202$)

Categories	Global rating of symptoms				Global impairment				Rating of overall deficit			
	Before		Follow-up		Before		Follow-up		Before		Follow-up	
	(*n*)	(%)	(*n*)	(%)	(*n*)	(%)	(*n*)	(%)	(*n*)	(%)	(*n*)	(%)
Type I ($n = 151$)												
None	1	0.8	44	30.8	4	3.7	35	23.2	1	0.8	24	15.9
Slight	1	0.8	45	31.5	27	25.2	49	32.5	5	4.0	62	41.1
Marked	11	9.3	17	11.9	39	36.5	26	17.2	43	34.1	28	18.5
Severe	50	42.4	12	8.4	28	26.2	10	6.6	62	49.2	13	8.6
Extreme	55	46.7	2	1.4	9	8.4	8	5.3	15	11.9	1	0.7
Deceased	0	0.0	23	16.0	0	0.0	23	15.2	0	0.0	23	15.2
Total	118	100	143	100	107	100	151	100	126	100	151	100
Type II ($n = 19$)												
None	0	0.0	4	22.2	1	6.2	2	10.5	1	5.3	3	15.8
Slight	0	0.0	5	27.8	2	12.5	12	63.2	2	10.5	10	52.6
Marked	6	33.3	7	38.9	7	43.8	5	26.3	4	21.1	5	26.3
Severe	6	33.3	2	11.1	5	31.3	0	0.0	10	52.2	1	5.3
Extreme	6	33.3	0	0.0	1	6.2	0	0.0	2	10.5	0	0.0
Deceased	0	0.0	0	0.0	0	0.0	0	0.0	0	0.0	0	0.0
Total	18	100	18	100	16	100	19	100	19	100	19	100
Diagnosis only ($n = 32$)												
None	0	0.0	3	10.0	4	21.1	8	25.0	1	4.0	3	9.3
Slight	0	0.0	8	26.7	6	31.5	11	34.4	5	20.0	12	37.5
Marked	5	26.3	6	20.0	4	21.1	1	3.1	9	36.0	6	18.7
Severe	9	47.4	3	10.0	3	15.8	3	9.4	6	24.0	2	6.3
Extreme	5	26.3	3	10.0	2	10.5	2	6.2	4	16.0	2	6.3
Deceased	0	0.0	7	23.3	0	0.0	7	21.9	0	0.0	7	21.9
Total	19	100	30	100	19	100	32	100	25	100	32	100

(55.7%, 73.7%, and 59.4%) do not differ significantly. The rating of overall deficit at follow-up (57%, 68.4%, and 46.8%) still mirrors the lower position of global symptoms of diagnosis only, but the overall deficit quotas do not attain statistical significance.

Demographic Status of the Combined Sample at Follow-up

Our global rating of psychosocial impairment (see Table 5) only compared data within and between our three samples. We also assessed the socio-demographic position of the sample within the general female population. Our combined sample (n = 172) had a mean age of 36 years. Hence, demographic variables were compared with the general female population of Germany in the age range of 30–35 years (n = 3 279 619; according to data from the Federal Statistical Office).

As expected, patients in the follow-up sample were more frequently unmarried (47.4% vs 11%; $p < .001$) than the corresponding general female population. There is no statistically significant difference concerning separated, divorced, or widowed individuals. This already implies a scarcity of housewives or part-time workers which is borne out (33.1% vs 50.5%; $p < .001$). Accordingly, we find an excess (39.6% vs 25.9%; $p < .001$) of full-time workers. However, we also find an overproportional number of unemployed (7.65% vs 3.6%; $p < .01$). In view of the clinically well known overzeal and hyperactivity of anorexia nervosa patients, unemployment is probably a consequence of disability caused by the

Table 6. Demographic status at follow-up (n = 172)

	Patient sample at follow-up		General female population[a] (%)	Significance
	(n)	(%)		
Family status				
Unmarried	82	47.4	11	$p < .001$
Married/remarried	72	41.9	81	$p < .001$
Separated/divorced/widowed	17	9.8	7.6	$p > .05$
No information	1	0.6	–	
Occupational status				
Working full-time	68	39.6	25.9	$p < .001$
Working in household/part-time	57	33.1	50.5	$p < .001$
Undergoing education	25	14.5	19.9	$p > .10$
Unemployed	13	7.6	3.6	$p < .01$
No information	9	5.2	–	

[a] Percentages for the female population between 30 and 35 years in the FRG in 1983 (n = 3 279 619), according to the Federal Census Bureau.

Table 7. Causes of death during follow-up interval

Cause	No. of patients (n = 218)	Percentage[a]
Cachexia	13	6
Suicide	7	3.2
Neoplasm	3	1.4
Other[b]	2	0.9
Unknown	5	2.3
Total	30	13.8

[a] Of total sample (n = 218).
[b] One death due to gastric hemorrhage and one after cholecystectomy.

disease. These figures show that our follow-up sample sociodemographically still deviates from the normal.

Mortality

Thirty (13.8%) of our 218 anorexia nervosa patients had died during the follow-up interval at the average age of 28.8 years. The causes of death – as far as could be ascertained – are listed in Table 7. If we add suicide to cachexia as probably being associated with anorectic psychopathology, mortality is due to anorexia nervosa in two thirds (20 of 30) of patients.

We were able to obtain additional information about 25 of these deceased patients. Their weight before death averaged 32.6 kg (−41.6% of their ideal weight). At discharge from our hospital they weighed 39.6 kg, which can be compared with the discharge weights of survivors of the type I and diagnosis only groups together (mean = 47.07 kg; n = 153). Thus, at the moment of death they had on average lost 7 kg of their discharge weight.

In Table 8 the social situation of the deceased patients is compared with the sample of survivors. In making this comparison we must be fully aware that the subsample of survivors already strongly deviates from the normal population (see Table 8).

The great majority of our deceased patients (82.3% vs 15.8%; $p < .001$) were living with their parents and (reciprocally) had no family or family equivalent of their own (0% vs 48.5%; $p < .001$). This points to a dependence on the family of origin. In consonance with this, not a single deceased woman was married (0% vs 42.1%; $p < .001$). Finally, our subsample of deceased patients contains significantly more unemployed/retired/incapacitated women than the group of survivors (71.6% vs 8%; $p < .001$). The lower quota of productive activity in deceased patients is

Table 8. Socioeconomic situation of living vs deceased patients

	Deceased		Living		Significance
	(*n*)	(%)	(*n*)	(%)	
Living situation					
With parents	14	82.3	27	15.8	$p < .001$
With partner or with partner and children or children only	0	0	88	48.5	$p < .001$
With friends, acquaintances	1	5.9	9	8.2	n.s.
Alone	2	11.8	42	24.6	n.s.
Other	0	0	5	2.9	
Total	17	100	171	100	
Marital status					
Unmarried	14	82.3	82	48.0	$p < .01$
Married	0	0	72	42.1	$p < .001$
Separated/divorced/widowed	3	17.7	17	9.9	n.s.
Total	17	100	171	100	
Occupational situation					$p < .001$[a]
Working full-time	1	7.1	68	41.8	
Working in own household	1	7.1	33	20.2	
Working part-time	1	7.1	19	11.6	
Helping family or husband	0	0	5	3.1	
Undergoing education	1	7.1	25	15.3	
Unemployed	5	35.8	13	8.0	
Prematurely pensioned	3	21.5	–	–	
Incapable of earning own living	2	14.3	–	–	
Total	14	100	163	100	

[a] All working or being educated vs unemployed/pensioned/disabled: df 1.

very probably not so much the result of dependence on the family of origin, but the consequence of pronounced and chronic cachexia.

Discussion

When we try to evaluate our results in comparison to those published by other authors, this obviously poses formidable problems due to selection factors and differences in follow-up procedures. The influence of these factors can be illustrated by the following findings. We employed a follow-up strategy with three levels of patient commitment in descending order: full follow-up, then (if refused) only questionnaires, then (if refused) phone interviews. Those consenting to full follow-up appeared to be more impaired than those only consenting to questionnaires, whereas the most pathologic mean rates were found in the group only providing phone information. If this can be extrapolated, authors who employ only one of these follow-up methods will obtain results which are biased accordingly.

Mortality illustrates the influence of both retrieval quota and length of follow-up. As the dead cannot provide information, follow-ups with a low retrieval quota probably mean part of the mortality will go undetected. We have also found that mortality dramatically rises with longer follow-up: 4.5% in the first 5 years, 23.8% in the range of 6–17 years after discharge. Among the outcome criteria mortality is the most unequivocal or "hard" datum. Its arithmetic means are surprisingly similar in other follow-up studies: referring to recent overviews we found 6.2% in Hsu (1980), 6.8% in Fichter (1985), and 5.7% in Steinhausen and Glanville (1982). The mortality in our present sample (mean = 13.8%; Table 7) appears catastrophically higher. However, this very plausibly is a consequence of our longer follow-up interval and our high retrieval quota. Consonantly, Theander (1983) in his long-term study found a mortality of 6% in his 5-year follow-up, 12% after 10 years, and 15% after more than 20 years.

Our results furnish additional information on mortality. There is a subgroup which can be identified by its compliance to gain weight under some pressure (type II). Hence, psychotherapy clinics or other institutions which select anorexia nervosa patients for outcome studies according to their motivation for therapy (e.g., Potreck-Rose 1987) will have minimal mortality in their follow-ups. Furthermore, mortality is strongly associated with remaining in the family of origin (82%) or living alone (11%) and thus not having a psychosexual partner or children (see Table 8). This is a trend already significantly present in our total sample (n = 172), when the latter is compared to the age-related female population (see Table 6). However, our deceased subsample strongly surpasses this deviation in the same direction. Although we obviously cannot discern cause from effect, we speculate that psychotherapy should encourage the forming of new relationships – preferably heterosexual ones (or friendships) – while relinquishing old ties to the family of origin.

Concerning somatic variables we think that remenorrhea is more telling than weight, because in a psychoendocrinological study of a subset of our present sample (Meyer et al. 1986) we found that remenorrhea is a functional indicator for biopsychosocial adaptataion. At follow-up 56% (types I and II) and 76% (diagnosis only; no significant difference; see Table 4) were free of menstrual disorders. Here again, our findings point to the importance of the follow-up interval. In the aforementioned study (Meyer et al. 1986) we found that the sample with remenorrhea versus the one which still (or again) showed amenorrhea had a significantly longer mean of follow-up interval (10 vs 6 years).

Conclusion

To the many varying factors already discussed which influence long-term outcome we must add the lack of standardization. All these make our comparisons with other authors quite vague. Nevertheless, we have gained the impression that the outcome of our two therapies is situated in the upper region of this spectrum of success. It may even be preferable for anorexia patients with a low weight attained in a comparatively short time (acute course).

As all our between sample differences are discrete, we think that the natural course of anorexia nervosa is more dominant than (at least our two) forms of therapy. Within this natural course psychosocial adaption seems important as we were able to demonstrate in the investigation of prognostic factors of anorexia nervosa (Engel 1988) and in the psychoendocrinological study (Meyer et al. 1986). In the future we will try to focus on these areas.

Acknowledgements. We gratefully acknowledge support from the Deutsche Forschungsgemeinschaft. The authors thank their patients for their cooperation, Ute Grünthal and Ilse Prockl-Pfeiffer for their dedicated effort to retrieve and motivate the former. Christa Flebbe and Ruth Limpinsel unweariedly typed several versions of this manuscript, Feodora Hinz the last one.

References

Dally P, Sargant W (1966) Treatment and outcome of anorexia nervosa. Br Med J 2:793–795

Deter H-C (1981) Methodik von katamnestischen Untersuchungen bei psychosomatischen Patienten. Psychother Med Psychol 31:48–52

Engel K (1988) Prognostic factors in anorexia nervosa. Psychother Psychosom 49: 137–144

Engel K, Meyer A-E (1982) Theorie und Empirie einer mehrfraktoriellen stationären Anorexie-Therapie für schwer erkrankte Patientinnen. Med Welt 33:1812–1816

Feighner JP, Robins E, Guze SB, Woodruff RA, Winokur G, Munor R (1972) Diagnostic criteria for use in psychiatric research. Arch Gen Psychiatry 26:57–62

Fichter MM (1985) Magersucht und Bulimia. Springer, Berlin Heidelberg New York

Fichter MM, Elton M, Engel K, Meyer A-E, Poustka F, Mall H, von der Heydte S (1989) The "Structured Interview for Anorexia and Bulimia Nervosa" (SIAB): development and characteristics of a (semi-)standardized interview. In: Fichter MM (ed) Bulimia nervosa: basic research, diagnosis and therapy. Wiley, Chichester, pp 57–70

Frahm H (1966) Beschreibung und Ergebnisse einer somatisch orientierten Behandlung von Kranken mit Anorexia nervosa. Med Welt 39:2001–2003, 2068–2072

Hentze M (1983) Katamnestische Langzeituntersuchungen an 105 Anorexia nervosa-Patientinnen. Thesis, University of Hamburg

Hsu LKG (1980) Outcome of anorexia nervosa: a review of the literature (1974–1978). Arch Gen Psychiatry 37:1041–1046

Meyer A-E, Holtzapfel von B, Deffner G, Engel K, Klick M (1986) Psychoendocrinology of remenorrhea in the late outcome of anorexia nervosa. Psychother Psychosom 45:174–185

Pierloot RA, Wellens W, Houben M (1975) Elements of resistance to a combined medical and psychotherapeutic program in anorexia nervosa: an overview. Psychother Psychosom 26:101–117

Potreck-Rose F (1987) Anorexia nervosa und Bulimia – Therapieerfolg und Therapieprozeß bei stationärer verhaltenstherapeutischer Behandlung. Deutscher Studienverlag, Weinheim

Schütze G (1980) Anorexia nervosa. Huber, Bern

Selvini Palazzoli M (1974) Self-starvation. From the intrapsychic to the transpersonal approach in anorexia nervosa. Chaucer, London

Steinhausen HC, Glanville K (1983) Follow-up studies of anorexia nervosa: a review of research findings. Psychol Med 13:239–249

Theander S (1983) Research on outcome and prognosis of anorexia nervosa and some results from a Swedish longterm study. Int J Eating Dis 2(4):167–174

The German Longitudinal Bulimia Nervosa Study I

M.M. Fichter, N. Quadflieg, and W. Rief

Introduction

Little is known about the long-term course and outcome of bulimia nervosa since the diagnostic criteria for this disorder have been formulated quite recently (Russell 1979; American Psychiatric Association 1980, 1987). About a dozen studies regarding outcome data on bulimia nervosa have recently been published (Abraham et al. 1983; Brotman et al. 1988; Fairburn et al. 1986; Herzog 1990; Herzog et al. 1990; Hsu and Holder 1986; Lacey 1983; Mitchell et al. 1986, 1989; Norman and Herzog 1986; Pope et al. 1985; Swift et al. 1987; Yager et al. 1987). Because of differences in research design, instruments of assessment, varying follow-up intervals, method of assessment (questionnaire, interview in person, or interview by telephone) and different diagnostic and outcome criteria, a direct comparison of these studies is hardly possible.

The "German Longitudinal Bulimia Nervosa Study" was designed on the basis of our clinical experience and the preliminary results of the above cited longitudinal studies on bulimia nervosa with relatively small patient samples and short follow-up intervals. The aims of our study were: (1) The assessment of the natural course of bulimia nervosa in a very large sample of severely bulimic inpatients during inpatient treatment and a 2-year follow-up interval; (2) the assessment of the course of illness in a prospective design over the years after discharge from inpatient treatment; (3) in order to avoid selection biases, we aimed for a very high completion rate; (4) it was intended to assess eating behavior and attitudes towards food and eating as well as the general psychopathology; (5) the study was designed to identify risk factors of relapse and unfavorable course and outcome; and (6) the study was designed to assess comorbidity with other mental disorders (affective illness, substance abuse, anxiety disorders, personality disorders).

The relationship between eating and affective disorders has been widely debated. A longitudinal study assessing eating pathology as well as general psychopathology may by its very nature (course over time, investigator-initiated contact, serial determinations) provide an opportunity to elucidate the relationship between bulimia nervosa and depression. While some have contended that depression is mainly a secondary feature of bulimia nervosa (Fairburn 1984; Fichter 1989; Sabine et al. 1983; Swift et al. 1987; Russell 1979), others have seen bulimia nervosa as a variant expression of a primary mood disturbance (Hudson et al. 1982; Pope and Hudson 1984).

Method

Sample Characteristics

Inpatient Sample. Some 250 patients (9 males, 241 females) with bulimia nervosa according to the DSM-III-R criteria were treated as inpatients in a hospital for behavioral medicine, Klinik Roseneck in Prien, Upper Bavaria, between May 1985 and June 1988. Of these, 86 patients (34.4%) had a history of anorexia nervosa. The age on admission was on the average 25 years 8 months with a standard deviation of 6 years 6 months. The age of onset of the bulimic disorder was 17 years 9 months (±SD 5 years). In comparison to other studies, patients had a very long duration of illness (7 years 11 months ±4 years 10 months). The inpatient treatment was on the average 100 (±45) days. The duration of the follow-up interval was 2 years 1 month (±9 months). Because it has been shown in previous studies on anorexia and bulimia nervosa that patients who refused to participate in follow-up assessments have rather high degrees of psychopathological illness (Fichter 1985; Swift et al. 1987) we made strong efforts to obtain an approximation of a 100% completion rate for the follow-up assessments. Of the total sample of 250 inpatients, we were able to reassess 247 patients (98.8%) at the 2-year follow-up: 222 patients had complete follow-up assessments (comprehensive questionnaire and subsequent interview by therapists); 10 patients participated in the detailed interview only; 1 patient filled out the comprehensive questionnaire, but did not participate in the interview; and in 14 cases we were able to hold at least a short interview at follow-up; 1 patient had died at age 24 years, $1\frac{1}{4}$ years after treatment in the Klinik Roseneck. After discharge from the hospital she had two further inpatient treatments in other hospitals and possible reasons for death were low body weight (32 kg; body mass index, BMI = 12.2), low plasma potassium levels, and severe alcohol abuse.

Untreated Sample for Comparison. In all, 1624 women with bulimic symptoms responded to a press survey in West Germany. In 1984–1985 they filled out a comprehensive questionnaire which was almost identical to the questionnaire used on admission and discharge in the inpatient sample. Based on this questionnaire those individuals of the press survey sample who fulfilled the criteria for bulimia nervosa according to DSM-III-R were selected; 4 months before the scheduled follow-up assessments these people received a brief questionnaire assessing the use of medical and psychotherapeutic services during the follow-up interview. From the press survey sample 79 women with bulimia nervosa according to DSM-III-R criteria in 1984–1985 were selected for comprehensive follow-up assessments; these women had neither had inpatient psychotherapy nor outpatient therapy of 50 sessions or more before the initial contact in 1984 or in the follow-up interval. Thus, this group of bulimic women does not constitute a control group to the inpatient sample. It is rather a quite differently selected group. Although women in this "untreated comparison sample" did fulfill the diagnostic criteria for bulimia nervosa according to DSM-III-R at initial assessment, their eating disturbances were less severe and of shorter duration, and they generally showed a lesser degree of psychopathological illness. The "untreated sample" was used for comparison and for reference, and we hypothesized that course and outcome in this group would be more favorable than in the severely disturbed bulimic inpatients. The average age of the 79 selected untreated bulimic women from the press survey sample at initial assessment was 26 years 11 months (±7 years 3 months).

Design and Instruments

Design. Treated inpatients were assessed at three points in time: at admission, discharge from hospital, and 2-year follow-up. The follow-up assessments were performed in two steps: All patients received a comprehensive 27-page questionnaire which contained the scales mentioned below and several additional items. After the patient had completed the questionnaire, he or she was contacted and a detailed 23-page interview, covering largely the same areas as the questionnaire, was conducted. For methodological reasons, the interview was conducted by a physician or clinical psychologist during a visit at the patient's home in a regional (Bavarian) sample of 62 subjects (home visit interview). In 184 subjects a physician or clinical psychologist (therapist) made a detailed interview by telephone (therapist interview by phone). In both types of interviews the interviewer had knowledge of details in the patient's medical chart and of the comprehensive follow-up questionnaire which the patient had previously filled out. At the beginning of the study, patients had given us permission to also contact their family physician and obtain data on diagnoses and use of medical services from their insurance companies. Some 62 of the 69 patients of the regional sample participated in the home visit interview and in 184 cases a therapist interview by phone was conducted.

The untreated sample for comparison was initially assessed in 1984 and was again assessed by questionnaire after an interval of 5 years 1 month (±3 months). In 72 of these 79 cases (91.14%) a therapist interview by phone was conducted.

Instruments. Table 1 shows the instruments used for assessment of the inpatient sample at admission, discharge, and follow-up. The scales used at the initial and the follow-up assessment in the press survey sample were practically identical with those of the inpatient sample. The instruments contained self-ratings and expert ratings and covered disturbed eating behavior and attitudes towards body shape, food, and eating as well as general psychopathology. For the follow-up, the Munich Diagnostic Check List for DSM-III-R (MDCL; Hiller et al. 1987) was used for assessment of psychiatric comorbidity. The Structured Interview for Anorexia and Bulimia Nervosa (SIAB; Fichter et al. 1990) was used to assess symptomatology related to eating disorder. In addition, the Morgan Russell-Scale (Morgan and Russell 1975) and the Psychiatric Status Rating Scale for Bulimia (PSRSB; Herzog et al. 1990) were used to assess the status at follow-up and the course over time. Nutritional status and somatic and psychiatric treatment in the interval since discharge were also documented.

Treatment

All patients of the inpatient sample participated in an intensive broad-spectrum behavioral therapy as described by Fichter (1990b). Special focus of the treatment was on improvement of interoceptive and emotional perception, expression of emotions, training of social communicative skills, and nutritional management in an "anti-diet group." Each patient was assigned to a therapist (physician or clinical psychologist) and the greater part of therapy was done in groups of patients homogeneous for eating disorder, excluding patients who had primarily other diagnoses.

Results

At admission 48.0% and at the 2-year follow-up 59.9% of the patients reported having a partnership (see Table 2).

Significant improvement was observed during inpatient therapy (from admission to discharge) in all measures used. Patients not only improved in attitudes towards eating and in eating behavior as measured in the subscales of the Eating Disorder Inventory (EDI) and the Anorexia Nervosa Inventory for Self-rating (ANIS) but also in measures of general health and psychopathology such as the Complaint List (CL) and the Hopkins Symptom Check List (SCL-90). Figure 1 shows the results for three relevant subscales of the Eating Disorder Inventory (EDI). The drive for thinness, feelings of ineffectiveness, and bulimic behaviors showed a marked improvement during inpatient therapy (Wilcoxon Test; $z > -9.98$; $p < 0.01$). During the time from discharge to follow-up, these scales showed a slight, statistically significant increase of symptomatology. A

Table 1. Instruments for assessment of the inpatient sample

			Admission	Discharge	Follow-up
Eating disturbance					
ANIS (Self-rating)	Fichter and Keeser (1980)	SR	X	X	X
EDI	Garner et al. (1983)	SR	X	X	X
SIAB (Self-rating)	Fichter et al. (1990)	SR	X	X	X
SIAB (Interview)	Fichter et al. (1990)	ER	(X)	–	X
Psychopathology					
Complaint List	von Zerssen (1976)	SR	X	X	X
Feeling of Wellbeing Scale	von Zerssen (1976)	SR	X	–	–
BDI	Beck et al. (1961)	SR	X	X	–
SCL-90	Derogatis et al. (1974)	SR	X	X	X
Demoralization Scale	Dohrenwend et al. (1980)	SR	X	X	–
PDS	von Zerssen (1976)	SR	X	X	–
MDCL (DSM-III-R Interview)	Hiller et al. (1989)	ER	–	–	X
Personality					
FPI-R (revised)	Fahrenberg et al. (1984)	SR	X	X	–
Locus of Control	Krampen (1979b)	SR	X	X	–
MMPI	Spreen (1963)	SR	X	–	–
Other					
Coping Scale	Fichter (1990a)		X	X	–
Sex-Role Stereotypes (GRO)	Krampen (1979a)	SR	X	X	–
Parental Bonding Instrument	Parker et al. (1979)	SR	X	–	–
Therapist's Documentation	Unpublished.	ER	X	–	–

SR, self-rating; ER, expert rating on the basis of interview; X, scale was used; (X), part of the scale was used; ANIS, Anorexia Nervosa Inventory for Self-rating; EDI, Eating Disorder Inventory; SIAB, Structured Interview for Anorexia and Bulimia Nervosa; BDI, Beck Depression Inventory; SCL, Johns Hopkins Symptom Checklist; PDS, Paranoid-Depression Scale; MDCL, Munich Diagnostic Check List for DSM-III-R; FPI-R, Freiburg Personality Inventory; MMPI, Minnesota Multiphasic Personality Inventory.

Table 2. Diagnosis, eating pathology, and marital status of bulimic inpatients

	Patients (*n*)	Admission (%)	Discharge (%)	2-Year follow-up (%)
DSM-III-R diagnosis	247			
Bulimia nervosa (total)		100.0		40.9
Bulimia nervosa only		85.0		36.0
Bulimia/Anorexia nervosa		15.0		4.9
Anorexia nervosa only		0		4.0
Eating disorder not otherwise specified		0		1.6
No major eating disorder (no bulimia nervosa, no anorexia nervosa)		0		53.5
Eating pathology				
Binge eating ≥2 times/week	207	100.0	45.4	41.5
Vomiting ≥2 times/week	207	84.1	45.0	41.1
Marked laxative abuse	207	32.9	5.8	6.2
Diuretics	207	6.3	0.5	2.4
Appetite suppressants	207	7.7	1.0	2.4
Intensive physical exercise	232	13.8	–	9.5
Major reduction in sexual drive	194	35.6	15.4	15.5
Amenorrhea	184	46.2	51.0	27.2
Status of personal relationships	227			
Single without partner		48.5		34.8
Divorced/separated without partner		3.5		5.3
Single with partner		29.5		36.1
Married		16.3		20.3
Divorced/separated with new partner		2.2		3.5

similar pattern was observed for other scales measuring areas directly or indirectly relating to eating disorders. All statistical tests for self-rating scales comparing admission with discharge data or comparing admission with follow-up data concerning general and eating specific psychopathology showed significant improvement ($p < 0.01$; Wilcoxon test admission vs. follow-up).

Figure 2 shows the results for the General Symptomatic Index of the SCL-90 as a general measure of degree of psychopathological illness. This and all other subscales of the SCL-90 showed a substantial improvement during inpatient therapy. From discharge to follow-up, these subscales showed a slight increase in symptoms (Wilcoxon Test; $p < 0.01$). When admission data are compared with those at the 2-year follow-up, the mental status at follow-up was significantly improved on all scales of the SCL-90.

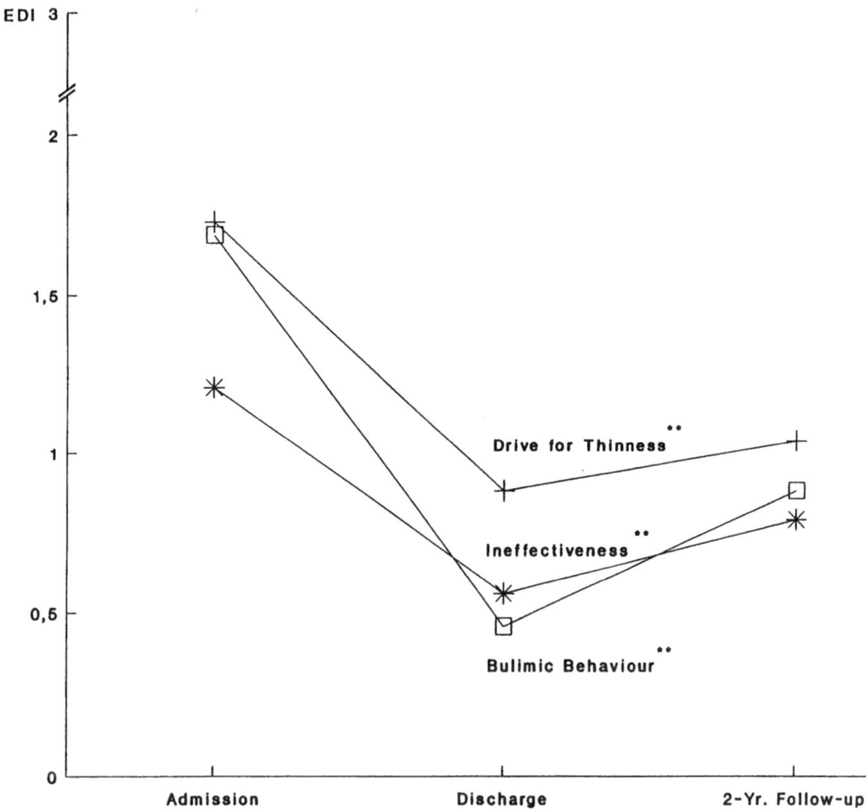

Fig. 1. Results of subscales of the Eating Disorder Inventory (*EDI*) over time in bulimia nervosa inpatients ($n = 196$). $^{**}p < .01$ (Wilcoxon test initial vs. follow-up score)

Figure 3 shows the results for the course of bulimic symptoms over the time of the follow-up period as assessed in the PSRSB. From the 4th to the 12th month after discharge, the bulimic symptoms on average were more pronounced than in the first 3 months after discharge and at follow-up. According to our data, the lowest expression of bulimic symptoms as measured in the PSRSB was at the time of the 2-year follow-up. Bulimic symptoms showed an increase after discharge with a peak at months 4 to 6 after discharge, which was maintained for months 7 to 12, and it then showed improvement. At the 2-year follow-up bulimic symptoms were more improved than at the first 3 months after discharge (Wilcoxon Test; $z = -3.63$; $p < 0.01$).

For different subgroups of patients as assessed on admission the course of symptoms over the time of the follow-up period as measured in the PSRSB generally had the same curvilinear pattern as shown in Fig. 3. The course was less favorable for those bulimic patients who had an additional

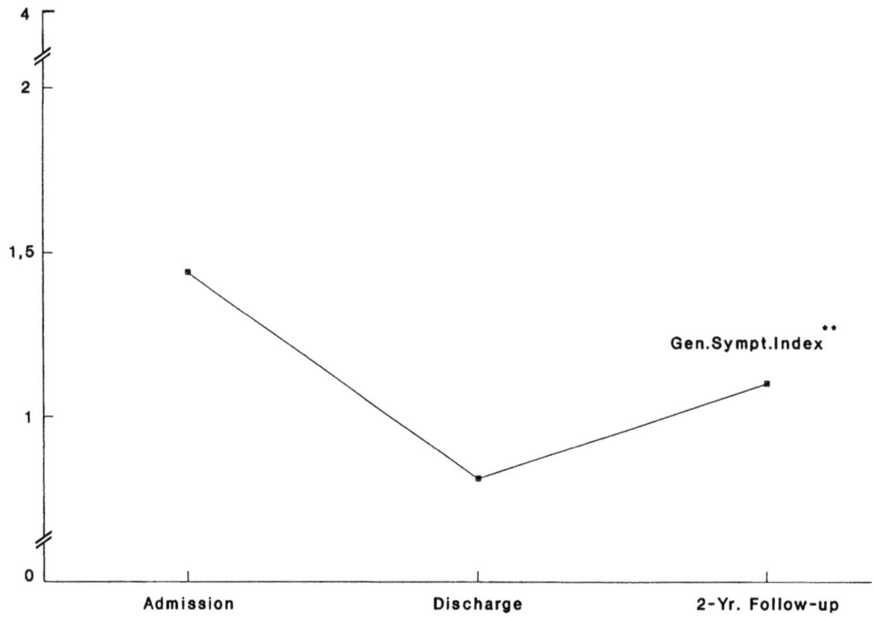

Fig. 2. General Symptomatic Index of the SCL-90 in bulimia nervosa inpatients over time (n = 186). **$p < .01$ (Wilcoxon test initial vs. follow-up score)

diagnosis of affective disorder, anxiety disorder, or personality disorder on admission (U-test for the time of follow-up; $z = -2.77$; $p < 0.01$). Also patients with a low body weight on admission (mainly patients with both anorexia and bulimia nervosa) had a less favorable course of bulimic symptoms over time than bulimic patients with higher body weight (U-test at follow-up; $z = -3.07$; $p < 0.01$).

The course over time, as measured in the total score of the EDI was less favorable for bulimic patients with a *higher expression of depression* on admission as measured in the Beck Depression Inventory (BDI; outcome at 2-year follow-up: U-test; EDI total score: $z = -4.08$; $p < 0.01$) for the EDI subscale bulimic behavior this difference was statistically not significant ($z = -1.37$; n.s.). Depression at the time of admission, as measured in the SCL-90 depression subscale and the BDI, indicated a slightly poorer prognosis as rated by the PSRSB; this, however, was statistically not significant (U-test; SCL 90 depression scale: $z = -1.22$; n.s.; and BDI: $z = -0.70$; n.s.).

At the time of initial assessment inpatients scored higher in the subscales bulimic behavior (EDI), anxiety (SCL-90), somatization (SCL-90), and the general symptomatic index (SCL-90) than persons in the nontreated comparison sample, while they scored about the same in depression (SCL-90) and the drive for thinness (EDI): At the time of the follow-up assessment both samples (treated and untreated) had improved

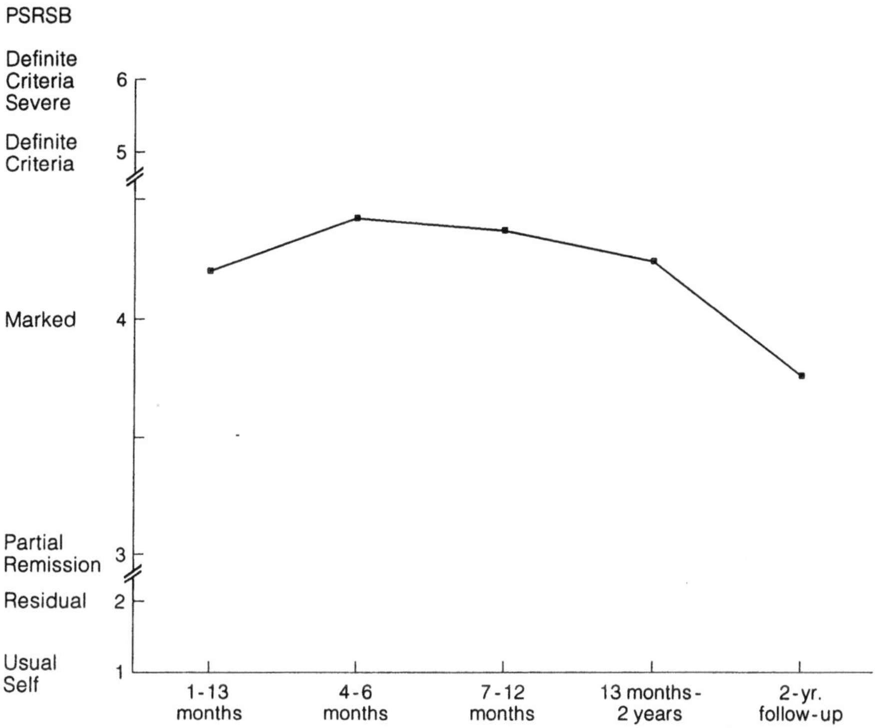

Fig. 3. Two-year course of bulimic symptoms, assessed by expert rating (Psychiatric Status Rating Scale for Bulimia, *PSRSB*)

in these scales. For bulimic inpatients the absolute score at follow-up was still slightly or markedly higher for anxiety, somatization, depression, and the general symptomatic index. Inpatients also had higher scores initially in these scales than subjects from the untreated sample.

For the inpatient sample an analysis was made of possible differences in course and outcome between those patients who agreed with no hesitation to participate in the follow-up interview (cooperative) and those who had been hard to persuade to participate (uncooperative). The upside-down U-shaped course of bulimic symptoms in the follow-up interval, as measured by the PSRSB, was seen for both groups; uncooperative patients had higher scores in PSRSB in the first year and thereafter showed improvement to the point of the cooperative patients. Thus, in our sample of bulimic inpatients, patients uncooperative at the follow-up did not show signs for substantially worse outcome, as has been reported for anorexia nervosa (Fichter 1985).

For a methodological comparison between results of home visit interviews and telephone interviews an inpatient sample of 61 matched pairs was analyzed. There were no significant differences of the expert assessments of binge eating, vomiting, laxative abuse, other bulimic symptoms, and amenorrhea (χ^2 tests). Comorbidity (affective, anxiety, personality disorders, and substance abuse) was more often diagnosed in telephone interviews. There were no significant differences of frequency in the use of DSM-III-R diagnoses. According to the telephone sample 34 persons (55.7%) had no major eating disorder, while there were 30 persons (49.2%) with no major eating disorder in the home interview sample. Anorexia nervosa was diagnosed twice (3.3%) in the telephone interview sample and in four cases (6.6%) in the personal interview sample. Both diagnoses – anorexia and bulimia nervosa – were found in two cases (3.3%) in the telephone interview sample and in four cases (6.6%) in the personal interview sample. Bulimia nervosa as the only eating disorder was diagnosed in 23 cases (37.7%) of the telephone interview sample and in 21 cases (34.4%) of the personal interview sample. Eating disorder not otherwise specified (DSM-III-R 307.50) was diagnosed in no case (0.0%) of the telephone interview sample and in two cases (3.3%) in the personal interview sample. In the Psychiatric Status Rating Scale for Bulimia home visit interviews resulted in higher ratings than telephone interviews only in the ratings for the first and second 3-month period of the follow-up interval (U-test; $z = -2.10$; $p < .05$; $z = -2.27$; $p < .05$, respectively) and not for the remaining three time periods of the follow-up. Generally, differences in personal versus telephone interviews were minor, with a tendency for more symptoms being reported in the telephone interviews.

Discussion

The German Longitudinal Bulimia Study is singular in the following respects. (1) The patient sample consisted of adult inpatients. (2) A very large patient sample was studied longitudinally ($n = 250$) in a prospective design. (3) In addition to meeting DSM-III-R criteria for bulimia nervosa, patients showed a severe, specific pathological eating disorder as well as a severe, general psychopathological condition, and the duration of illness previous to inpatient treatment (almost 8 years) was very long. Thus, the patient sample probably constitutes a "negative" selection of patients; these patients are therefore likely to have a poorer prognosis than patients with a less severe eating disturbance and general psychopathological state. All patients received intensive, broad-spectrum behavioral psychotherapy focusing on perception and expression of emotions, social skills and nutritional management. (4) Since a matched, untreated control group was not possible for ethical and practical reasons, we assessed an untreated "comparison sample" of respondents to a press survey, who never had received any inpatient or any noteworthy outpatient treatment (<50 sessions) for problems related to their eating disorder.

While several studies cited above have reported on the course of bulimia in outpatients, there is at present only one major published study on the course of bulimic symptoms (and depression) in bulimia nervosa inpatients (Swift et al. 1985, 1987). The sample size in this study was much lower; patients were evaluated over a period varying from 2 to 5 years. Swift et al. reported "surprisingly low" depression scores in their inpatient sample at follow-up (BDI 7.7 ± 7.2; SCL-90 depression scale 1.12 ± 0.82); the patients also had a much shorter duration of bulimia before admission (less than 3 years = 32.7 months) than patients in our study (7; 11 years). The patients in their sample were also younger at the initial evaluation (19; 3 years) than our sample (25 years 8 months), were treated on a pediatric eating disorder unit and conditions diagnosed according to DSM-III criteria (and not DSM-III-R as in our study). There was also a higher drop-out rate throughout the study by Swift et al. and follow-up data were obtained only on 30 out of 38 consecutively admitted bulimic patients. In the study by Swift et al. the mean frequency of bulimic episodes per month decreased substantially, although most patients still regularly exhibited bulimic behavior. As in our study, both bingeing and vomiting showed a substantial decline over time. The study by Swift et al. was not prospective. Based on the status over a 6-month period before follow-up, Swift et al. identified three outcome groups: 27% showed a good outcome (bingeing and vomiting less than monthly and no laxative-diuretic abuse), 40% showed an intermediate outcome (bingeing and/or vomiting more than once per month, but less than daily and no laxative-diuretic abuse at follow-up) and 33% had a poor outcome (bingeing/vomiting daily or ongoing laxative-diuretic abuse). Two to five years after discharge from the hospital 26 out of 30 patients (87%) still met the DSM-III criteria for bulimia (50% met DSM-III-R criteria). In the German Longitudinal Bulimia Study I only 36.0% of the patients met DSM-III-R criteria for bulimia nervosa alone and an additional 4.9% met criteria for both anorexia and bulimia nervosa at the 2-year follow-up (total: 40.9%).

Most longitudinal studies with bulimia (nervosa) patients reported a decline of depressive symptoms parallel to improvement in eating pathology (Norman and Herzog 1986; Fairburn et al. 1986; Pope et al. 1985; Brotman et al. 1988; Hsu and Holder 1986; Swift et al. 1987). In our study we also found a decrease in depression, measured by different scales (BDI, SCL-90 depression subscale, ANIS insufficiency scale) over the course of time parallel to the improvement in eating pathology. Improvement was also seen for anxiety and somatization (SCL-90)

The identification of risk factors is of theoretical and practical importance. Identifying patients with a high risk for relapse and unfavorable course can be used for designing more effective treatment and maintenance programs for patients at risk. In agreement with Swift et al. (1987), we found that bulimic inpatients with comorbidity of depression on admission had a poorer prognosis concerning bulimic symptoms. We did confirm their finding concerning the presence or absence of an affective illness on admission but found only a trend for a predictive value for self-rating scores in depression (SCL 90, BDI). In addition we found that those bulimic patients who had an additional diagnosis of affective, anxiety, or personality disorder according to DSM-III-R at admission had a poorer prognosis concerning their bulimic symptoms. Of special interest is our finding that the best outcome was seen in bulimic patients with a higher body weight on admission (BMI > 26) and the worst outcome was seen in bulimic patients with a lower body weight at time of admission (BMI < 16). Of 34 bulimia nervosa patients in the low weight group 27 patients (79.4%) also fulfilled the DSM-III-R criteria for anorexia nervosa on admission.

A relatively high comorbidity with affective disorders in patients with bulimia nervosa has been reported in several studies; Herzog et al. (1990) reported that 25 of their 30 bulimic outpatients (83%) had a lifetime diagnosis of affective illness. At the time of the initial assessment 17 patients (57%) had an affective illness according to RDC (Spitzer et al. 1978). Of these 17 patients, 5 (29%) had a diagnosis of major depression, 3 (18%) had dysthymia and 5 (29%) had minor depression. The remaining four patients had major depression in association with another affective disturbance. A relatively high comorbidity with affcetive illness in bulimia (nervosa) has also been reported by Hatsukami et al. (1984) and Walsh et al. (1985). In the patient sample (n = 232 for whom these data were obtained) we found a lifetime rate according to DSM-III-R for affective illness of 34.1% (n = 79), for anxiety disorders of 16.8% (n = 39), for borderline personality disorder of 8.2% (n = 19), and for substance abuse disorder of 30.2% (n = 70). Swift et al. had reported a high positive correlation between depression and bulimic symptoms.

In our study drive for thinness (EDI) and laxative abuse were not associated with a poorer prognosis in the PSRSB-expert rating.

Our press survey sample of untreated patients is of interest not as a control group but as a comparison and reference group of women who can be expected to have a positive prognosis. Although these women met DSM-III-R criteria for bulimia nervosa at initial assessment, the specific pathology of their eating disorder and especially their general psychopathological condition was much less severe than in our inpatient sample. In addition, the duration of illness was much shorter in the press survey sample, and it is quite likely that they did not seek professional help, because there was little need for it. The outcome of this untreated comparison group concerning specific eating pathology and general psychopathology was indeed quite positive.

There is one study assessing a similar press survey sample. Yager et al. (1987) studied longitudinally a national sample of 628 women with an eating disorder. Of these, 392 had normal weight bulimia according to DSM-III criteria at initial assessment. The follow-up period in this study was 20 months, which was much shorter than the 5-year follow-up of our untreated press survey sample. At follow-up 43% of the sample of Yager et al. with bulimia had "improved enough to be classified as having a subdiagnostic disorder" (p. 1172). Table 3 shows the scores on the subscale of the EDI for our inpatient sample and both press survey samples at initial assessment and at follow-up. Overall the German press survey sample of untreated bulimics showed fewer symptoms than were measured with the EDI in the sample of Yager et al. Except for maturity fears at initial assessment, the German sample had lower scores in every scale of the EDI than did the American sample. This comparison shows that the women in our sample – being additionally selected on the basis of having had no treatment – had a very mild degree of bulimia nervosa. In this untreated press survey sample, symptoms, as measured by the EDI, decreased significantly from initial assessment to follow-up except for the subscale perfectionism, for which the decrease was not significant. The inpatient sample generally had higher values at initial assessment than our press survey sample. The total EDI score of Yager and coworkers' press survey sample was substantially higher than in the German press survey sample and (surprisingly) slightly higher than in our inpatient sample.

Conclusion

In this longitudinal study of a very large sample of bulimia nervosa inpatients ($n = 250$) a 98.8% completion rate was obtained at 2-year follow-up assessments of specific eating pathology as well as general psychopathology. A methodological comparison between results of home visit interviews and telephone interviews confirmed that the latter was a valid method of assessment in our particular context. Comorbidity with other mental disorders (affective disorders, substance abuse disorders, anxiety disorders, and personality disorders) was shown to be a risk factor for unfavorable course. A sample of untreated women with a mild expression of bulimia nervosa and general psychopathology (untreated comparison sample) showed a very favorable course of illness. Compared to admission, inpatients with severe bulimia nervosa were significantly improved at follow-up. During the first year after discharge the inpatient sample showed more bulimic symptoms than in the later periods of the follow-up interval. When compared to the other published study about bulimic inpatients (Swift et al. 1985, 1987), our 2-year follow-up data shows a more favorable course and outcome.

In future follow-up research of bulimic eating disorders, studies are needed measuring the course over time in short intervals. A study of this type is presently being conducted by Herzog (1990). In addition, as time elapses, studies following samples of bulimic patients over long periods of time (5, 10, and 20 years) are needed. The German Longitudinal Bulimia

Table 3. Eating Disorder Inventory mean scores (0–3) at initial assessment and at follow-up in two press survey samples of women with bulimia (nervosa) according to DSM-III-(R)

Eating Disorder Inventory Scale	Mean			SD		
	A German inpatients	B Yager et al. survey	C German survey	A German inpatients	B Yager et al. survey	C German survey
Drive for thinness						
Initial	1.73[c]	2.39[a]	1.86	0.78	0.54	0.79
Follow-up	1.04	1.99[a,b]	1.06[c]	0.87	0.80	0.93
Interoceptive awareness						
Initial	1.20[c,d]	1.1[a]	0.69	0.59	0.69	0.41
Follow-up	0.71[d]	0.8[a,b]	0.45[c]	0.66	0.62	0.45
Bulimia						
Initial	1.69[c,d]	1.86[a]	1.41	0.69	0.64	0.81
Follow-up	0.88	1.24[a,b]	0.69[c]	0.88	0.83	0.79
Body dissatisfaction						
Initial	1.70[c]	2.19[a]	1.65	0.94	0.73	0.91
Follow-up	1.23	2.03[a,b]	1.31[c]	0.98	0.83	0.94
Ineffectiveness						
Initial	1.21[c,d]	1.11[a]	0.72	0.70	0.75	0.63
Follow-up	0.79[d]	0.91[a,b]	0.51[c]	0.75	0.75	0.54

Maturity fears						
Initial	0.87[c,d]	0.56[a]	0.58	0.66	0.69	0.52
Follow-up	0.59	0.63	0.43[c]	0.56	0.49	0.37
Perfectionism						
Initial	1.16[c,d]	1.65[a]	0.85	0.72	0.80	0.66
Follow-up	1.01	1.45[a]	0.94	0.70	0.83	0.73
Interpersonal distrust						
Initial	1.02[c,d]	0.87[a]	0.66	0.64	0.69	0.64
Follow-up	0.75[d]	0.69[a,b]	0.53[c]	0.67	0.66	0.58
Total inventory						
Initial	1.32[c,d]	1.44[a]	1.03	0.50	0.42	0.44
Follow-up	0.87	1.19[a,b]	0.73[c]	0.57	0.50	0.45

A, German Inpatients on Admission (German Longitudinal Study I) (N = 196); B, American Press Survey: Yager et al. $n = 392$; DSM-III; 1.6 year follow-up interval (1987); C, German Press Survey: German Longitudinal Bulimia Study; $n = 73$; DSM-III-R; 5-year follow-up interval.
[a] $p < .05$ (unbalanced ANOVA for differences)
[b] $p < .05$ (Student's t-test for change score)
[c] $p < .05$ (Wilcoxon test initial vs. follow-up score)
[d] $p < .05$ (U test A vs. C)

Nervosa Study was designed as a long term follow-up study. A 7-year follow-up study of the same sample is in preparation.

References

Abraham SF, Mira M, Llewellyn-Jones D (1983) Bulimia: a study of outcome. Int J Eating Dis 2(4):175–180

American Psychiatric Association (1980) Diagnostic and statistical manual for mental disorders, 3rd edn. American Psychiatric Association, Washington DC

American Psychiatric Association (1987) Diagnostic and statistical manual for mental disorders, 3rd edn: revised. American Psychiatric Association, Washington DC

Beck AT, Ward CH, Mendelson M, Mock J, Erbaugh J (1961) An inventory for measuring depression. Arch Gen Psychiatry 4:561–571

Brotman AW, Herzog DB, Hamburg P (1988) Long-term course of bulimic patients treated with psychotherapy. Clin Psychiatry 49:157

Derogatis LR, Liberman RS, Rickels K, Uhlenhutz EH, Cori L (1974) The Hopkins Symptom Checklist (HSCL). A self-report-symptom inventory. Behav Sci 19:1–4

Dohrenwend BP, Shrout PE, Egri G, Mendelsohn FS (1980) Nonspecific psychological distress and other dimensions of psychopathology. Arch Gen Psychiatry 37: 1129–1236

Fahrenberg J, Hampel R, Selg H (1984) Das Freiburger Persönlichkeitsinventar, Revidierte Fassung FPI-R. Hogrefe, Göttingen

Fairburn CG (1984) Bulimia: its epidemiology and management. In: Stunkard AJ, Stellar E (eds) Eating and its disorders. Raven, New York, pp 235–258

Fairburn CG, Kirk J, O'Connor M, Cooper PJ (1986) A comparison of two psychological treatments for bulimia nervosa. Behav Res The 24:629–643

Fichter MM (1985) Magersucht und Bulimia: empirische Untersuchungen zur Epidemiologie, Symptomatologie, Nosologie und zum Verlauf. Springer, Berlin Heidelberg New York

Fichter MM (1989) Bulimia nervosa und bulimisches Verhalten. In: Fichter MM (ed) Bulimia nervosa. Grundlagen und Behandlung. Enke, Stuttgart, pp 1–9

Fichter MM (1990a) Verlauf psychischer Erkrankungen in der Bevölkerung. Springer, Berlin Heidelberg New York

Fichter MM (1990b) Bulimia nervosa: basic research, diagnosis and therapy. Wiley, Chichester

Fichter MM, Keeser W (1980) Das Anorexia Inventar zur Selbstbeurteilung (ANIS). Arch Psychiatr Nervenkr 228:67–89

Fichter MM, Elton M, Engel K, et al. (1990) The Structured Interview for Anorexia and Bulimia Nervosa (SIAB): development and characteristics of a (semi-)standardized instrument. In: Fichter MM (ed) Bulimia nervosa: basic research, diagnosis and therapy. Wiley, Chichester, pp 57–70

Garner DM, Olmstead MP, Polivy J (1983) Development and validation of a multidimensional Eating Disorder Inventory of anorexia nervosa and bulimia. Int J Eating Dis 2:14–34

Hatsukami DK, Eckert E, Mitchell JE, Pyle R (1984) Affective disorder and substance abuse in women with bulimia. Psychol Med 14:701–704

Herzog DB (1990) Predictors in eating disorders. Paper presented at the 4th international conference on eating disorders, New York, 27–29 April

Herzog DB, Keller MB, Lavori PhW, Bradburn IS, Ott IL (1990) Course and outcome of bulimia nervosa. In: Fichter MM (ed) Bulimia nervosa: basic research, diagnosis and therapy. Wiley, Chichester, pp 126–141

Hiller W, Zaudig M, Mombour W (1989) Münchner Diagnosen Checklisten für DSM-III-R. Logomed, Munich
Hsu LKG, Holder D (1986) Bulimia nervosa: treatment and short-term outcome. Psychol Med 16:65–70
Hudson J, Laffer P, Pope H (1982) Bulimia related to affective disorder by family history and response to dexamethasone suppression rest. Am J Psychiatry 139: 685–687
Krampen G (1979a) Eine Skala zur Messung der normativen Geschlechterrollen-orientierung (GRO-Skala). Z Soziol 8:254–266
Krampen G (1979b) IPC-Fragebogen zu Kontrollüberzeugungen: Handanweisung. Hogrefe, Göttingen
Lacey H (1983) Bulimia nervosa, binge eating and psychogenic vomiting: a controlled treatment study and long term outcome. Br Med J 286:1609–1613
Mitchell JE, Davis L, Goff G, Pyle R (1986) A follow-up study of patients with bulimia. Int J Eating Dis 5:441–450
Mitchell JE, Pyle RL, Hatsukami D, Goff G, Glotter D, Harper J (1989) A two year follow-up study of patients treated for bulimia. Int J Eating Dis 8:157–165
Morgan HG, Russell GFM (1975) Value of family background and clinical features as predictors of long-term outcome in anorexia nervosa: four-year follow-up study of 41 patients. Psychol Med 5:355–371
Norman DK, Herzog DB (1986) A three-year outcome study in normal weight bulimia. Assessment of psychosocial functioning and eating attitudes. Psychiatry Res 19:199–205
Parker G, Tulping H, Brown LB (1979) A parental bonding instrument. Br J Med Psychol 52:1–10
Pope H, Hudson J (1984) New hope for binge eaters. Harper, New York
Pope HG, Hudson JI, Jonas JM, Yurgelun-Todd D (1985) Antidepressant treatment of bulimia: a two-year follow-up study. Clin Psychopharmacol 5:320–327
Russell GFM (1979) Bulimia nervosa: an ominous variant of anorexia nervosa. Psychol Med 9:429–448
Sabine E, Yonace A, Farrington K, Barrat K, Wakeling A (1983) Bulimia nervosa: a placebo controlled double-blind therapeutic trail of mianserin. Br J Clin Pharmacol 15:195S–202S
Spitzer RL, Endicott J, Robins L (1978) Research diagnostic criteria: rationale and reliability. Arch Gen Psychiatry 25:773–780
Spreen O (1963) MMPI-Saarbrücken. Handbuch zur deutschen Ausgabe des MMPI von SR Hathaway und JC McKinley. Huber, Bern
Swift WJ, Kalin NH, Wamboldt FS, Kaslow N, Ritholz M (1985) Depression in bulimia at 2 to 5 year follow-up. Psychiatry Res 16:111–122
Swift WJ, Ritholz M, Kalin NH, Kaslow N (1987) A follow-up study of thirty hospitalized bulimics. Psychosom Med 49:45–55
Walsh BT, Roose SP, Glassman AH, Gladis M, Sadik C (1985) Bulimia and depression. Psychosom Med 47:123–131
Yager J, Landsverk J, Edelstein CK (1987) A 20-month follow-up study of 628 women with eating disorders, I: course and severity. Am J Psychiatry 144:1172–1177
von Zerssen D (1976) Klinische Selbstbeurteilungs-Skalen (KS) aus dem Münchner Psychiatrischen Informationssystem. Beltz, Weinheim

Long-Term Follow-up of Bulimic Patients Treated in Integrated Behavioural and Psychodynamic Treatment Programmes

J.H. Lacey

Introduction

When analysing the efficacy of treatments aimed at the amelioration of bulimia nervosa two cardinal features need to be borne in mind: bulimia is both common and heterogeneous. There can therefore never be a single treatment: a holistic answer, as it were; rather, there must be different programmes aimed at particular subgroups, each of which must be tested on the relevant population. With such a large-scale problem there is little advantage in developing treatments which are long-term, expensive and only capable of being conducted by charismatic psychotherapists using sophisticated therapies. Treatment must be short, on an outpatient basis, effective and highly cost-effective (Lacey 1983). It is against this background that the reader must judge the capacity, benefits and efficacy of the treatment programmes developed by my research team.

This chapter will examine both the immediate effect of treatment and also the long-term outcome. It will examine the individual effects of certain components of the treatment programme, and especially the impact on "drop-out". Treatment packages aimed at certain subgroups will be studied, particularly treatment for multi-impulsive bulimics (Lacey and Evans 1986) and pregnant bulimics. Finally, the need for family therapy and the treatment needs of the sexual partners of bulimics will be reported.

Methods

Subjects

Diagnosis

All patients were interviewed by the author and were under his consultant care. To be considered an index case, all patients were diagnosed as having bulimia according to the criteria of DSM-III and DSM-III-R but excluding patients who have anorexia nervosa, by Crisp's criteria (according to Crisp 1980, a weight phobia is the pathognomic psychological feature of anorexia nervosa), or who had massive obesity, defined as being over 50% mean-matched population weight (MMPW). In addition to the aberrant eating behaviour, the expressed psychological feelings were always a profound sense of lack of control and low self-esteem, often anger and depression, but never a phobic avoidance of normal body weight. The author's views on diagnosis can be read elsewhere (Lacey 1982).

The Catchment Area

All the treatment programmes were tested on representative populations. The authors strongly reject the use of advertisement or special selection of any kind to collect study groups particularly using predominantly students or the middle classes. Treatment must work in the rough-and-tumble of general psychiatry. All patients came from the same circumscribed area. The catchment area studied was the London Boroughs of Wandsworth and Merton. These urban boroughs, which have a population of 350 000 persons, form a triangular wedge of South-West London, with Battersea at the apex, spreading out through Clapham, Balham, Tooting, Mitcham and Wimbledon. There are some areas of deprivation and others of great opulence but, in the main, they are typical city suburbs with terrace, villa and semi-detached family dwellings or "flat" accommodation with a stable population. There is a substantial ethnic minority with 10% of heads of household being Asian or black (Lacey and Dolan 1988). The Bulimia Clinic is at St George's Hospital which is centrally situated with good transport links.

Pretreatment Selection

Only patients referred by a doctor were offered an appointment. Patients who did not have a medical referral letter and who wrote to or telephoned the author were encouraged to approach their family doctors or a staff-health clinic. It was the practice to see patients whose doctors reported that their patients binged with food and were within a normal range of weight. Bulimic patients seen by other psychiatrists in the area would be re-referred to the clinic: there was no financial or other incentive to do otherwise.

The studies mirror normal referral practices in the United Kingdom; that is, a patient is referred without prompting by a local family doctor on the basis of his/her judgement alone. Advertisement, even if only a letter to a general practitioner requesting referrals, would have two effects the authors wished to avoid. First, referral rates, in any one year at least, would be artificial. Second, as bulimia fluctuates particularly in its symptoms, elicited referrals may present differently from those prompted by medical judgement in the natural course of the illness. There was no attempt to accept patients who were "severe" or "interesting". The sample is therefore representative of a clinic population unselected by advertisement, payment or "special centre" bias and the author would reasonably claim that it is representative of the total spectrum of the disorder in the community.

Procedure

All patients were asked to attend for a 2-h assessment interview. On presentation, weight, height and a physical examination was done. All patients completed extensive demographic and clinical history questionnaires and standardised eating disorder questionnaires. They were also accompanied by an informant, usually their mothers, who were also interviewed.

Demographics

The mean age of the patients (all females) was 24.8 years (SD = 5.0) with a range from 16 years to 40 years. The modal 5-year cohort was 20 to 24 years. The women attending the clinic were, therefore, young: whilst the age of presentation ranged from 16 years to 40 years, this was so for only 42% of all women living in Wandsworth and Merton. Three-quarters of the women in the series were single, 21.5% were married, 3.6% were separated and a similar percentage were divorced. Women in the clinic were less likely to be married than their sisters (42.8%) or similarly aged women in the district (51%; Office of Population Census and Surveys, OPCS, 1982).

The mean weight was 60.9 kg (SD = 8.7) and was not normally distributed. The average MMPW was 105.5% (SD = 14.3). Statistical comparisons for weight and height with the gener population are inappropriate. A proportional survey based on OPCS (1982) data by age, social class and region suggest that the clinic population is slightly taller, heavier and had a fuller figure than the general population; and this was strongly supported by the express views of the patients themselves. No patient was below 85% and 60% of the sample were over the MMPW. The mean desired weight was 53.2 kg (SD = 5.6). No patient wished to be less than 86% MMPW and 87% of patients wished to be between 86% and 100% MMPW; 13% wished to be over their MMPW.

Over 95% of patients were either employed or in full-time education, or caring for their children. Over three-quarters of the clinic were from social classes I, II and III, with 44.6% from social class III. Only 9% were students. The clinic population therefore comes from a broad range of social classes generally similar to the pattern of social class in the catchment area itself (Lacey 1984). The absence of social class V is probably a function of few black and Asian referrals (Lacey and Dolan 1988). The percentage of students (9%) suggests that the various community surveys have been done fortuitously on a population in which bulimia is particularly prevalent (Dolan et al. 1989).

In summary, bulimia would appear to be normally distributed by social class, although with a slight upward shift because, although bulimia is a common pathway for both anorexia nervosa and obesity, the clinic sample is probably more heavily weighted with patients with a history of anorexia than previous obesity.

Clinical Features – The Disordered Eating

The average age of onset of binge eating was 18.8 years (SD = 3.7). The mode was 17 years, median 18 years and the range 12–31 years. The average weight at onset was 59.2 kg (SD = 9.3); 104.2% MMPW (SD = 14.1). Binge eating was at its most severe at 23.2 years (SD = 4.6) at a mean weight of 105 MMPW (SD = 16.2). Over the 4 weeks prior to assessment 63% had been binge eating at least daily and 23% reported binge eating at least twice a day. Half the sample binge ate only in their home. The most common time for binge eating was the early evening (79%). Carbohydrates were the most commonly used binge foods. At presentation all patients claimed that their appetite was out-of-control, 82% craved particular foods, but only 15% felt hungry. Before binge eating the most common emotions were unhappiness (62%), anger (51%) and anxiety (46%). Afterwards, humiliation (56%), disgust (66%) and guilt (64%) were commonly reported.

Clinical Features – Weight Control

Vomiting began at an average age of 20.3 years (SD = 4.1). The mode was 18 years, minimum age 13 years and maximum 33 years. On average vomiting began 1.7 years after binge eating; one-fifth of the patients began vomiting before the onset of binge eating. The significant and moderate-sized negative correlations between age of onset of binge eating and the *interval* before onset of vomiting ($F = -0.25$; $p = 0.01$) and the *interval* before the problem was most severe ($F = -0.32$; $p = 0.001$) indicate a tendency for later-onset cases to progress more rapidly. Nearly 86% of patients induced vomiting in the 4 weeks prior to assessment, over two-thirds were vomiting at least once and 44% were vomiting twice or more per day. Over half (57%) were using laxatives ostensibly to control weight although more detailed enquiry indicated that they were used as diuretics and to give intestinal hurry so that the stomach could become more speedily "flattened" (Lacey and Gibson 1985). Less than a third (31%) gave a history of laxative abuse but all patients reported that they had, at times, used exercise as a means of weight control but only a quarter (27%) had done this in the 4 weeks before assessment.

Just less than a quarter (24%) reported alcohol abuse, 28% reported a history of regularly taking "street drugs" particularly amphetamines and marijuana; 41% gave a history of stealing, of which 21% had stolen on two or more occasions; 23% reported having taken at least one overdose.

Treatment

The Theory of Treatment

Bulimia nervosa is more than a behavioural illness surrounding binge eating. It is a complex syndrome beneath which there is a mixed affective state and a subjective feeling of lack of control (Lacey 1982). The patient's fear that this lack of control may break out into other maladaptive behaviours of a social or sexual kind must be dealt with by any successful treatment programme. Further, treatment must recognise that the total syndrome has its roots in and is driven by a dynamic conflict derived not only by individual experiences, but also familial characteristics and pathology as well as cultural factors (Lacey et al. 1986). Particularly relevant among the latter are the social and sexual roles which are imposed upon women in our society.

The patient must not be allowed to escape into other neurotic disorders such as agoraphobia, obesity or alcohol abuse. She must, therefore, gain during treatment, a thorough understanding of the factors which are maintaining her bizarre behaviour and develop other ways of dealing with her emotional needs. The clinical treatment cannot be expected to deal with all the problems. It should be enough if it leads the patient to the point where she can make her own way, learning, as we all do, by broadening her relationships and experiences. Further, although the patient does not have a weight phobia (if she did, she would be suffering from anorexia nervosa according to Crisp 1980), she is wary of weight *gain* and wishes to remain within a normal range of weight; dietary and weight controls, therefore, must be adequate.

The theoretical basis of treatment is shown in Table 1. The eating disorder and the emotional and social problems attendant on it need to be dealt with by different approaches in different settings, but by the same therapist. This consistency of therapist (rather than therapy) is important and without it the treatment would disintegrate. The eating disorder itself can be controlled by simple behavioural techniques and counselling in one-to-one therapy. As emotional and interpersonal difficulties become exposed, the behavioural therapy needs to change to insight-directed

Table 1. The theoretical basis of treatment

Problem	Treatment	Settings
Eating disorder	Behaviour therapy, counselling	Individual sessions
Emotional conflicts	Insight-directed therapy	Individual sessions, group sessions
Social problems	Counselling, insight-directed therapy	Individual sessions, group sessions

psychotherapy, although counselling is still needed. The movement from behavioural to insight therapy should be determined by the patient and her progress. Patients who are slow to give up symptoms need more counselling and behavioural help, whilst those who give it up speedily can move onto psychodynamic therapy. Undoubtedly, the latter forms a part of the reward for eliminating the bulimic behaviour and in a sense dynamic therapy can be seen as a reinforcer within the behavioural treatment.

General Details of Treatment Programmes

The treatment programmes fall into two parts. The first part is a concentrated treatment involving a weekly attendance at the clinic and lasting for 10 weeks. This controls the overt symptoms and gives the patient an understanding of the superficial dynamics. The second phase of treatment is much less frequent but is nonetheless regular. It involves clinic attendance at 1 month and then 3-monthly after treatment's end. This allows the patient to touch a reassuring base so as to discuss current experiences and to test her ideas. During this time the patient will practise the controls learnt during the first phase of treatment, develop a fuller appreciation of her own emotional needs and have a growing sense of safety in her own society.

The programme consists of a graduated contract, a series of structured controls and formal opportunities for therapy. The contract is made between patient and therapist. It is developed and agreed during each individual session and written down by the patient herself in her dietary diary (see below) and a copy made by the therapist in the notes. Before the onset of treatment the patient contracts, firstly, to attend the ten sessions and their related groups. Secondly, she agrees to maintain her presenting weight throughout the ten weeks of the programme. Thirdly, she contracts to eat the prescribed diet (see below). This means eating three meals a day at "set" times, though the patient is not told by the therapist when these times must be but decides them herself, taking into account her own prearranged social events. The therapist must therefore take great care that initiating these "behavioural" procedures does not demean or patronise the patient.

From the beginning of the treatment programme, the patient is encouraged to stop eating in binges and vomiting. Though this is rarely possible, the patient is encouraged to reduce, then stop, the binge eating, each stage being marked by its incorporation in the contract. If binge eating occurs, she is encouraged, then contracted, not to vomit. Most importantly, she contracts to eat the next prescribed meal irrespective of the proximity of the previous binge. This final "contract" is of course the core element which switches the patient out of her behavioural disorder and forces her to face the emotional underpinnings.

Five controls are built into the programme. Most importantly, the patient uses the rigid structure of the treatment programme itself and the transference relationship between herself and her therapist. The relationship between patient and therapist is not confrontatory or authoritarian, but it does put the responsibility for change – that is, getting better – on the patient herself. It cultivates the patient's undoubted motivation. In analytic terms, an alliance is formed between the therapist and the rational self of the patient against the dietary and emotional chaos.

Three other stratagems are also used. Firstly, the therapist gives the patient a dietary diary each week. This dietary diary is perhaps the single most important measure which works against the need for hospital admission. It becomes an intensely personal document, symbolising the relationship between therapist and patient: it may, perhaps, be described as a transitional object. The diary provides control and discipline throughout the week and is a constantly available outlet for emotional feelings.

The written instructions on the diary state that the patient must carry it on her person everywhere (including the lavatory) until it is handed back and kept by the therapist at the beginning of the following individual session. It is emphasised that the diary *belongs to the therapist not the patient* and that the therapist must be addressed *by name* within it. Each page of the diary is divided into columns and the patient records all food and drink consumed, the time taken, and the incidence of binge eating, vomiting or laxative abuse. She also records her feelings and thoughts in temporal sequence to the details of food consumed or eating symptoms displayed. She is encouraged to explain why she has binged or vomited at that particular moment. If the diary does indeed become a transitional object and if the transferential relationship is powerfully formed, it becomes virtually impossible for the patient to vomit with the diary open before her.

The second stratagem of control is a diet sheet, which outlines a prescribed diet structured into meals. The quantity of bread, potatoes and other "carbohydrate" foods which the patient contracted to eat is clearly marked. The patient has a choice, however, in both the variety and quantity of other foods. The aim is to control her carbohydrate intake, not the energy value of all food eaten. The patient is told that she cannot change her diet without reference to her therapist. The diet sheet emphasises that bulimia can be thwarted by adequate carbohydrates when taken regularly. By giving a choice of other foods, battles of authority between therapist and patient are reduced. It informs the patient of the amount of food needed to maintain her weight, a matter about which she has little knowledge.

The last control used is that the patient is weighed by the therapist (never the clinic nurse) on the same scales just prior to the therapy session. It is, perhaps, the therapist's insistence on the patient maintaining her presenting weight (thereby temporarily removing the drive to diet) which allows the binge eating to stop under the impact of the programme.

The Four Studies

The four studies to be described all follow the style, philosophy and general structure described in the above two sections. They differ, however, in aims, scale and in the opportunities built into each programme for therapy. The first study and the follow-up study are essentially of a pilot nature. The later two test the findings on large populations.

The First Study

The first study was a controlled evaluation of a mixed individual and group treatment programme. Thirty patients were treated in six groups of five, one patient of which had "recovered" from previous anorexia nervosa but was now at normal body weight. The patients were allocated alternatively to the treatment programme or to a nontreatment control programme, both of which lasted 10 weeks. Those who were initially placed within the control programme then entered the treatment programme. The eating behaviour and mood state of the patients were monitored by a daily symptom diary and mood analogue scale. This was done not only during the 10-week treatment programme but also during the 2-week assessment periods before and immediately after both the treatment and control programmes. The thirty patients were consecutive and unselected and acted as their own control. Their ages ranged from 21 to 37 years and they gave a history of both bulimia and self-induced vomiting from 3 to 18 years. The vomiting (mean number of episodes 3.78/day) and binge eating (3.32/day) were severe. The mean presenting weight was 3.2% above MMPW.

Three formal treatments were built into this study: firstly, the diary which was "available" throughout the day; secondly, the individual sessions; and thirdly, the groups. The aim of therapy was the recognition and delineation of emotional and social factors which are associated with, or generated by, the patient's symptoms. The patient is encouraged to experiment with new ways of dealing with her feelings and interpersonal difficulties. The aims of the therapy are formulated into a handout which is given to the patient before the treatment programme. Most emphasis is given to exploration of current feelings, redefinition of symptoms, exploration of maintaining factors and preparation for future difficulties; less emphasis on exploration of origin, experimentation with feelings, "termination" and problem solving.

The insight-directed therapy, but not the behaviour therapy, is provided in both the individual sessions and in a psychotherapy group. The reasons for this will be discussed later, but the cost-effectiveness of conducting psychotherapy within a group setting is self-evident, particularly when the number of women seeking treatment is formidable.

Each of the five patients in each group attended for half an hour's individual session with one therapist before both therapists and patients meet together for a group session lasting $1\frac{1}{2}$ h.

Follow-up Study

The second study was a follow-up project of the 30 patients who entered study 1. They were assessed 1 month after treatment and then at 3-monthly intervals by a psychiatrist using a standard symptom check list in formal follow-up clinics. After 2 years this study was extended for a further 2 years to test whether the 3-monthly appointments themselves influence the long-term results. In other words, this study was aimed at determining the length of contact that the patients need in the clinic, frequency of drop-out and complications of treatment.

Catchment Area Individual Study

The same principles and schema of treatment apply as described in the section "General Details of Treatment Programmes". The design of the programme was the same as study 1 except that the individual sessions lasted 55 min and there was no group component.

The aim of this study was to measure the efficacy of the original treatment programme translated into the needs of National Health Service (NHS) psychiatry. A standard treatment package based on study 1, containing all its main features, was designed (see "The First Study"). Each patient completed clinical and demographic questionnaires. Following diagnosis, the patient met her therapist for 30 min, and the times and date of the treatment programme itself were determined.

Each treatment package consisted of ten 1-h sessions after which patients were seen in follow-up for 1 h after 1 month, and then 3-monthly intervals for 18 months. In addition, they would see the assessing psychiatrist immediately after the main 10-week treatment package and again at the end of the follow-up period of 18 months. A total of 250 patients entered the study. The demographic and clinical features are given in the section "Subjects" as are the diagnostic criteria.

Catchment Area Individual/Group Treatment

The same schema of treatment and principles apply to this study as already described in the section on general details. The aims were similar to the large catchment area study, that is, to measure the efficacy of the treatment translated into NHS practice. The difference between this study and the preceding one is that it combines both individual and group treatment in the same way as the first study. Five patients were treated in each group. Each patient attended for a half-an-hour individual session with one therapist before both therapists and the five patients met together for a group session lasting $1\frac{1}{2}$ h. In addition, they saw the assessing psychiatrist immediately after the main 10-week treatment package and again at the end of the follow-up period of 18 months. All patients were seen individually at 1 month, and then 3-monthly intervals for 18 months. They were interviewed by their therapist. A total of 50 patients entered study.

A subgroup was asked to complete the "most important event" questionnaire. Bloch et al. This tool is used to explore what the patients found helpful in the experience of the group. Each patient was asked to complete the questionnaire at the end of sessions 3, 6 and 9 of the 10-week group (details of procedure are to be found in Hobbs et al. 1989).

Summary

Two treatment programmes have been devised in these four studies. One uses both individual and group techniques, and the other uses only individual sessions. Both are set in the outpatient department and are time-limited, lasting for 10 weeks. These outpatient programmes are aimed at the majority of bulimic patients. When the condition is more complicated and the patient presents with a multi-impulsive disorder (Lacey and Evans 1986) embracing alcohol or drug dependence, sexual disinhibition, stealing or self-harm, a third programme – an inpatient treatment – is preferred. Details of this and its testing are given in the section "Multi-impulsive Bulimia".

Results

Results of First Study

The control programme had no demonstrable effect on the incidence of bulimia or self-induced vomiting. Although individual fluctuations occurred, the patients reported a mean of 45 binge-eating episodes during the 2-week assessment period before the control period and just under a mean of 47 episodes afterwards.

The treatment programme, however, had a significant impact on the incidence of both binge eating and vomiting. A mean of 47 binge-eating episodes were reported during the 14 days before the treatment programme. This fell to a mean of 2.3 episodes during the 2 weeks after assessment ($p > 0.01$, Mann-Whitney). The incidence of self-induced

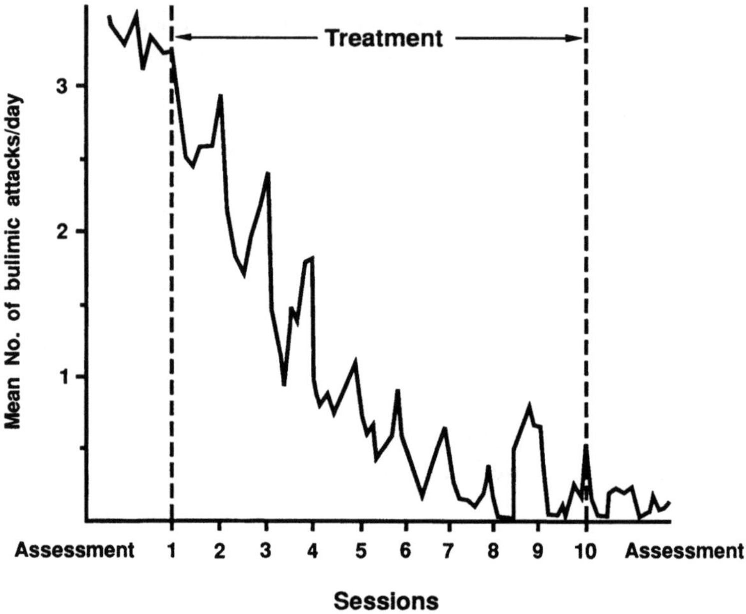

Fig. 1. Mean daily incidence of bulimia in patients before, during and after treatment

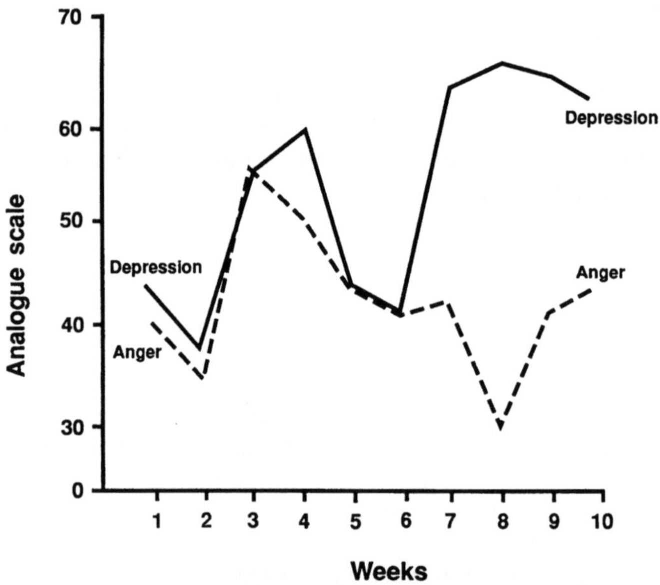

Fig. 2. Mean analogue scale scores for depression and anger during each week of treatment

vomiting similarly fell from 53 episodes to a mean of just over two episodes during the assessment period ($p > 0.01$). After the treatment all the women had fewer symptoms; 24 (80%) stopped binge eating and vomiting completely at the end of the ten sessions. During the treatment programme, all patients reduced the frequency of their binge eating. Figure 1 shows this graphically and shows the tendency of symptoms to exacerbate in the latter part of the week as the impact of the treatment session waned (Lacey 1983).

For the beginning of treatment the patients were sad and angry as judged clinically and as measured on self-rated analogue scales (Fig. 2). Initial control of the eating disorder was associated with profound mood fluctuations and a feeling of being "disturbed". All patients who are sufficiently motivated brought their eating under control between the 4th and the 6th week of treatment. At just this point their sense of sadness became profound. I see it as the duty of the psychiatrist dealing with the eating disorder to "create" depression not to relieve it! The treatment programme must translate the behavioural disorder of binge eating into an emotional disorder of sadness and anger which can then be dealt with psychotherapeutically.

Follow-up Study

The success of the initial treatment begs the question: what are the longer-term results? To determine this, the original 30 patients were followed up for 4 years. Over the first 2-year period, the patients were seen 1 month after the end of the 10-week treatment package and then at 3-monthly intervals. Although one patient refused follow-up and one needed an inpatient admission, the remaining 28 patients judged the treatment to be a success. Twenty had no bulimic episodes, whilst the remainder had occasional episodes roughly three to four times a year. Between 22 and 27 patients were not binge eating in any 3-month assessment period. There was no "drop-out" other than the one patient mentioned above.

The question arises whether the 3-monthly appointments themselves influence the longer-term results. The appointments consisted of a formal interview using a standard format. In no sense were these interviews psychodynamic in the way that the initial weekly appointments were, but they were certainly supportive. To test this, the population was divided into two groups of 14 patients during the 3rd year of follow-up. One group continued to have 3-monthly appointments in exactly the same way as they had in the previous 2 years. The other 14 patients had no follow-up. The results are shown in Table 2. Of the 14 who continued their follow-up appointments little change occurred in the 3rd year. However, in those who had no follow-up appointments, two developed severe bulimia, six had occasional bulimic episodes and only six remained clear of dietary

Table 2. Effective follow-up appointments on longer-term follow-up

Patients	1st 2 Years			3rd Year			
	No binge eating (*n*)	Occasional binge eating (*n*)	Severe binge eating (*n*)	No binge eating (*n*)	Occasional binge eating (*n*)	Severe binge eating (*n*)	Follow-up
Group 1	10	4	0	9	5	0	Yes
Group 2	10	4	0	6	6	2	No
Total	20	8	0				

Table 3. The impact of follow-up appointments on 14 bulimic women binge eating

	None	Occasional	Severe	Follow-up
1st 2 Years	10	4	0	Yes
3rd Year	6	6	2	No
4th Year	8	6	0	Yes

abuse. Thus, some of those patients who maintained their improvement over 2 years significantly deteriorated in their 3rd year when they had no follow-up appointments. Those who had continued to have only minimal contact – in some cases by letter alone – maintained their improvement. During the 4th year of follow-up the groups were reversed, that is, those who had no follow-up appointments in the 3rd year resumed their appointments and vice versa. Those patients who had 3 continuous years of follow-up appointments showed no change in the 4th year. The results of the remainder are shown in Table 3. With the resumption of follow-up appointments in the 4th year, an improvement occurred, though not to a level noted in the first 2 years of follow-up.

Body weight did not change significantly during follow-up. The presenting weight of the sample was 3.2% above MMPW; at the end of treatment average weights were 3.2% above MMPW and 2.8% above after 1 year.

Catchment Area Individual Study

Table 4 shows the distribution of binge eating before and after the 10-week initial treatment package. The majority of patients were binge eating between one and three times a day. After treatment 82% were not binge eating, 10% were bingeing up to once per day and 5% showed no im-

Table 4. Effect of current or past history of alcohol abuse on response to treatment

	Before treatment (binges/week)	After treatment (binges/week)
Bulimia (n = 250)	17	0.7
Bulimia with alcohol abuse (n = 89)	22	4

Table 5. Long-term outcome of treatment (number of patients who binge eat in each assessment period)

Follow-up period (months)	Periodic total		Cumulative total	
	(n)	(%)	(n)	(%)
End of treatment	37	15	37	15
1	44	18	44	18
2–3	56	22	68	27
4–6	62	24	74	29
7–9	58	23	82	33
10–12	61	24	85	34
13–15	67	27	87	35
16–18	60	24	89	36

provement. Those patients with a history of current or past alcohol abuse did less well (see "Multi-impulsive Bulimia").

Table 5 shows the long-term outcome of treatment. It indicates the number and percentage of patients who were binge eating in each assessment period, that is, during the week at the end of treatment; during the 4 weeks immediately following treatment (1-month assessment); during the 2nd and 3rd month after treatment (3-month assessment) and during each 3-month period up to 18 months. During the last week of treatment 85% were not abusing food. However, during the 4 weeks immediately after, 18% had at least one binge episode. Thereafter between 22% and 27% of the total sample were binge eating during any 3-month period; in other words, three-quarters of this unselected sample, treated not with the intense resources of a research programme but with the limited facilities available on the NHS, were in control of their eating. Table 5 also shows the accumulative total: two-thirds, after a 10-week outpatient treatment programme with follow-up at regular and at frequent intervals, reported full control of their eating pattern.

Body weights did not change significantly. The average MMPW was 105.5%; at the end of the 10 weeks of initial treatment, the mean weight was 107.2% and at 1 year, after treatment, 106.2% MMPW. Of the fifth of the sample who were over 115% MMPW, most (16%) wished to lower their weights after treatment for cosmetic reasons; all were offered guidance (briefly, that they should maintain "carbohydrate" levels but reduce protein and fatty foods). Most, 8% felt "bingey" and in some panic stopped the reducing diet. Some (6%) achieved weight loss and a few (2%) returned briefly to bulimia.

Many of the sexual partners of bulimics experienced difficulties during or shortly after the end of treatment. In each case there was a shift in the relationship, generated by anger, when the patient gave up her eating symptoms, leading to the marriage being questioned. Full details of the psychopathological status of the spouses is beyond the scope of this chapter, but readers who are interested should contact the author. For details of the clinical psychiatric state, physical health and psychosocial functioning as well as changes in menstrual histories and fertility reference should be made to the author.

The pattern of improvement took three forms. First, the majority (161 patients: 63%) had no binge-eating or vomiting episodes at all. Most of these patients felt the impulse to binge eat but resisted the compulsion. Most described feeling intermittently emotionally "disturbed"; only 15% pursued further psychotherapy after the end of treatment. The second group had occasional bulimic episodes. Before treatment, these patients judged their diet to be in daily "chaos". After treatment these chaotic periods were reduced and most limited their disordered eating to discreet circumscribed periods usually lasting about 6 days. They had been instructed that if the symptoms recurred they were to place themselves back on the diet and dietary diary and speak to their therapist on the telephone. Some found this enough, some brought their next appointment forward and some, interestingly, decided not to contact the clinic. These latter patients remained with their symptoms for a longer period of time; it appeared that they lacked the motivation to change temporarily, putting off contact with the clinic until they felt that they could again face underlying conflicts. The third group (5%) were those who did not benefit from therapy.

Catchment Area Individual/Group Study

Fifty patients were treated in this study. Only one patient dropped out (2%). Of the 250 treated in the catchment area study (2.33) 18 patients dropped out (6%).

The findings on the "most important event" questionnaire suggested that patients regard "nonspecific" factors such as realising shared experience, learning from the experience of others and gaining hope from observing positive change in others as important as the more "specific" factors such as gaining understanding into the nature of their difficulties.

Multi-impulsive Bulimia

The first study (see "Results") indicated that a history of alcohol abuse was associated with poorer outcome. The patient who refused follow-up was alcoholic; of the eight who had occasional mild bulimic episodes during follow-up, four had a history of alcohol abuse (including two who were alcoholic). Only one of the twenty without any binge-eating or vomiting episodes during follow-up had a history of alcohol abuse, and that was mild (see "Results"). Similarly, Table 4 shows the differential impact of a current or previous history of alcohol abuse in a catchment area study.

These patients may also show a variety of other impulsive self-damaging behaviours such as drug abuse, stealing, repeated overdoses or mutilation or sexual disinhibition. They present to alcohol and drug-abuse clinics (Lacey and Mourelli 1986; Beary et al. 1986) and place large demands on psychiatric and emergency services. Such patients have been labelled as multi-impulsive (Lacey and Evans 1986) and they should be considered as a unitary problem with special needs. One potential treatment problem is the interchangeable nature of the symptoms so that if, for example, the eating disorder is addressed in the bulimia treatment clinic, the patient may stop binge eating but move to self-harm or drug abuse. Such symptom substitution tends to thwart focusing therapies, and only by focusing on all symptoms can the underlying psychopathology be tackled. As these patients need strict behavioural limits and are usually more disturbed than uni-impulsive bulimics, treatment is typically more intensive and on an inpatient basis. The diagnostic criteria of the multi-impulsive form of bulimia is shown in Table 6.

Methods

The Sample

The sample was drawn from consecutive admissions to the author's inpatient unit who fulfilled the operational criteria for multi-impulsive personality disorder (see Table 6) and who had been inpatients for at least 1 month. These ten patients gave informed consent.

The average age of admission was 26.2 years, the average age of onset of symptoms of binge eating 17.5 years, and the average duration of symptoms 8.7 years. Five patients had a definite history of previous anorexia nervosa, and two a possible history of anorexia nervosa. The length of admission varied between 17 and 30 weeks.

Treatment

The treatment setting is a 40-bedded unit based on therapeutic community principles and with special experience in the field of eating disorders.

Table 6. Diagnostic criteria for the multi-impulsive form of bulimia

1. Bulimia is associated with one or more of the following:
   Gross alcohol abuse
   "Street drug abuse"
   Multiple overdoses
   Repeated mutilation
   Sexual disinhibition
   Shoplifting
2. Each behaviour is associated with a similar sense of being out-of-control.
3. Each of these patterns of behaviour may fluctuate, and they are interchangeable and impulsive.
4. The patients' affects are of depression and intense anger which are declared when the behaviours are controlled.

Patients are usually referred from general practitioners or other psychiatrists and are assessed in the outpatient department. They come from a wide area of the United Kingdom.

After outpatient assessment and diagnosis patients are invited to visit the unit. They are then recommended to participate in the inpatient programme and are told of the likely length and structure. They are initially admitted for a 2-week period of mutual assessment during which time all permanent members of staff interview the patient. At the end of the 2 weeks the aims of admission are documented. It is expected that the aims will include not only behavioural control but changes at psychological, psychodynamic and social levels. The patients are recommended not to leave the unit for the first 2 weeks. The admission itself then begins. On average, it lasts 3–4 months, followed by a period of 2 months as a day-patient before outpatient appointments for up to 9–12 months further.

On admission the patient agrees to a verbal contract: that is, she will not binge eat, vomit or attempt weight loss; she will not be violent to herself, to others or to hospital property; she will not use unprescribed drugs; she will not drink alcohol on the unit (and if alcohol is a problem, off the unit also); she will not form any sexual relationships on the unit; and she will become actively involved in the treatment programmes.

The patient has a nurse-counsellor who is responsible for helping her with control of food and weight (along exactly the same lines as described in the outpatient programmes). In addition, the patient has an individual therapist; again, the principles of psychotherapy are as described above. Patients are also assessed for family therapy either with their parents or sexual partner: the decision as to whether family therapy is appropriate is based on the principles that normally justify such treatments. The patient also attends a series of psychotherapy groups with other multi-impulsive bulimic patients. The groups deal particularly with issues of sexuality, family matters and interpersonal relationships both on and off the unit.

In addition the patient engages in creative therapies particularly art therapy, psychodrama, dance/movement therapy, creative writing, assertion training, interactional skill training and relaxation programmes. Reality-orientated therapies include domestic skills, practical living and leisure skills, and body image work.

Once symptoms have been controlled, intense affects of depression and anger become declared (Lacey 1983) and interpsychic conflicts become more accessible. A transference may develop in relation to the unit and the staff reflecting the patient's internal dynamics. The unit provides a holding empathic environment with firm behavioural limits. Pitfalls are staff splitting, power struggles with patients, over-responding to manipulative behaviour and failure to provide sufficient clear and constant boundaries and structure. The ethos and personal contract emphasise that all aberrant and self-destructive behaviours must cease immediately. The patient may request a period of voluntary restriction before responsibility is gradually transferred back to her. Meals may initially be "plated", but the expectation is that she will choose and serve her own meals speedily within the context of weight stability. If any bulimic or self-damaging behaviours take place, it is put to the patient that she is putting her emotions into behaviour rather than into the wealth of psychotherapy on offer to her. She is then asked to leave the unit for 1 week to consider whether or not she is "ready" to pursue this treatment programme. It is emphasised that there is no sense of failure if she decides to stop it and, indeed, she is told that outpatient support will be provided. Usually, however, after the week away from the unit, the patient returns determined to maintain control. This period of rustication is insisted upon even if the patient is suicidal. Repeated breaches of behavioural limits may lead to formal discharge but, again, outpatient support is provided.

Follow-up

The follow-up interview was performed face-to-face and lasted 90 min. The first part of the interview was unstructured and aimed at evaluating current circumstances, lifestyle and mental state. The second part dealt with patterns of impulsive behaviour over the previous 6 months and attitudes towards the treatment regime. This was assessed by a semi-structured interview schedule devised by the author. Finally the patients were interviewed with the Morgan-Russell Outcome Assessment Schedule (MRAS), which has been used extensively in outcome studies of anorexia nervosa (Morgan and Hayward 1988).

Results

Tables 7 and 8 list the self-damaging behaviours of the nine patients before treatment and at 18 months after the end of admission. The severity of

Table 7. Past behaviours on admission

Patient	Frequent bingeing	Vomiting	Laxative abuse	Alcohol abuse	Drug abuse	DSH	Sexual disinhibition	Shop-lifting	Score
1	++	+	+	+	–	++	+	+	9
2	++++	++	–	+++	+	–	–	–	10
3	++++	++	+	+++	+	++	++	+	16
4	++++	++	–	+++	–	+	+	–	11
5	++++	++	–	+	+	–	++	–	10
6	+++	+	–	+++	++	–	++	—	11
7	++++	+	+	+++	++	++	+	–	14
8	++++	++	+	+	+	–	+	+	11
9	++++	++	–	+++	–	+	–	–	10
								Average score	11.3

Scores are for the highest frequency of behaviours. Total score calculated by adding the number of +.

Binge eating:	+,	occasional;
	++,	>weekly;
	+++,	daily;
	++++,	>daily
Alcohol abuse:	+,	occasional;
	++,	>weekly;
	+++,	daily
DSH:	+,	overdoses or cutting/burning etc.;
	++,	both OD and cutting etc. Or >1 OD per week
Vomiting:	+,	occasional;
	++,	after each binge
Drug abuse:	+,	occasional;
	++,	frequent, i.e. >weekly
Sexual disinhibition:	+,	socially;
	++,	prostitution

Table 8. Behaviours at follow-up (previous 6 months)

Patient	Frequent bingeing	Vomiting	Laxative abuse	Alcohol abuse	Drug abuse	DSH	Sexual disinhibition	Shop-lifting	Score
1	+	–	–	+	+	++	–	–	5
2	++	++	–	++	–	–	–	–	6
3	++++	++	+	+	+	+	–	+	11
4	–	–	–	–	–	–	–	–	0
5	+	+	–	+	–	–	–	–	3
6	–	–	–	++	++	–	–	–	4
7	++++	+	+	++	+	++	+	–	12
8	–	–	–	+	–	–	++	–	3
9	–	–	–	+++	–	+	–	–	4
								Average score	5.3

For definitions see Table 7.

symptoms was less than half the rate prior to admission, and if we exclude two patients (three and seven) where little improvement occurred, then the overall results of the remaining eight are substantial.

Discussion

This chapter indicates the development of a powerful series of treatment programmes. They have not just been tested on research groups, but on an unselected general population of bulimic women. No other treatment programme yet published has been tested in a like manner and none have been proved to be so robust on such large numbers. Further, no other psychological treatment is as cost-effective.

Great store has been placed on the demography of the population. In social class terms, this sample is very different from the Cooper and Fairburn study. In the latter, social class was determined by parental occupation as only 21.8% were in paid employment and nearly 60% were students. Although Cooper and Fairburn claim their sample is "representative of patients with bulimia nervosa seen by general psychiatrists in Britain" there are grounds for suspecting that, on social class terms at least, their sample is atypical of the average British urban area. The average age of the author's patients was slightly older than reported by Cooper and Fairburn (1983) which might be a reflection of the way the samples were collected; that is, elicited referrals would tend to arrive at the clinic somewhat earlier than "spontaneous" referrals.

Bulimia is a fluctuating symptom occurring in often suggestible patients. The first study was in part designed to show whether any improvement resulted from the treatment programme or from other factors, such as the intense assessment, monitoring procedures, knowledge of impending treatment or the simple passage of time. The control programme, however, had no effect on symptoms. The patients, who acted as their own controls, showed a rapid improvement in symptoms during the treatment programme.

These results demonstrate that the initial concentrated weekly treatment, although short-term, can effectively interrupt a disordered eating pattern by behavioural self-monitoring. The results are good. Although they are the product of the whole programme, certain features are noteworthy. The treatment is eclectic, drawing on both behavioural and psychodynamic techniques. Both are conducted by the same therapist.

It is important to recognise that behavioural treatment only leads to an interruption of symptoms and that, without other therapy, they will most likely reappear. The aim therefore must be to achieve a period of stability during which time other therapies can attempt a more permanent resolution of the patient's problems. A patient who is duly motivated can usually stop her binge eating between the 4th and 6th week of treatment.

No longer hindered by food manipulation, underlying emotional issues become exposed and sadness and anger are more appropriately felt. In other words, although the patient's eating pattern achieves stability, the converse is true of her emotional state. It appears logical therefore that successful treatment requires a forum to examine these emotional feelings and their underlying conflicts. Two have been tested. First, a combined group and individual treatments and second, just individual sessions (see "The Four Studies"). The aim of both is focal, short-term and interpretive. Similar overall results are found by both methods, but it is clear that the addition of a group prevents drop-out. We know that the patients who meet in a group continue as a self-help group after the formal end of treatment (Deeble et al. 1990).

The treatment's power to alter, within psychotherapy, entrenched and core symptoms has been further shown in a study of the disturbed body perception of these patients (Birtchnell et al. 1985). Prior to treatment the bulimic with normal body weight overestimates various (usually emotionally charged) areas of her body. This overestimation is more than normal-weight controls but not significantly so and less than that shown in anorectic populations. An examination of 29 bulimic patients before and after their treatment showed a significant decrease in overestimation of body weight which appeared to be an effect of treatment in that no similar reduction of body-image distortion was reported in control subjects. In that study we argued that emotional conflicts – particularly concerned with a woman's role, femininity, or problems associated with interpersonal relationships – which are ordinarily displaced, are rechannelled by psychodynamic psychotherapy. This leads to a retreat of the overestimation towards the norm.

The first controlled study demonstrated that sadness increased at just the point where the eating disorder tended to come under control, a translation, as it were, of the eating problem into emotional terms. Generally the middle part of the 10-week treatment programme is a time of major and fluctuating emotional feelings. The patients feel significantly more guilty, frightened, anxious and isolated, and it is essential that the treatment emphasise the temporary nature of these feelings and the need to tackle them by determining the underlying problems which fuel them. Generally these results indicate that group psychotherapy sessions in conjunction with individual therapy sessions provide an efficient means of exploiting the control of eating gained by behavioural techniques.

The follow-up study demonstrates that regular but infrequent attendance at the clinic following the initial stage of treatment is essential. Patients who did well tended to be ambitious, hard-working women whose predominant clinical symptoms, apart from their eating disorder, were sadness and anger – the latter being initially denied. In general they appeared to be resourceful yet "neurotic" (Lacey 1982, 1984). The long-term results of bulimics who have had a previous history of anorexia nervosa are similar to those who have not. However, they tend to respond more slowly to treatment and there is the occasional danger that they will return to anorexia when the binge eating is removed in treatment. On the other hand, a poorer prognosis is significantly associated with abuse of alcohol. These multi-impulsive patients (Lacey and Evans 1986) in whom the disorder is associated with alcohol and drug abuse, and perhaps, sexual disinhibition and self-mutilation, need longer and more intensive treatment, the more severe cases needing the inpatient programme described above. What is impressive is how the same basic principles of treatment, if set in a disciplined treatment structure marked by a contract to give up the aberrant behaviours, are so successful. When these indicators are taken into account, the prognosis of bulimic patients is not as poor as was originally judged.

The reader should not forget the large number of women who seek treatment because they intend shortly to have a child (Lacey and Smith 1987). Treatment for the pregnant bulimic patient should be concentrated during the weeks of the third trimester when the eating behaviour is quiescent and deeper issues not thereby obscured. The 3-monthly follow-up sessions of the standard treatment take place during the puerperium, fitting well with the schedule of a busy young mother.

These programmes are brief, effective and cost-effective and have the advantage that, with the exception of the inpatient programme, they can be conducted by paramedical staff under medical supervision. All can be set up as part of a general psychiatric service, and therefore they provide perhaps the best opportunity for the large number of patients at present seeking help. Certainly the cardinal features of these programmes – a contract through which the patient agrees to maintain a stabilised weight and a prescribed diet whilst being held in therapeutic control and given supportive, yet insight-directed psychotherapy – are appreciated by patients and are successful. Perhaps most importantly, these programmes, when set in the NHS clinics, have been shown to be as effective as when first they were developed over 12 years ago, and the results hold with time.

It is no longer acceptable to state that the answer to bulimia lies in cognitive therapy and medication (Fairburn 1990). Successes on the scale described here have never been shown by the use of medication, and our studies clearly indicate that brief insight-directed therapy can deliver, cheaply, to large numbers of patients a cure of their unhappy disorder.

Acknowledgement. The author acknowledges the help and assistance of those who have from time to time been members of his research-team, particularly Drs F Mourelli, C Evans, B Dolan, S Birtchnell, S McCluskey, A Kent, T Read, Ms F Hume and Ms M Kramers.

References

Beary MD, Lacey JH, Merry J (1986) Alcoholism and eating disorders in women of fertile age. Br J Addict 81:769–774

Birtchnell SA, Lacey JH, Harte A (1985) Body image distortion in bulimia nervosa. Br J Psychiatry 147:408–412

Cooper PJ, Fairburn CG (1983) Binge-eating and self-induced vomiting in the community: a preliminary study. Br J Psychiatry 142:139–141

Crisp AH (1980) Anorexia nervosa: Let me be. Academic Press, London

Deeble EA, Crisp AH, Lacey JH, Bhat AV (1990) A comparison between women seeking self-help and psychiatric treatment in anorexia nervosa and bulimia. Br J Med Psychol 63:65–72

Dolan BM, Evans CDH, Lacey JH (1989) Family composition and social class in bulimia: a catchment area study of a clinical and comparison group. J Nerv Ment Dis 177:267–272

Fairburn CG (1990) Bulimia nervosa. Br Med J 300:485–486

Fairburn CG, Cooper PJ (1983) The clinical features of Bulimia Nervosa. Br J Psychiatry 144:238–246

Hobbs M, Harte A, Birtchnell SA, Lacey JH (1989) Therapeutic factors in short-term group therapy for women with bulimia. Int J Eating Dis 8:623–633

Lacey JH (1982) The bulimic syndrome at normal body weight: reflections of pathogenesis and clinical features. Int J Eating Dis 2(1): 59–66

Lacey JH (1983) Bulimia nervosa, binge-eating and psychogenic vomiting: a controlled treatment study and long-term outcome. Br Med J 286:1609–1613

Lacey JH (1984) Moderation of bulimia. J Psychosom Res 28:355–360

Lacey JH, Dolan B (1988) Bulimia in British blacks and Asians: a catchment area study. Br J Psychiatry 52:73–79

Lacey JH, Evans CDH (1986) The impulsivist: a multi-impulsive personality disorder. Br J Addict 81:715–723

Lacey JH, Gibson E (1985) Does laxative abuse control body weight? A comparative study of purging and vomiting bulimics. J Human Nutr (Appl Nutr) 39A 36

Lacey JH, Mourelli E (1986) Bulimic alcoholics: some features of a clinical sub-group. Br J Addict 81:389–393

Lacey JH, Smith J (1987) Bulimia nervosa: the impact of pregnancy on mother and child. Br J Psychiatry 150:777–781

Lacey JH, Coker S, Birtchnell SA (1986) Bulimia: factors associated with its etiology and maintenance. Int J Eating Dis 5:475–487

Morgan HG, Hayward AE (1988) Clinical assessment of anorexia nervosa: the Morgan-Russell outcome assessment schedules. Br J Psychiatry 152:367–372

Office of Population Censuses and Surveys (1982) Census. 1981. Sex, age and marital status. SCEN 81. SAM. HMSO, London

A Long-Term Follow-up Study of Outpatients with Bulimia Nervosa Treated in a Structured Group Psychotherapy Program*

J.E. Mitchell and R.L. Pyle

Introduction

Data on the longitudinal course of bulimia nervosa are limited and difficult to synthesize into a clear picture. Researchers have not studied this disorder in a large enough number of subjects over a sufficient period of time to come to any firm conclusions about its course and the impact of various treatments on the long-term outcome. Our research group at the University of Minnesota has been interested in studying the clinical characteristics, course, and response to treatment for persons with bulimia nervosa since 1979.

Our first follow-up study involved a total of 83 patients with bulimia (DSM-III criteria) who we attempted to contact 12–15 months following outpatient evaluation in our eating disorders clinic (Mitchell et al. 1986). Seventy-five (90.4%) of the 83 patients were contacted and interviewed. Not all of these patients had participated in treatment in our clinic. The median age of this group of patients was 25. Seventy-five (95%) were female and all were Caucasian. Most (52, 69%) had never married. During the period of follow-up 62 (83%) had participated in some form of treatment or had been involved in self-help groups. Forty-three (57%) had participated in self-help groups, 40 (53%) had participated in weekly group or individual therapy, and 40 (53%) had participated in a structured outpatient intensive evening program for bulimia at the University of Minnesota. Seventeen (23%) had received antidepressant treatment, 14 (19%) had received outpatient counseling for other problems and 9 (12%) had attended Overeaters Anonymous. Overall, 25 (33%) were classified as free of bulimic symptoms at follow-up, 10 (13%) as much improved, 13 (17%) as improved, and 27 (36%) as minimally improved or unchanged.

We did a separate analysis on 30 bulimic patients who had originally gained control of their eating behavior for a minimum of 2 months but had subsequently relapsed. More than half of those who relapsed did so within

* This work was supported in part by Grants MH40377 and MH43296 from the NIMH and a grant from the Minnesota Medical Foundation.

2 months of finishing treatment, and one third relapsed while still involved in aftercare therapy. Situations associated with relapse included stressful situations and feeling anxious, nervous, or depressed (Mitchell et al. 1985a).

Our group subsequently completed and published some of the results from a 2- to 5-year follow-up study on a larger cohort of patients. The remainder of this chapter will be devoted to a presentation and discussion of data from that project (Mitchell et al. 1988).

Follow-up Study

The study involved a follow-up assessment of patients who had been referred for treatment to an outpatient intensive evening group psychotherapy program for bulimia nervosa offered in our clinic. The original sample consisted of 100 patients who had been evaluated in the outpatient Eating Disorders Clinic and had been referred by the evaluating staff for participation in this group psychotherapy treatment program. Patients who come to this clinic frequently come or are referred for participation in this program, which involves a total of 45 h of group work over a 10-week period. This program is held in the evening so as to allow students or employed persons to participate with little interference with scheduled activities. All patients satisfied DSM-III criteria for bulimia with the additional criterion of binge eating coupled with self-induced vomiting and/or laxative abuse at a minimum frequency of three times a week for a minimum duration of 6 months. Most, if not all, would have satisfied DSM-III-R criteria for bulimia nervosa. At evaluation the patients underwent a routine psychiatric evaluation, had a physical examination performed unless one had recently been completed elsewhere, and had screening laboratory tests performed. They also completed a psychiatric data base and an extensive instrument designed to obtain detailed information about eating and weight history, the Eating Disorders Questionnaire (EDQ; Mitchell et al. 1985b).

An attempt was made to follow up all patients regardless of whether or not they completed the group program or had been successful during treatment. Of the original 100 patients, 91 were contacted and interviewed. Five could not be located, one had died by suicide (approximately 15 months after finishing the group), and three refused to be interviewed. All had given written informed consent for follow-up at the time of their initial evaluation and treatment.

At initial evaluation the median age of this group of subjects was 22 (range 17–44) with a median age of onset of the eating disorder of 17 (range 11 to 27). Social class distribution (Hollingshead and Redlich Two Factor Index of Social Position) was: I = 14 (17%), II = 14 (17%), III = 26 (31%), IV = 4 (25%), V = 8 (10%; data not available on eight subjects). Sixteen (18%) had prior inpatient treatment for bulimia, 5 (5%) had inpatient treatment for anorexia nervosa, 15 (17%) had had inpatient treatment for depression, and 6 (7%) had inpatient treatment for alcohol or drug abuse problems. Forty (44%) had outpatient treatment for bulimia, 7 (8%) had outpatient treatment for anorexia nervosa, 17 (19%) had outpatient treatment for depression, and 3 (3%) had outpatient treatment for alcohol or drug abuse. Sixteen (17%) were married, 5 (5%) were separated or divorced, and 70 (77%) had never married.

The outpatient evening treatment program (referred to as the intensive program) was specifically developed to treat patients with bulimia and bulimia nervosa. The program has been in operation for nearly 10 years in various forms and has been described in the literature previously (Mitchell et al. 1984). This is a manual-based approach that is divided into three phases. During the first 2 weeks of treatment (phase I) patients come into the clinic two nights a week for 2 h each time. This phase, which is led by the program dietitian, focuses on nutritional issues and the development of meal planning skills. The second phase (phase II) of treatment is 4 weeks in duration and involves an initial high intensity with a tapering schedule over the period of 4 weeks. During the 1st week patients come into the clinic 5 nights a week, during the 2nd week, 4 nights and so on in a tapering schedule. Each evening consists of a lecture with discussion and a review of homework, a meal eaten in the hospital cafeteria with one of the therapists, and group therapy, which again is highly structured and manual based. Phase II emphasizes the interruption of bulimic symptoms with a strong expectation that during the intensive phase patients will try to gain control of their eating behavior and begin to eat regular balanced meals. The third portion of treatment (phase III), which involves weekly therapy sessions, focuses on relapse prevention through the reintroduction of feared foods and exposure to high-risk situations. Following treatment, patients are offered participation in a weekly aftercare group.

The mean duration to follow-up was 42 months (range 20–60 months). All subjects were interviewed by phone using a semi-structured interview designed for the purpose. Interviews were approximately 30–45 min in length. Detailed information was obtained concerning treatment and course of eating symptoms. The interviews were conducted by members of the Eating Disorder Treatment team, including psychiatrists, psychologists and social workers. Interviews were scheduled so that therapists did not contact patients with whom they had worked previously. Training sessions were held to improve interrater reliability on the instrument.

Of the 91 subjects who were followed up, 79 (87%) had completed the intensive treatment program and 11 (12%) had dropped out, while 1 (1%) had been asked to discontinue participation because of noncompliance with attendance. Fifty-six (62%) were between 90% and 110% of ideal body weight (Metropolitan Life Insurance Tables 1959), eighteen (20%) were less than 90% of ideal body weight and 22 (24%) were greater than 110%

Table 1. Outcome of 91 patients followed-up at 2–5 years

Outcome	Patients	
	(*n*)	(%)
Group 1		
Continually abstinent	17	19
Abstinent at follow-up but history of isolated lapses	8	9
Abstinent at follow-up but relapse <1 month in duration, no further treatment necessary	6	7
Group 2		
Abstinent at follow-up relapse >1 month in duration, no further treatment necessary	8	9
Group 3		
Abstinent at follow-up relapse >1 month in duration, further treatment necessary	21	23
Group 4		
Improved but still symptomatic at follow-up	8	9
Group 5		
Treatment failures	23	25
Total	91	100

ideal body weight. At follow-up there was somewhat less variability in weight, with fewer patients being overweight or underweight: 11 (12%) less than 90% ideal body weight and 18 (18%) greater than 110% ideal body weight.

Based on examination of the self-reported eating behavior over the period of time to follow-up, the patients were classified into five groups as to outcome. These data are summarized in Table 1. Overall, 60 (66%) of these patients were doing well and reported freedom from bulimic behavior at follow-up although some had achieved remission through additional treatment, whereas 8 (9%) were improved but still symptomatic and 23 (25%) were classified as treatment failures.

In order to further examine predictors of success and failure we compared patients in group 1 (the best outcome group) with those in group 5 (the worst outcome group).

The comparison of the two outcome groups on demographic variables is summarized in Table 2. The two groups did not differ significantly either at baseline or at follow-up as to age or percent ideal body weight. There were also no significant differences in marital status or social class. It is of note, however, that the patients in the best outcome group weighed on

Table 2. Demographic data on patients

Variable	Best outcome group	Worst outcome group	f	χ^2	p
Age at follow-up	26.7 ± 5.6	26.8 ± 5.2	1.14		.733
% Ideal body weight	100.2 ± 10.9	104.2 ± 11.5	1.10		.824
Married	17 (55%)	8 (35%)		1.406	.236
Social class					
I	5 (17%)	1 (6%)		2.758	.236
II	3 (10%)	3 (18%)			
III	12 (40%)	6 (36%)			
IV	9 (30%)	5 (29%)			
V	1 (3%)	2 (12%)			

Table 3. Frequency of bulimic behaviors prior to treatment

Variable	Best outcome group		Worst outcome group		χ^2	p
	(n)	(%)	(n)	(%)		
Binge eating						
> Once/day	8	26	13	56	8.067	.153
Once/day	7	23	4	17		
< Once/day	16	52	6	26		
Vomiting						
> Once/day	10	32	26	70	9.028	.108
Once/day	3	10	2	9		
< Once/day	18	58	5	22		
Laxatives						
< Once/day	0	0	0	0	1.961	.743
Once/day	1	3	0	0		
Several/week	3	10	3	14		
< Several/week	27	87	19	36		
Diet pills						
> Once/day	0	0	0	0	3.131	.536
Once/day	3	10	0	0		
Several/week	2	7	1	5		
< Several/week	25	83	21	95		

average slightly less than those in the worst outcome group at baseline yet were thinner at follow-up, although this difference was not significant.

Data on the frequency of various bulimic behaviors at baseline are summarized in Table 3. As can be seen there were no significant difference between groups. However, if the comparison is changed to those binge eating more than once a day versus those binge eating once a day or less, the difference is significant, and it does appear overall that the high frequency binge eating and vomiting patients tended to cluster in the worst

Table 4. Treatment histories of patients

Variable	Best outcome group		Worst outcome group		χ^2	p
	(n)	(%)	(n)	(%)		
Inpatient						
Bulimia nervosa	5	16	5	22	.029	.864
Chemical dependency	1	3	1	4	.000	1.000
Depression	6	19	2	9	.494	.482
Outpatient						
Bulimia nervosa	17	55	8	35	1.406	.236
Chemical dependency	0	0	0	0	–	–
Depression	8	26	2	9	1.553	.213

outcome group. Laxative use and diet pill usage were uncommon and did not appear to have predictive value.

Data are summarized in Table 4 relative to the treatment histories of the two groups of patients. Again there were no significant differences between the two groups on these variables. However, it is clear that the worst outcome group did not have an exaggerated history of poor prior treatment response, since numerically patients in the best outcome group were more likely to have been treated previously for bulimia nervosa and for depression as outpatients.

We also examined attitudes towards weight and shape at follow-up. These data are summarized in Table 5. As can be seen, there were no significant differences between the two groups in terms of perceived weight. However, there were significant differences in terms of fear of becoming fat and dissatisfaction with body proportions, with those in the worst outcome group generally scoring higher on both of these measures.

Discussion

There are several important limitations to the current study. First, not all subjects were followed up, and it is certainly likely that some of the patients who could not be located or who refused to be interviewed were doing poorly at follow-up. Second, there are significant questions concerning the validity of the data obtained. Information was based on patients' self-report, gathered by phone interview. It is certainly possible that a direct interview technique may have allowed for a more complete and reliable assessment. However, the number of patients willing to cooperate in such an interview probably would have decreased significantly and some patients may have been more open when not directly encountering the interviewer. Third, the breadth of information obtained was

Table 5. Attitudes toward weight and shape at follow-up

Variable	Best outcome group		Worst outcome group		χ^2	p
	(n)	(%)	(n)	(%)		
Weight						
Somewhat thin	4	13	0	0	3.127	.372
Normal	19	61	3	13		
Somewhat overweight	0	0	0	0		
Moderately overweight	7	23	10	42		
Extremely overweight	1	3	1	4		
Fear of becoming fat						
Extremely	3	10	11	48	14.742	.005
Very	7	24	7	30		
Moderately	4	14	2	9		
Slightly	6	21	3	12		
Not at all	9	31	0	0		
Dissatisfaction with body proportions						
Extremely	0	0	2	9	11.132	.025
Very	4	13	5	22		
Moderately	3	10	6	26		
Slightly	8	26	7	30		
Not at all	16	52	3	13		

fairly limited in this study, focusing primarily on eating behavior and information on relapse. These patients were not assessed relative to psychosocial or physiological functioning.

Based on these results we have drawn certain tentative conclusions:

1. Many people who participate in this structured treatment program appear to be doing well at 2- to 5-year follow-up, although treatment has clearly failed in some and some require additional therapy after this initial program involvement.
2. There is some suggestion that the patients who do poorly over time are the ones with the worst symptoms at baseline, which parallels the results from treatment studies in many areas of psychiatry.
3. Most patients who normalize their eating behavior do not gain weight. Indeed, there was a nonsignificant finding that they tended to maintain a lower body weight at follow-up than did patients who were continuing to have bulimic problems.
4. Many patients who at long-term follow-up are doing well may have slips or lapses during the process of recovery. This suggests that a pure "abstinence" model that treats such slips as failures rather than as learning experiences may not necessarily be the most useful theoretical orientation in treating these patients.

In an attempt to further evaluate the efficacy of this structured treatment program, we have recently completed a four way comparison study wherein subjects were randomly assigned to one of four treatment groups: treatment with the antidepressant imipramine hydrochloride, treatment with placebo, treatment with the antidepressant imipramine hydrochloride combined with involvement in the intensive structured group psychotherapy program, treatment with placebo combined with involvement in the intensive structured group psychotherapy program. The stort-term results of this study and 6-month follow-up results have been published (Mitchell et al. 1990; Pyle et al. 1990). An 18-month follow-up is currently in progress. Our current plan is to examine the long-term outcome of these various interventions in outpatients with bulimia nervosa.

References

Metropolitan Life Insurance Company (1959) New height standards for men and women. Stat Bull Metrop Insur Co 40:1–4

Mitchell JE, Goff G, Eckert ED, Hatsukami D, Pyle RL, Davis LE (1984) Intensive outpatient group treatment for bulimia. In: Garner D, Garfinkel P (eds) Handbook of psychotherapy for anorexia nervosa and bulimia. Guilford, New York, pp 240–253

Mitchell JE, Davis L, Goff G (1985a) The process of relapse in patients with bulimia. Int J Eating Dis 4:457–463

Mitchell JE, Hatsukami D, Eckert ED, Pyle RL (1985b) The eating disorders questionnaire. Psychopharmacol Bull 21:1025–1043

Mitchell JE, Davis L, Goff G, Pyle RL (1986) A follow-up study of patients with bulimia. Int J Eating Dis 5:441–450

Mitchell JE, Pyle RL, Hatsukami D, Goff G, Glotter D, Harper J (1988) A 2–5 year follow-up study of patients treated for bulimia. Int J Eating Dis 8:157–165

Mitchell JE, Pyle RL, Eckert ED, Hatsukami D, Pomeroy C, Zimmerman R (1990) A comparison study of antidepressants and structured intensive group psychotherapy in the treatment of bulimia nervosa. Arch Gen Psychiatry 47:149–157

Pyle RL, Mitchell JE, Eckert ED, Hatsukami D, Pomeroy C, Zimmerman R (1990) Maintenance treatment and 6-month outcome for bulimia patients who respond to initial treatment. Am J Psychiatry 147:871–875

A Large-Scale Longitudinal Follow-up Study of Patients with Eating Disorders: Methodological Issues and Preliminary Results

W. Vandereycken and G. Pieters

Background

The University Psychiatric Center in Kortenberg has been especially interested in treatment and research of eating disorders for about a quarter of a century (for an overview see, for example, Vandereycken 1985; Vandereycken and Meermann 1984; Vandereycken et al. 1989). Follow-up investigations have become a part of the standard research program.

Previous Follow-up Studies

Our *first* and often quoted study (Pierloot et al. 1975) concerned a group of 32 female anorexia nervosa patients who had been treated on an inpatient basis between 1967 and 1973. The follow-up ranged from 1 to 6 years after treatment. The status of the patients at follow-up was assessed by the principal therapist who based his judgment on a direct interview with the patients. The outcome was correlated with clinical variables at admission (a retrospective analysis of medical records) as well as personality characteristics measured by psychodiagnostic tests (e.g., Minnesota Multiphasic Personality Inventory, MMPI).

In 1980 the first author (WV) undertook a *second* study: all the files of patients admitted to the hospital (1967–1979) with some kind of an "anorectic" problem were screened. Using the diagnostic criteria of Feighner et al. (1972), except for the age limit (some patients were older than 25 years at onset), we collected a total sample of 141 female anorexia nervosa patients for whom we had sufficient information at admission. It included the sample of the previous study, except for two patients who did not meet the Feighner criteria for anorexia nervosa. Follow-up information was only gathered in patients who at the time of the study had been discharged from our hospital at least 1 year previously (a total of 128 patients). All patients were invited by letter to a personal interview. If they objected or did not react after a second invitation, they received a follow-up questionnaire concerning their medical and psychosocial history since discharge. In cases where the patient did not return the questionnaire, we tried to obtain information from the general practitioner. For 40 patients we obtained insufficient or no information at all, except for the fact that they were all alive at the time of the study. Results of this research project have been scattered in different publications focusing on particular issues: diagnostic subclassification (Vandereycken and Pierloot 1983c), treatment program (Pierloot et al. 1982), drop-outs during treatment (Vandereycken and Pierloot 1983a), methodological problems in follow-up studies (Vandereycken and Pierloot 1983b).

In 1985 a *third* research project was started, which due to its longitudinal nature is still going on. It is essentially a combined retrospective and prospective study with repeated measurements at admission, after 6 months, 1 year, 2 years, and 5 years. This design was based on two fundamental (methodological and clinical) questions:

1. Is there a difference between retrospective (unannounced beforehand to the patients) and prospective follow-up (announced to the patients at the beginning of the program)?
2. What patterns of course do patients show after treatment, and is there a relationship between short-term (from 6 months to 2 years) and long-term effects (after 5 years)?

Before we report some preliminary results of this study, we describe in detail the concrete procedure, with respect especially to the methodological issues discussed elsewhere in this book (see Vandereycken and Meermann, this volume).

Present Follow-up

Study Design

The study sample consists of all patients who have been admitted to our hospital with a DSM-III-R diagnosis of eating disorder since 1980. For those patients hospitalized between January 1980 and December 1984 the study could be considered as partially retrospective and not preprogrammed (we call this the "retrospective" sample), whereas for all patients admitted from 1985 on, and up until today, the study was/is announced at the start of the program (the "prospective" sample). Since January 1985 we proceed as follows. Each month we send a letter of invitation to patients admitted 6 months, 1 year, and 2 and 5 years before. So between January and July of 1985 invitations were sent only to the retrospective sample and from July 1985 the first patients of the prospective sample received an invitation. In January 1990 we finished the study of the retrospective sample (5 years follow-up for those admitted in December 1989) which implies that for this sample a complete set of four repeated follow-up assessments (from 6 months to 5 years) could only be obtained for the patients who entered the hospital between July and December 1984. For the rest of the retrospective sample the number of repeated assessments depended on the time which had passed since their admission. The prospective study, on the other hand, will be finished (with four repeated measurements) for the whole sample in 1995.

Admission

Our treatment center has built up a certain reputation in the treatment of eating disorders. Although we have an outpatient department, most of our outpatient activities are devoted to the aftercare of previously hospitalized patients. The main thrust of our work lies in the inpatient treatment of "difficult" cases. Patients are referred to us from all over the Flemish-speaking part of Belgium (those from the French-speaking part can only be accepted if they speak Dutch). We also see many patients from the Netherlands where up to a few years ago only limited possibilities for specialized treatment existed. Due to our health care system, up to 90% of the treatment costs (unlimited time) are covered by the state health insurance (so there are no financial barriers involved in the selection). The whole year around we have a waiting list for admission (our total capacity is 18 beds), but we like to limit the waiting time to a maximum of 3 months. Indication for inpatient treatment is usually based on a combination of medical, psychotherapeutic, and psychosocial criteria (see Vandereycken 1987b). The overall sample admitted may be considered a *"negative" selection*. If we think that patients can be treated on an out-patient basis or in another hospital (with less intensive treatment), we refer them elsewhere. This means that we adopted the rule that inpatient treatment is indicated solely in patients in whom other treatments have failed (except for those living in the region for which our hospital is obliged to offer health care services). In principle, we only admit patients from the age of 15 years on (with a few occasional exceptions) and there is no upper age limit. The severity or duration of the disorder is not an exclusion criterion (in fact, "complex" cases are an indication instead) and we do not request specific motivation from the patients: if the parents of minors or the spouse of a married patient agree, we accept involuntary admissions.

Once a patient is admitted to the eating disorder unit, the principal therapist (one experienced clinical psychologist for each of the two groups of nine inpatients) fills in a *standardized research file*, including the usual demographic variables and a series of clinical characteristics: diagnosis according to DSM-III(-R), presence and severity (frequency) of behaviors and attitudes known to be linked to eating disorders, age at onset and duration of illness, number and type of previous therapies, attitude of patient and parents (or spouse) towards admission, reason for entering the inpatient program, data on family of origin (and on partner if living together), judgement of psychosocial and sexual functioning. All research files are screened by the supervising psychiatrist (WV) to ensure accuracy and consistency in reporting. On the first day of admission all patients fill out the Eating Disorder Inventory (EDI; Garner et al. 1983). Patients and parents (spouses) also complete other tests in the context of particular research projects, especially regarding family and marital interaction (see Vandereycken et al. 1989) and body experience (e.g., Vandereycken et al. 1988).

Treatment

Our inpatient treatment approach has been changed several times, mostly guided by research findings (see Vandereycken 1985). From 1982 on, the program was established along the conceptual lines which are still valid today (though some practical variations on the same theme took place over time, of course). First of all, the program is *time limited* with a maximum stay of 6 months but we emphasize the importance of well-planned aftercare including follow-up assessments (a rule explained to the patients and the family before the start of the program). Major emphasis is placed on a "constructive" *family-oriented* approach (Vandereycken 1987a) including the flexible use of a variety of methods to involve the family (spouse) as actively as possible in the treatment (for details and examples see Vandereycken et al. 1989). All formal therapies are conducted in two groups of nine patients each (mixing bulimics and anorectics, adolescents and adults); the setting of this *group approach* (no individual therapies) can be characterized as highly structured, directive and confrontational with repeated evaluations (see Vandereycken 1988).

The use of psychotropic drugs is a great exception and, when prescribed in a few cases, limited to a minor tranquilizer or hypnotic for a maximum of 1 week. The *multifaceted* psychotherapeutic elements of the program are as insofar as possible integrated according to a formal treatment plan (including short-term and long-term goals) proposed by the patient and approved by the group members and the staff. The central integrating figure, the "principal therapist," is the clinical psychologist who conducts the group psychotherapy and organizes regular sessions with family members (parents, siblings, spouse). Four times a week the practical execution of the plan (translated into concrete steps in the "here and now") is discussed in a special "work session" of the patient group and nurses. With regard to specifically anorectic or bulimic behavior (e.g., weight loss, abnormal eating habits, vomiting, laxative abuse) a contract system is used based on behavior therapy principles (Vandereycken and Meermann 1984). With regard to weight restoration our goal (to be reached before discharge) is at least 90% of the ideal body weight and we very seldom apply tube feeding, although its use as a last resort is stipulated in the treatment contract. In the art/occupational therapy (five sessions per week) patients are asked to express and elaborate specific themes with, alternately, a personal importance as to their own treatment process and a significance within the group interaction. Three times a week a special body-oriented therapy session takes place combining ingredients from psychomotor and dance-movement therapy, with particular emphasis on video confrontation (see Vandereycken et al. 1987; Vandereycken 1990). Group psychotherapy sessions (three times a week) are usually quite structured by the rather directive therapists who stress both cognitive and experiential explorations of a particular theme brought in by each group member in turn (Vandereycken et al. 1986).

Patients have to stop work or school during the whole inpatient treatment and are not allowed to leave the hospital in the first month (visitors only permitted on weekends). Then progressively, and depending on the individual's progress, contacts with the outside world are stimulated and more weekends spent elsewhere. In the final weeks (if practical) patients only come to the "day program" within the same group and sleep elsewhere (either at home or in a new environment). At the end of inpatient treatment the *aftercare* is planned. For those who can come to our clinic (traveling distance less than 2h) – and this constitutes the majority of our population – we are responsible for further outpatient treatment: patients have to come at least once every 2 weeks for 6 months to one of our "aftercare groups" (following the same basic principles as in the inpatient treatment), usually alternated with more or less frequent family sessions. This implies that the patient is followed up by the clinical psychologist who was his or her "principal therapist" while hospitalized. If patients live too far away, contacts with another clinician are always arranged before the patient is discharged. Moreover, in these cases during the first months we act as consultants to the aftercare therapist (by telephone) and invite the patient back for at least two meetings with the former "principal therapist." In each form of aftercare a contract is made which stipulates when the patient must return for more intensive outpatient therapy or, if necessary, be readmitted (limited to 2 weeks, usually for quick weight restoration above a certain minimum). If patients drop out prematurely during the therapeutic program, they are always given a second, but not a third, chance to reenter treatment if they so desire.

Follow-up Assessments

As explained above, follow-up measurements are repeated four times (starting from the admission date): after 6 months, 1 year, and 2 and 5 years. Patients receive a written invitation to particate in the study each time. If they do not reply within a month, a reminder is sent. Since we had no special research budget for this project, neither patients nor investigators could be paid. In the first 2 years of the project we had a part-time research assistant coordinating the work, but later on we had to rely on volunteer trainees (clinical psychologists) to help us. This explains why the intensity of the follow-up work was greater during the first 2 years: patients would receive a third reminder if they had not replied; if their address had changed, we tried to locate them insofar as was possible; and all patients were invited for an interview. Afterwards, only one reminder was sent, no effort was made to locate patients, and interviews were organized only for patients who were still coming to the center for their aftercare. If an interview took place, the patient was also weighed at our unit.

The interviews and the analysis of the questionnaires were always done by an assessor who previously had not been in contact with the patients. The items covered in the interview and the follow-up questionnaire were the same as those assessed at admission (including the completion of the EDI, see above). In order to ensure accuracy and consistency in scoring, all final results were checked by the first author (WV, supervisor of the research and treatment program). In case of discrepancy in judgement/interpretation of the findings between assessor and supervisor a consensus was made with one major rule in mind: since the supervisor could be biased as to patients who fully completed the treatment, the "worst" judgement was chosen in doubtful cases (which in turn might be viewed as a *"negative" bias*, of course). In a pilot study of the cases sampled in the first 2 years, a randomly selected group of patients was asked to fill in the follow-up questionnaire just before the interview took place. The interviewer assessed the same items independently, and afterwards both scorings were compared. As expected from another study comparing evaluations by patients and by staff members (Vanderlinden and Vandereycken 1988), patients tended to be generally more optimistic in evaluating their status than the assessors, particularly with respect to their actual body weight (estimation 5%–10% higher than actual weight) and the judgement of their relationship with their family of origin. This finding reinforced our previously mentioned choice of a "negative" bias, especially with regard to questionnaire data (see also Vandereycken and Vanderlinden 1983).

Part of the clinical assessment (at admission and consecutive follow-ups) was standardized into a global scoring system we called the *Eating Disorder Evaluation Scale* (EDES). The different variables and their scoring are summarized in Table 1. As reported elsewhere (Vandereycken 1987b) the design of the scale – reflecting general clinical impressions – was inspired by similar instruments described by Garfinkel et al. (1977) and Hall and Crisp (1983). Provisionally we differentiate the severity of the eating disorder, according to the total EDES score, into five categories: within "normal" range (above 80), "slightly to moderately" disturbed (61–80), "clearly" disturbed (41–60), "seriously" disturbed (21–40) and

Table 1. Eating Disorder Evaluation Scale (EDES)

Actual body weight (in percentage of matched mean population weight)			
90–109	= 6		
80–89 or 110–119	= 4		
70–79 or 120–130	= 2		
<70 or >130	= 0		a = . . .
Body weight fluctuations (during the last 6 months)			
<1 kg	= 6		
1–2 kg	= 4		
2–5 kg	= 2		
>5 kg	= 0		b = . . .
Eating behavior			
Never	= 6		
Sometimes	= 4		
Often (weekly)	= 2		
Very often (daily)	= 0		
		Food restriction	c = . . .
		Bulimia	d = . . .
		Vomiting	e = . . .
		Use of laxatives	f = . . .
Preoccupations			
None	= 6		
Moderate	= 4		
Strong	= 2		
Extreme	= 0		
		With food/eating	g = . . .
		With body shape	h = . . .
Menstrual function[a]			
Regular cycles (pregnancy)	= 6		
Irregular or artificial cycles	= 4		
Sporadic menstrual bleeding	= 2		
Absent	= 0		i = . . .
Sexual adjustment			
Pleasure = 6 – 4 – 2 – 0 = avoidance/dislike			
		Attitude toward sexuality	j = . . .
		Active sexual behavior	k = . . .
Social adjustment			
Good/satisfactory	= 6		
Fair/variable	= 4		
Inadequate/poor	= 2		
Bad	= 0		
		Attitude toward family[b]	l = . . .
		Contacts outside family	m = . . .
		Occupational functioning	n = . . .
Mental state			
6 = Normal			
4 = Mildly abnormal (e.g. slight mood swings or psychosomatic complaints for which no treatment is necessary)			
2 = Markedly abnormal (e.g. anxiety disorder, affective disorder, and other problems for which out-patient psychiatric/psychotherapeutic treatment is needed)			
0 = Grossly abnormal (e.g. serious suicide attempt, addiction, psychotic behavior and problems for which hospitalization is necessary)			
			o = . . .
Final score (a to o; maximum = 90)		Total	= . . .

[a] In case of male patients one could ask about the frequency of ejaculations (through spontaneous emission, masturbation or sexual contact).

[b] In married patients we mean the family of origin.

"extremely" disturbed (20 and less). The usefulness of the EDES and the (differential) value of the various subscores will be tested in a large patient sample (this might lead to a different scoring or subgrouping of variables). Currently the scale is also being tested in a normal population to obtain some normative data.

Preliminary Results

As explained above, only part of the study is finished. We will first present the data related to basic methodological issues (referring to the form of the present study). Next, some preliminary findings will be reported on the content of the project, namely, outcome after 5 years in a subsample. We distinguish between four diagnostic categories: restricting anorectics (pure anorexia nervosa), anorectics who also binge and/or purge (mixed anorexia nervosa), bulimia nervosa and atypical eating disorders. The total sample of which we present data here consists of 315 inpatients (309 females, 6 males). Table 2 summarizes some characteristics at the time of admission: acteristics at the time of admission:

- Age, duration of illness, weight, marital status, social class.
- By first admission we mean that the patient has not been hospitalized before being admitted to our inpatient unit.
- First treatment means that no regular outpatient therapy (for at least 2 months) had been tried out before admission.
- No comorbidity refers to the finding that no additional DSM-III-R diagnosis (on axis 1 or 2) was present.
- The EDES was developed in a later phase of the study, so that the percentages given in Table 2 refer to a smaller sample (total = 191): 66 restricting anorexia nervosa, 61 mixed anorexia nervosa, 47 bulimia nervosa and 15 atypical eating disorders; the highest score in this sample was 52.
- Because of its international recognition, we also present the mean raw scores on the EDI (Garner et al. 1983).

Methodological Issues

In a large-scale project such as this, taking into account the many practical restrictions already discussed, it is no wonder that for only part of the total sample are we able to obtain a complete set of four follow-up evaluations (after 6 months, 1 year, and 2 and 5 years). To assess the well-known bias of patient selection caused by the *attrition rate*, we compared the patients who did not respond to any of our repeated requests for information (n = 24) with the group of patients who, at least once during the follow-up period, were willing to cooperate (n = 291, including all six male patients).

Table 2. Characteristics of different diagnostic subgroups ($n = 315$)

	AN-R ($n = 92$)	AN-M ($n = 115$)	BN ($n = 78$)	AT ($n = 30$)
Age (years)	20.1 ± 5.4	23.8 ± 6.4	23.7 ± 4.5	25.8 ± 8.2
Duration of illness (years)	3.5 ± 3.4	5.4 ± 5.1	6.4 ± 4.6	4.8 ± 5.1
Weight (% IBW)	68.5 ± 9.9	75.0 ± 15.1	102.0 ± 19.7	88.5 ± 26.9
Unmarried	83 (90.3%)	91 (79.1%)	60 (76.9%)	20 (66.7%)
Lower class	23 (25.0%)	40 (34.8%)	28 (35.9%)	12 (40.0%)
Middle class	53 (57.6%)	56 (48.7%)	33 (42.3%)	15 (50.0%)
Upper class	16 (17.4%)	19 (16.5%)	17 (21.8%)	3 (10.0%)
First admission	56 (60.9%)	72 (62.6%)	46 (59.0%)	15 (50.0%)
First treatment	24 (26.1%)	25 (21.7%)	9 (11.5%)	9 (18.0%)
No comorbidity	80 (86.9%)	68 (59.1%)	40 (51.2%)	19 (63.3%)
EDES total				
41–60	18.2%	4.9%	17.0%	23.5%
21–40	78.8%	55.7%	66.0%	64.7%
0–20	3.0%	39.4%	17.0%	11.8%
EDI scores[a]				
Total	57.8 ± 29.5	68.8 ± 29.1	92.4 ± 26.4	75.4 ± 28.2
DT	8.9 ± 7.3	11.5 ± 6.3	14.6 ± 5.3	10.6 ± 6.8
B	0.9 ± 2.3	4.0 ± 4.9	11.3 ± 4.9	3.1 ± 4.3
BD	10.0 ± 6.9	10.9 ± 6.2	16.6 ± 7.3	13.3 ± 6.9
I	10.3 ± 7.5	12.0 ± 7.2	14.7 ± 6.2	13.4 ± 6.8
P	6.2 ± 4.0	6.4 ± 4.5	8.2 ± 4.7	7.0 ± 4.2
ID	6.8 ± 4.3	6.8 ± 4.4	6.8 ± 4.6	7.0 ± 4.8
IA	8.2 ± 6.0	10.7 ± 6.7	12.9 ± 5.9	11.4 ± 5.9
MF	6.6 ± 5.5	6.6 ± 5.2	7.0 ± 6.2	8.0 ± 6.2

IBW, ideal body weight (according to Metropolitan Life Insurance Tables); EDES, Eating Disorder Evaluation Scale; EDI, Eating Disorder Inventory; AN-R, restricting anorexia nervosa; AN-M, mixed anorexia nervosa; BN, bulimia nervosa; AT, atypical eating disorder; EDI, subscales: DT, dieting; B, bulimia; BD, body dissatisfaction; I, ineffectiveness; P, perfectionism; I, interpersonal distrust; IA, interoceptive awareness; MF, maturity fears.

[a] The higher the score, the more abnormal the symptom category is considered to be.

As can be seen in Table 3, the noncooperative patients seem to belong to a more seriously disturbed subgroup: they are older, more often married, show a longer duration of illness with a more pronounced clinical picture (lower weight, more bingeing and/or purging, more additional mental disturbances, poor functioning at school or work) and they have been hospitalized more often before the present admission. Following the ethical guidelines discussed in the first chapter of this book, we did not pursue our search for follow-up information any further if patients did not react or overtly refused cooperation. It is therefore possible that of this group one or more patients have died in the meantime.

To test whether a *retrospective versus prospective* follow-up procedure makes a difference is not that easy. One of the problems involved is that a

Table 3. Comparison between patients who did and those who did not cooperate with the follow-up assessment (characteristics at admission)

	Cooperative (n = 291)	Noncooperative (n = 24)	Statistical significance
Age (years)	22.5 ± 5.8	24.5 ± 6.6	*
Duration of illness (years)	4.5 ± 4.2	6.8 ± 5.4	***
Weight (% IBW)	80.3 ± 36.8	77.3 ± 20.0	*
Unmarried	242 (83.8%)	17 (70.8%)	**
Lower class	85 (29.2%)	8 (33.3%)	NS
Middle class	153 (52.6%)	13 (54.2%)	NS
Upper class	53 (18.2%)	3 (12.5%)	NS
First admission	198 (68.0%)	14 (58.3%)	**
First treatment	62 (21.3%)	6 (25.0%)	NS
No comorbidity	203 (69.8%)	13 (54.2%)	**
No bingeing	159 (54.6%)	11 (45.8%)	*
No purging	153 (52.6%)	10 (41.7%)	*
Poor at school or work[a]	55.9%	76.9%	**
EDES total	30.9 ± 10.4	25.9 ± 11.0	*
EDI total	70.7 ± 30.2	74.1 ± 37.6	NS
EDI bulimia[b]	4.3 ± 5.5	7.4 ± 6.8	*

EDES, Eating Disorder Evaluation Scale; EDI, Eating Disorder Inventory.
[a] Based on EDES (n = 167 and 14); no significant differences on other items.
[b] No significant differences on other EDI subscales.
* $p < .05$, ** $p < .01$; *** $p < .001$

comparison between our retrospective sample from 1980–1985 (group 1) and the prospective one from 1985–1989 (group 2) is difficult to interpret since the population referred to the hospital has changed over time. A global comparison of the two groups showed some significant differences on various characteristics, which we summarize as follows:

- No difference in age and duration of illness.
- More mixed anorectics and bulimia nervosa patients in group 2.
- Patients from group 2 had been hospitalized significantly more often before their admission to our center.

Moreover, some elements of the treatment program have been altered, including the aftercare (the average length of inpatient treatment was significantly longer for group 2). Hence, we need to carefully match patients from both groups so that they are comparable on some crucial variables. This part of the study has not yet been done.

A final methodological issue had to do with the *form of assessment*, the way of obtaining information from the patient: by direct interview or by mailed questionnaire. We have explained the practical difficulties in this respect and underlined already that for some variables (e.g., self-reported weight) the patient's information might be misleading. On the other hand,

Table 4. Outcome after 5 years: demographic and clinical features (n = 174)

	AN-R (n = 61)	AN-M (n = 54)	BN (n = 45)	AT (n = 14)
Demographic data				
Marital status				
Unmarried	57 (93.4%)	46 (85.2%)	34 (75.5%)	9 (64.3%)
No partner	55 (90.2%)	40 (74.1%)	28 (62.2%)	8 (57.2%)
Engaged	2 (3.2%)	6 (11.1%)	6 (13.3%)	1 (7.1%)
Living situation				
Alone	27 (44.3%)	11 (20.4%)	6 (13.3%)	5 (35.7%)
With family	13 (21.3%)	20 (37.0%)	11 (24.5%)	1 (7.2%)
With partner	5 (8.2%)	7 (13.0%)	10 (22.2%)	3 (21.4%)
Other[a]	16 (26.2%)	16 (29.6%)	18 (40.0%)	5 (35.7%)
Own child(ren)	1 (1.6%)	2 (3.7%)	5 (11.1%)	4 (28.6%)
Occupation				
Studying	34 (55.7%)	9 (16.7%)	9 (20.0%)	2 (14.3%)
Working	13 (21.3%)	25 (46.3%)	21 (46.7%)	3 (21.4%)
Other[b]	7 (11.5%)	10 (18.5%)	8 (17.8%)	0 (0.0%)
None	7 (11.5%)	10 (18.5%)	7 (15.5%)	9 (64.3%)
Clinical data				
Weeks of hospitalization[c]	1.0 ± 4.	1.3 ± 4.3	9.9 ± 20.5	35.7 ± 59.2
Months in regular therapy[d]	1.5 ± 4.4	3.9 ± 8.6	3.3 ± 9.3	20.0 ± 18.3
General outcome				
Good	34 (55.7%)	17 (31.5%)	15 (33.3%)	3 (21.4%)
Moderate	23 (37.7%)	24 (44.4%)	17 (37.8%)	8 (57.2%)
Bad	4 (6.6%)	13 (24.1%)	13 (28.9%)	3 (21.4%)
Death	0	0	5	0
Weight (% IBW)	93.2 ± 9.2	92.3 ± 11.2	100.5 ± 20.0	93.3 ± 3.5
EDI total	11.4 ± 9.7	33.7 ± 34.7	58.4 ± 47.1	91.7 ± 65.2
EDES total	72.5 ± 9.3	64.9 ± 14.1	56.1 ± 19.8	49.3 ± 19.2
EDES items[e]				
Weight/normal				
Normal-good	57 (93.4%)	47 (87.0%)	39 (86.7%)	11 (78.6%)
Poor-bad	4 (6.6%)	7 (13.0%)	6 (13.3%)	3 (21.4%)
Fluctuations				
Normal-good	9 (14.8%)	8 (14.8%)	3 (6.7%)	1 (7.1%)
Poor-bad	52 (85.2%)	46 (85.2%)	42 (93.3%)	13 (92.9%)
Dieting				
Normal-good	44 (72.1%)	34 (63.0%)	30 (66.7%)	7 (50.0%)
Poor-bad	17 (27.9%)	20 (37.0%)	15 (33.3%)	7 (50.0%)
Vomiting				
Normal-good	60 (98.4%)	43 (79.6%)	36 (80.0%)	12 (85.7%)
Poor-bad	1 (1.6%)	11 (20.4%)	9 (20.0%)	2 (14.3%)
Bingeing				
Normal-good	59 (96.7%)	44 (81.5%)	25 (55.6%)	13 (92.9%)
Poor-bad	2 (3.3%)	10 (18.5%)	20 (44.4%)	1 (7.1%)
Laxatives				
Normal-good	58 (95.1%)	50 (92.6%)	40 (88.9%)	12 (85.7%)
Poor-bad	3 (4.9%)	4 (7.4%)	5 (11.1%)	2 (14.3%)
Food preoccupation				
Normal-good	40 (65.6%)	21 (38.9%)	13 (28.9%)	7 (50.0%)
Poor-bad	21 (34.4%)	33 (61.1%)	32 (71.1%)	7 (50.0%)

Table 4 (*continued*)

	AN-R (n = 61)	AN-M (n = 54)	BN (n = 45)	AT (n = 14)
Weight preoccupation				
Normal-good	20 (32.8%)	19 (35.2%)	13 (28.9%)	7 (50.0%)
Poor-bad	41 (67.2%)	35 (64.8%)	32 (71.1%)	7 (50.0%)
Menses				
Normal-good	14 (23.0%)	25 (46.3%)	34 (75.6%)	9 (64.3%)
Poor-bad	47 (77.0%)	29 (53.7%)	11 (24.4%)	5 (35.7%)
Sex attitude				
Normal-good	40 (65.6%)	44 (81.5%)	35 (77.8%)	8 (57.1%)
Poor-bad	21 (34.4%)	10 (18.5%)	10 (22.2%)	6 (42.9%)
Sex behavior				
Normal-good	16 (26.2%)	19 (35.2%)	20 (44.4%)	5 (35.7%)
Poor-bad	45 (73.8%)	35 (64.8%)	25 (55.6%)	9 (64.3%)
Family relations				
Good	41 (67.2%)	33 (61.1%)	28 (62.2%)	9 (64.3%)
Bad	20 (32.8%)	21 (38.9%)	17 (37.8%)	5 (35.7%)
Social contacts				
Good	47 (77.0%)	35 (64.8%)	32 (71.1%)	9 (64.3%)
Bad	14 (23.0%)	19 (35.2%)	13 (28.9%)	5 (35.7%)
Study or work				
Good	49 (80.3%)	45 (83.3%)	28 (62.2%)	12 (85.7%)
Bad	12 (19.7%)	9 (16.7%)	17 (37.8%)	2 (14.3%)
Mental status				
Good	52 (85.2%)	37 (68.5%)	23 (51.1%)	7 (50.0%)
Bad	9 (14.8%)	17 (31.5%)	22 (48.9%)	7 (50.0%)

AN-R, restricting anorexia nervosa; AN-M, mixed anorexia nervosa; BN, bulimia nervosa; AT, atypical eating disorder; EDI, Eating Disorder Inventory; EDES, Eating Disorder Evaluation Scale; IBW, ideal body weight.

[a] Many were living in a student home, some in a halfway house or foster family, and a few in a hospital setting at the time of follow-up assessment.

[b] No regular studies or work but active, for example, in a job as volunteer, in a hobby club or partial training course.

[c] During the last 2 years because of eating or other mental disorder.

[d] Outpatient psychiatric treatment or psychotherapy at least once every 2 weeks over the last 2 years.

[e] For each item (see Table 1) we combined the score 0 and 2 as "poor/bad" and 4 and 6 as "normal/good", some of our criteria for "normality" are probably too strict and should be corrected once we have data on a normal control group.

the fact that after a while we were obliged to rely in most cases on questionnaire data might constitute another bias. The sample only assessed with questionnaires (group A; 76.5%) did not differ from the interviewed sample (group B; 23.5%) on most of the variables mentioned in Table 2 except for the following characteristics: patients of group B were less often married (12.5% vs. 21.6%), less often unemployed (9.4% vs. 26.7%) and came more frequently from the middle social classes (75% vs. 50%). Some elements of treatment may be linked to the patient's willingness to accept an invitation for an interview: patients from group B more often stayed in day hospital treatment as part of their whole therapeutic program at our center (but this can also be due to geographic differences, i.e., patients living closer to the hospital are more likely to come in for day treatment and accept more easily having to return for a follow-up interview). Whether this is connected with the fact that patients from group B were less likely to stop treatment prematurely (12.5% vs. 25.6%) is not clear. Hence, some of these findings seem to suggest that the interviewed patients had "better contact" with the treatment center!

Outcome After 5 Years

Of the total sample of 291 patients who complied to our follow-up request, up to now five have died (all from the bulimia nervosa subgroup): two within 6 months after admission (one starvation, one suicide), one suicide in the 2nd year, and two deaths between the 3rd and 5th year (one starvation, one suicide). As mentioned above, it is possible that among the noncooperative patients for whom we lack information there might be some patient(s) who have died. Table 4 summarizes our data on 174 patients (still living) 5 years after admission at our hospital.

Differences in demographic features (marital status, living situation, occupation) are to some extent related to the patients' age, which is somewhat lower for restricting anorectics (see Table 2). The figures of the atypical eating disorder group should be interpreted with caution due to its relatively small size (these patients show a very variable clinical picture). When we compare the three other groups, overall the restricting anorectics show the best outcome as reflected in the global evaluation, the total EDI and EDES scores. But many of them still do not have regular menstruations and show some reluctance towards sexuality. The mixed anorectics seem to resemble the bulimia nervosa patients, although the latter show a worse outcome (counting the five deaths, i.e., poor outcome): 30% good, 34% moderate, and 36% poor (several had to be rehospitalized).

Discussion

Inspired by the research literature (Vandereycken and Meermann, and Herzog et al., this volume) and our own previous follow-up studies, we are convinced that the impact of treatment on the course of eating disorders should be assessed in a prospective longitudinal investigation with repeated evaluations over fixed time intervals (see also Keller et al. 1987). But, because treatment effects will obviously diminish over time and life events will inevitably interfere, we are not directly interested in the outcome in the long term (after 5 years and more). Our approach is a compromise between short-term and long-term evaluations: starting at the patient's admission (T1), with most treatment efforts exerted within the following months, a second evaluation takes place after 6 months (T2) and a third one after another 6 months (T3). So, using three assessments within 1 year, we hope to find certain profiles of (treatment-induced) changes. Looking for predictive features, both the findings at T1 (before treatment) and T2 (after treatment) can be utilized as independent variables to be correlated with later assessments (the relationship between T1 and T2 is, of course, also an interesting one). For reasons of comparability, the interval between T1 and T3 is the same as between T3 and T4, namely, 1 year (T4 being 2 years after admission). Finally, since the literature suggests a follow-up of at least 4 years, we planned a fifth evaluation 5 years after the first assessment.

We wanted to analyze the course of differentiated forms of eating disorders in large samples (at the end of the project in 1995, more than 500 patients will have gone through the research procedure). Having arrived at the halfway mark, we must admit that this probably is too ambitious an endeavor with respect to the very limited potential we have to carry out such studies (besides many other current investigations at our center). We have explained already that the method suffered from the large scale of the enterprise: perhaps it would have been better to take a limited population (e.g., 100 patients) and study this in detail with a sound and consistent method. On the other hand, regardless of its limitations, we think that our research has already yielded a wealth of information, and some of it is offered in this chapter.

Having struggled ourselves with methodological problems in the past (see Vandereycken and Pierloot 1983b) the findings presented in the previous pages confirm the significance of specific elements of the *follow-up method*: patients who do not comply to requests for cooperation in this type of study probably constitute a "negatively" biased subgroup, and those who are willing to return for a direct interview most likely belong to a "better" subsample (see also Ringler et al. 1983). One should take these factors carefully into account before making general conclusions on the sample studied. We also have reason to doubt the reliability of patients' self-reported information through questionnaire surveys (see, e.g., Vandereycken and Vanderlinden 1983) or telephone interviews. A typical example is a so-called "hard" fact such as body weight: it has been shown that self-reports about weight in women (and a fortiori in anorectics) have a questionable validity (see, e.g., Cash et al. 1989).

Finally, the results we presented as to the outcome after 5 years confirmed our previously stated opinion (see Vandereycken and Pierloot 1983c) about the significance of *subclassification* in eating disorders: one should distinguish in the anorexia nervosa population the classical type of fasting patients (abstainers, restrictors) and the mixed type of patients whose eating behavior (binge eating, vomiting and/or laxative abuse) resemble more the bulimia nervosa picture. The first type of anorexia nervosa appears to have in general a better outcome. As we have argued elsewhere (Vandereycken and Meermann 1984), we are convinced that many patients pass through a certain eating disorder "career": beginning as dieters (restrained eaters) and in several cases passing through a really anorectic phase, the older they become and the longer their disturbed eating pattern lasts, the more likely they will develop in the "bulimic" direction (from mixed anorexia nervosa to typical bulimia nervosa). Treatment may influence this course in both positive and negative (iatrogenic) ways. When the rest of our research data is mature enough to be analyzed and presented, we hope to find some crucial determinants of this process. After all, research efforts should benefit everyday clinical practice!

Acknowledgements. We gratefully acknowledge the help from Dr. R. Stroobants, Dr. O. Van den Bergh and Ms. Ch. Vercauteren.

References

Cash TF, Counts B, Hangen J, Huffine CE (1989) How much do you weigh? Determinants of validity of self-reported body weight. Percept Motor Skills 69:248–250

Feighner JP, Robins E, Guze SB, Woodruff RA, Winokur G, Munoz R (1972) Diagnostic criteria for use in psychiatric research. Arch Gen Psychiatry 26:57–63

Garfinkel PE, Moldofsky D, Garner DM (1977) The outcome of anorexia nervosa. In: Vigersky RA (ed) Anorexia nervosa. Raven, New York, pp 315–329

Garner DM, Olmsted MP, Polivy J (1983) Development and validation of a multidimensional eating disorder inventory for anorexia nervosa and bulimia. Int J Eating Dis 2(2):15–34

Hall A, Crisp AH (1983) Brief psychotherapy in the treatment of anorexia nervosa. In: Darby PL, Garfinkel PE, Garner DM, Coscina DV (eds) Anorexia nervosa: recent developments in research. Liss, New York, pp 427–439

Keller MB, Lavori PW, Friedman B, Nielsen E, Endicott J, McDonald-Scott P, Andreasen NC (1987) The longitudinal interval follow-up evaluation: a comprehensive method for assessing outcome in prospective longitudinal studies. Arch Gen Psychiatry 44:540–548

Pierloot RA, Wellens W, Houben ME (1975) Elements of resistance to a combined medical and psychotherapeutic program in anorexia nervosa. An overview. Psychother Psychosom 26:101–117

Pierloot R, Vandereycken W, Verhaest S (1982) An inpatient treatment program for anorexia nervosa patients. Acta Psychiat Scand 66:1–8

Ringler M, Jandl-Jager E, Maritsch F (1983) Patientenkarrieren: Repräsentativität und Motivation zur Teilnahme an einem Katamnesegespräch nach 5 Jahren. Psychother Med Psychol 33:153–157

Vandereycken W (1985) Inpatient treatment of anorexia nervosa: some research-guided changes. J Psychiat Res 19:413–422

Vandereycken W (1987a) The constructive family approach to eating disorders: critical remarks on the use of family therapy in anorexia nervosa and bulimia. Int J Eating Dis 6:455–467

Vandereycken W (1987b) The management of patients with anorexia nervosa and bulimia: basic principles and general guidelines. In: Beumont PJV, Burrows GD, Casper RC (eds) Anorexia and bulimia nervosa. Elsevier, Amsterdam, pp 235–253 (Handbook of eating disorders, part 1)

Vandereycken W (1988) Organization and evaluation of an inpatient treatment program for eating disorders. Behav Residential Treatment 3:153–165

Vandereycken W (1990) The relevance of body-image disturbances for the treatment of bulimia. In: Fichter MM (ed) Bulimia nervosa: basic research, diagnosis and treatment. Wiley, Chichester, pp 320–330

Vandereycken W, Meermann R (1984) Anorexia nervosa: a clinician's guide to treatment. de Gruyter, Berlin

Vandereycken W, Pierloot R (1983a) Drop-out during inpatient treatment of anorexia nervosa: a clinical study of 133 patients. Br J Med Psychol 56:145–156

Vandereycken W, Pierloot R (1983b) Long-term outcome research in anorexia nervosa: the problem of patient selection and follow-up duration. Int J Eating Dis 2(4):237–242

Vandereycken W, Pierloot R (1983c) The significance of subclassification in anorexia nervosa: a comparative study of clinical features in 141 patients. Psychol Med 13:543–549

Vandereycken W, Vanderlinden J (1983) Denial of illness and the use of self-reporting measures in anorexia nervosa patients. Int J Eating Dis 2(4):101–107

Vandereycken W, Vanderlinden J, Van Werde D (1986) Directive group therapy for patients with anorexia nervosa and bulimia. In: Larocca FEF (ed) Eating disorders: effective care and treatment. Ishiyaku EuroAmerica, St Louis p 53–69

Vandereycken W, Depreitere L, Probst M (1987) Body-oriented therapy for anorexia nervosa patients. Am J Psychotherapy 61:252–259

Vandereycken W, Probst M, Meermann R (1988) An experimental videoconfrontation procedure as a therapeutic technique and a research tool in the treatment of eating disorder. In: Pirke KM, Vandereycken W, Ploog D (eds) The psychobiology of bulimia nervosa. Springer, Berlin Heidelberg New York, pp 172–178

Vandereycken W, Kog E, Vanderlinden J (1989) The family approach to eating disorders: assessment and treatment of anorexia nervosa and bulimia. PMA Publications, New York

Vanderlinden J, Vandereycken W (1988) Perception of changes in eating disorder patients during group treatment. Psychother Psychosom 49:160–163

The Prognosis of Eating Disorders: A Clinician's Approach

G.F.M. Russell

After his first assessment of a patient with anorexia nervosa, the clinician will be expected to convey to her relatives his opinion about the likely outcome of her illness. More specifically he will be asked to answer several crucial questions. First, is the patient likely to recover and, secondly, how long is she likely to remain ill? In grappling with these questions the clinician will rely on existing knowledge of those factors which are known to influence the course and outcome of the illness, for better or worse. These factors are known as the predictors of prognosis. If the clinician believes he can be reasonably optimistic with his prognosis, he will be asked further whether the recovery is likely to be permanent, or relapses are to be expected. Finally the relatives will turn to him for advice as to the treatment most likely to arrest the course of the illness and increase the chances of a complete recovery.

In trying to answer these questions the clinician will rely in part on his previous experience, but he will also turn to the vast literature containing the results of long-term follow-up studies of patients with anorexia nervosa. These studies have shed more light on some of these questions than on others. This chapter is concerned with an appraisal of the value and limitations of follow-up studies: have they succeeded or failed to illumine these questions which are crucial to the practising clinician, his patient and her relatives?

To be more precise about definitions, a "clinical prognosis" is the onerous task of forecasting the future course and final outcome of a disease or illness in an individual patient. The term is sometimes used to signify in broad terms the likely course and outcome of a given disease. It is better, however, to restrict the usage of the term "clinical prognosis" to the predictions that can be made for an individual patient afflicted with the illness which has been specifically diagnosed in him or her.

The opportunity to provide a chapter on the clinician's perspective of this important subject is welcomed. It is the clinician who bears the heavy burden of breaking to the patient's relatives the import of the diagnosis of anorexia nervosa. His prognosis will also guide the patient and her family with regard to the efforts that will be needed in cooperating with an appropriate therapeutic programme. For example, the clinician may consider the prognosis to be so serious that only a vigorous and sustained course of treatment can be expected to lead to a full recovery. On the other hand, if the clinician is optimistic about the natural outcome of the illness, he may be inclined to let nature take its course and only offer the minimum intervention. Unfortunately such optimism is rarely experienced by the clinical psychiatrist who first meets a patient with an eating disorder in a hospital clinic: he will often observe that she has already suffered from anorexia nervosa or bulimia nervosa for several years. The position may be somewhat different in patients with "partial syndromes" who present themselves in the setting of general or family practice (King 1986). This chapter is written, however, from the perspective of the clinical psychiatrist based in a hospital clinic offering specialist treatment to patients with entrenched eating disorders.

The clinical prognosis in the individual patient with anorexia nervosa is a complex issue. This is in spite of the numerous thorough long-term follow-up studies whose findings should serve to guide the clinician. The reason for this gap between research findings and clinical practice will soon become apparent. The reader of this book will already have ascertained from the studies reported therein and from earlier studies (Theander 1970; Morgan and Russell 1975; Hsu et al. 1979; Morgan et al. 1983) that the course of anorexia nervosa is extremely variable. Furthermore, the long-term effects of treatment have hitherto remained uncertain in spite of short-term benefits from established treatments. This difficulty is compounded by the fact that in the majority of patients the illness lasts a number of years, only 15% of patients recovering in less than 2 years in Theander's series (Theander 1970). These difficulties led me, in 1977, to observe that one of the mysteries of anorexia nervosa is the unpredictability of the prognosis in an individual patient. It seemed then impossible to promise more than a short-term improvement to most patients. "The long-term course appears to be a function more of the individual patient's illness rather than the treatment she receives" (Russell 1977). A similar kind of cautious scepticism was expressed by Theander in 1985. Again, it was stated in another review that "the prediction of outcome in an individual case is generally believed to remain something of a guess" (Szmukler and Russell 1986).

One of the aims of the present review is to assess whether it remains impossible to predict the long-term outcome of anorexia nervosa in the individual patient. Difficulties in determining the prognosis in patients with bulimia nervosa are even greater, partly because of the absence of any long-term outcome studies, and in spite of the excellent review of the subject by Manfred Fichter in this book. This problem will be further discussed in a later section of this chapter.

Another aim is to examine the relationship between treatments and long-term outcome. This, too, is a complex issue. The clinician is apprehensive that his painstaking treatments may fail to carry any long-term benefits. On the other hand, the clinical scientist is concerned lest natural outcome studies are contaminated by wide variations in the nature and intensity of the treatment previously administered to the patients followed up. Finally, an attempt is made to help the clinician interpret conflicting findings in published follow-up studies in order to guide him when making a clinical prognosis on patients with eating disorders.

Value and Limitations of Follow-up Studies in Anorexia Nervosa

Follow-up studies in anorexia nervosa and dissertations on their methodology hold a special fascination for clinical researchers in this field. In turn these studies have been comprehensively reviewed and appraised (Steinhausen and Glanville 1983; Swift 1982; Hsu 1988; Steinhausen et al. 1991). Over 60 follow-up studies have been brought together by these authors, and a further number are included in this book. The reviewers have rendered a further service through their careful critique of the methodology employed, including its defects. They have also put forward criteria for the near-perfect follow-up study of anorexia nervosa. For example, they recommend that the diagnostic features of patients studied should be fully documented. Not only should syndrome-specific features be recorded (e.g. low weight, amenorrhoea and a morbid dread of fatness) but data should also be presented on the more general psychiatric disturbances and on the patients' psychosocial adjustment. The patients should be re-examined personally, using standardized interviews. The success rate in tracing patients should approximate to 100%. Most important, in view of the commonly prolonged course of the illness, the follow-up should be sufficiently long to permit a determination of its eventual outcome. A somewhat arbitrary follow-up interval of at least 4 years was set by Morgan and Russell (1975) and most recent studies have adhered to this recommendation. Thus there has been no lack of good advice on the design and execution of follow-up studies in anorexia nervosa, and this has been accompanied by a hive of industry in this field.

In spite of the volume and quality of this research there remains a wide variation in the pattern of outcome described between different follow-up studies and within them. These variations pose serious problems for the practising clinician, hence the necessity to explain their origins.

Variations in Outcome Between Studies

Reviewers of outcome studies in anorexia nervosa have attempted to evaluate the overall rates of outcome in terms of general categories. The estimates of the illness running a chronic course fall within a relatively narrow range (about 20%), whereas the rates of good outcomes or recoveries vary more widely (from 40% to 60%), with the remaining patients (from 20% to 40%) having intermediate outcomes (Hsu 1988; Steinhausen et al. 1991). These figures ignore deaths, the most ominous of all outcomes, which become more clearly apparent with long follow-ups such as those spanning 20 years (Theander 1985; Ratnasuriya et al. 1991). The relatively wide range found for these basic data can readily be illustrated by comparing three studies which were carried out in England in three different centres, but using the same methods of assessment (Morgan and Russell 1975; Hsu et al. 1979; Morgan et al. 1983). In all three studies the duration of the follow-up was at least 4 years, and the main measures used were derived from the "Morgan-Russell" scales (Morgan and Hayward 1988). The first study emanated from the Maudsley Hospital and the second from St. George's Hospital, both centres based in London and accepting patients referred from afar. In contrast, the third study was based on a clinical unit in Bristol which provided a more local service.

The "general outcome categories" at follow-up favoured the Bristol series of patients in contrast with the Maudsley series, the St. George's patients occupying an intermediate place but closer to the Maudsley series. Thus a "good outcome" (weight maintained within 15% of average and regular menstruation) occurred in 58% of the Bristol patients but in only 38% of the Maudsley series (48% at St. George's). On the other hand, a "poor outcome" (weight never approaching 85% of average and absent menstruation) occurred in only 19% of the Bristol patients and 29% of the Maudsley series (20% at St. George's). Other measures tended to confirm these relative positions of patient outcome in these three series of patients. The authors of the Bristol study (Morgan et al. 1983) tentatively suggested that the better outcomes obtained in their series might be due to the less frequent hospitalisation of their patients in comparison with the Maudsley and St. George's patients. An important lesson to stress in comparing the outcome results of different clinical centres is that differences in the quality of the available treatment are hardly ever the main reason for differences in outcome. This theme will be further developed later on. A much more likely explanation is that powerful selection biases determine the nature of the patient population under study, even when the researchers are careful to admit consecutively referred patients to their study. In support of this argument, a comparison of the Maudsley and Bristol patient populations shows that the latter had a shorter duration of illness and a higher mean body weight than the Maudsley patients. In particular, only 12% of Bristol patients had received previous psychiatric treatment for anorexia nervosa, whereas the comparable figure for the Maudsley series was 39%. The failure of previous psychiatric treatment is probably the most telling index of the severity of illness in a series of patients and a high percentage figure is in keeping with the selective referral pattern prevailing in hospital clinics which accept referrals of patients from beyond the local population.

Much doubt remains, therefore, regarding the general outcome of anorexia nervosa since the course of the illness will differ from one clinic population to another. Moreover, clinical populations will differ even more from patients in the general population who never seek treatment at a clinic. Such are the cases of "partial syndromes" identified through the screening of patients receiving primary care in general practice (King 1986). A study of anorexia nervosa in a representative sample of the population has not been conducted and would in any case present considerable difficulties.

Variations in Outcome Within Studies

It has been the main justification of follow-up research that, within any given series of patients, differences in outcome might be associated with personal or clinical factors which predate the follow-up assessments and thus serve as "prognostic predictors".

The two London studies previously quoted (Morgan and Russell 1975; Hsu et al. 1979) agreed with each other in identifying factors "predictive" of a poorer outcome:

1. An older age of onset
2. A longer duration of illness
3. A poorer adjustment in childhood
4. Disturbed family relationships
5. A history of previous psychiatric treatment
6. A lower body weight when referred for treatment

Various other studies have tended to confirm these prognostic factors, especially the first four listed above, but there have also been exceptions. The relationship between an early onset and a good outcome has tended to be sustained when the age of the patient population ranges from adolescence upwards. Among very young patients, however, the relationship between age and outcome remains uncertain (Swift 1982). Bryant-Waugh et al. (1988) thought that the prognosis was more serious in very young patients. It has been suggested that the relationship between age of onset and outcome might follow a U-shaped curve, the worst outcomes occurring when the illness begins in mature women or in prepubertal children (Russell 1991). Additional evidence for associating a short interval between onset and hospitalization with a good outcome has come from the studies of Hall et al. (1984), and in the case of male patients from Burns and Crisp (1984).

Further support for the consistency of the first four prognostic indicators listed above has come from following-up patients from the Maudsley Hospital, London, after a mean of 20 years (Ratnasuriya et al. 1991). As this was the same series of patients as was reported on in 1975 by Morgan and Russell, it was also possible to test the consistency of the earlier findings in this respect. The age at the onset of the illness remained a strong predictor of outcome even after 20 years. Of 18 patients with an onset of illness between 11 and 15 years only two had a poor outcome. In contrast, among 22 patients with an age of onset of 16 or more, 15 had a poor outcome. The duration of illness at the time of the key admission had a weaker but still significant association with outcome at 20 years. Among the premorbid factors associated with a poor outcome, high levels of significance were obtained with a history of a poor individual adjustment (neurotic traits in childhood and personality difficulties) and disturbed relationships within the family.

Nevertheless, the recognition of these few prognostic indicators is a slender reward for much work which still leaves too many questions unanswered. For example, a proportion of anorexic patients experience bulimia and vomiting during their illness, but it remains uncertain what proportion of them later develop bulimia nervosa. In some patients these features are benign, as suggested by long-term follow-up studies in which it was found that about one third of anorexic patients who recovered developed bulimia as a temporary feature (Theander 1970; Dally et al. 1979). Several studies discounted the occurrence of bulimia or vomiting as in any way presaging an unfavourable outcome (Morgan and Russell 1975; Hsu et al. 1979; Morgan et al. 1983). On the other hand, the 10-year follow-up study by Eckert and Halmi (1990) revealed that 33% of anorexic restrictors developed bulimia (possibly a disturbance which falls short of bulimia nervosa). Hsu (1988) re-examined the data from the St. George's study and estimated that at least 19% of his patients had bulimia nervosa at follow-up. Ratnasuriya et al. (1991) in their 20-year follow-up reported slightly lower rates: 15% who developed bulimia nervosa and a further 18% who exhibited bulimic behaviour short of the full syndrome.

The clinician must therefore remain cautious before applying the few prognostic predictors that have emanated from follow-up studies to the task of reaching a prognosis in an individual patient. These prognostic predictors may also depend on the nature of the population under study, though they are probably less vulnerable to sampling biases than the outcome categories themselves. In practice, the search for prognostic indicators has been judged disappointing (Theander 1985) and their application deemed of limited usefulness (Szmukler and Russell 1986).

Variations of Outcome with Time

Anorexia nervosa is often a prolonged illness even in patients who subsequently make a full recovery. It follows that no pronouncement on the outcome of anorexia nervosa in a population of patients (as opposed to an individual patient) should be made until sufficient time has been allowed for the full clinical picture to be displayed. An arbitrary minimum period of 4 years for follow-up was recommended by Morgan and Russell (1975), but it should now be questioned whether this is sufficiently long.

It is worth underlining the need for a prolonged follow-up. Vandereycken and Pierloot (1983) found that only 13% of their patients had recovered within 2 years, compared with a 56% recovery rate after 5 years. Similarly, Theander (1985) reported a 15% recovery within 2 years which rose to 33% by three years. In the series of patients studied by Dally et al. (1979) a return to a normal weight was achieved by 38% at the end of the 2nd year, 67% after 4 years and 70% after 10 years. It is clear therefore that sufficient time should elapse for the recovery curve to level off before reaching valid conclusions about the outcome of anorexia nervosa. The long-term study by Ratnasuriya et al. (1991) specifically examined whether a follow-up after 20 years (on average) confirmed the initial pattern of outcome found after a mean of 5 years, both calculated from the time of admission to hospital. Reference has already been made to this study in the discussion on predictors of prognosis, but fuller details of the methodology and results will now be presented.

A 20-year Follow-up on Anorexia Nervosa Subjects

The series of patients was the one originally studied by Morgan and Russell (1975) and consisted of 41 patients (38 female and 3 male) who were admitted consecutively to the Maudsley Hospital between November 1959 and October 1966. They all satisfied the diagnostic criteria defined by Russell (1970) which correspond closely to those of ICD-10 (WHO 1987). The mean age of onset was 18.0 years (±6.7) and the mean age on admission was 21.5 years (±8.6). The mean duration of the illness was 3.7 years (±3.7). On admission the mean body weight was 64.27% (±9.2) of average body weight (A.B.W.; Society of Actuaries 1959). The patients were therefore severely undernourished with prolonged illnesses. Most of them displayed neurotic or abnormal personality traits and almost half had previously received psychiatric treatment. At that time the main emphasis of treatment at the Maudsley Hospital was on rapid weight gain achieved by trained nursing staff who developed a therapeutic relationship with their patients. The mean maximum weight reached in hospital was 84.3% A.B.W. After discharge from hospital they were given a limited number of out-patient supportive interviews but no systematic psychotherapy or family therapy.

Methods

The first follow-up took place at least 4 years after the key admission (mean of 5 years). The second follow-up occurred at least 18 years after admission (mean of 20 years). Of the 41 patients traced on the first occasion, 40 were traced on the second. Among the surviving patients, 70% were interviewed by a psychiatrist. For the remainder reliance was placed on a detailed questionnaire completed by the patient (18%) or by her general practitioner (12%).

The clinical data were rated using the Morgan and Russell Scales which provided scores on nutritional status, menstrual function, mental state, psychosexual adjustment and socioeconomic functioning (Morgan and Russell 1975; Morgan and Hayward 1988). From these scales a final "average outcome score" was computed. An overall rating of general outcome was also made by placing each surviving individual into one of the three following categories:

- Good outcome: normal weight (100% ± 15% A.B.W.) with normal menstruation
- Intermediate outcome: normal or near normal weight (75%–100% A.B.W.) and/or menstrual abnormalities
- Poor outcome: low weight and absent or scanty menstruation

Results

The average age of the patients at the time of follow-up was 40.9 years (SD = 9.5).

Deaths

Seven patients had died during the 20-year follow-up, but only six deaths were attributable to anorexia nervosa. Two had died by the 5-year follow-up (one suicide and another from combined nutritional deficiencies). The five further deaths were from suicide (2), nutritional complications and electrolyte imbalance (2) and one unlawful killing by a stranger. Excluding this last patient, the crude death rate (i.e. without a correction for expected deaths) was 15% after a 20-year follow-up.

Comparison of the 5-Year and the 20-Year Follow-up

Table 1 compares the general outcome of the 40 patients at 5 years and 20 years. It will be noted that some patients were "misclassified" at the 5-year follow-up in comparison with the 20-year outcome. Table 1 can be simplified by adding together the patients in the good and intermediate outcome groups. This was readily justified by comparing the patients' average outcome scores in the good outcome group (10.1, SD ± 1.6) and those in the intermediate outcome group (9.6, SD ± 2.3), and observing that they did not differ significantly from each other (using a one-way analysis of variance). On the other hand, both these average outcome

Table 1. Comparison of outcome at the 5-year and 20-year follow-up

		20 Year follow-up				
		Good	Intermediate	Poor		Total
				Ill	Dead	
5-Year follow-up	Good	6	8	1	1	16
	Intermediate	4	3	2	2	11
	Poor					
	ill	2	2	5	2	11
	dead	–	–	–	2	2
	Total	12	13	8	7	40

scores were significantly higher (and better) than those in the poor outcome group (3.8, SD ± 1.9). When the good and intermediate outcome groups are collapsed, a clearer pattern emerges. Twenty-one of the 27 patients in the good or intermediate outcome group at 5 years continued to do well at 20 years, but 3 patients had deteriorated into the poor outcome group and 3 had died. On the other hand, 7 of the 11 patients with a poor outcome at 5 years remained seriously ill or had died, but 4 had improved by the 20-year follow-up.

Another way of comparing patient outcome at 20 years with the earlier findings at 5 years is to calculate the correlation between individual average outcome scores at the two intervals. A strong correlation was found between the two follow-ups ($r = 0.66$; $p < 0.01$). Surprisingly, there was little association found between the body weights recorded at 5 years and 20 years. This was partly due to the fact that five patients had developed bulimia nervosa and their weights had fluctuated between emaciation and obesity. When these patients were excluded from the analysis a statistically significant correlation was found, but its level remained rather modest ($r = 0.36$; $p < 0.05$).

Duration of Illness and Outcome

The disconcertingly high death rate at the end of 20 years has already been presented, but it is appropriate to redress the balance and point to the finding that recovery from anorexia nervosa can occur even after many years' duration. In a long-term study such as this the most appropriate statistical techniques of examining the relationship between the duration of illness and the probability of a given eventual outcome is that of "survival methods" (Isager et al. 1985; Kalbfleisch and Prentice 1980). They permit the analysis to take into account two important facts: the duration of follow-up varies between patients, and the size of the population also diminishes with time. Once again, it was considered justified to group together the good and intermediate outcome categories as they consisted of patients who had made at least a partial "recovery". The survival analysis indicated that at first there was a fairly constant rate of recovery – approximately 15% of those still ill in each successive 3-year period. After 12 years the "recovery" rate dropped sharply so that it was less than 7% after 15 years. Nevertheless, even after this time, a few patients continued to "recover".

Conclusions from the 20-Year Follow-up

1. On the one hand, the results of the 20-year follow-up were less favourable than those anticipated at the 5-year follow-up: the deaths attributable to anorexia nervosa rose to 15% by the end of the 20 years. On the other hand, at least partial recoveries were still possible even after illnesses lasting for over 15 years, a finding which is in accord with Theander's (1985) study. This observation contrasts with the views expressed in review articles (Hsu 1980; Schwartz and Thompson 1981; Steinhausen and Glanville 1983) and with the assertion that recovery is not possible after 7 years (Dally 1969).
2. Seen together, the two Maudsley studies provide for the first time a detailed comparison of the outcome in the same cohort of patients followed up 5 and 20 years after a key admission. Although there was a fair degree of consistency for the group as a whole, a number of individual patients changed considerably. The majority of the patients with a poor outcome at 5 years were still faring badly after 20 years, but about one third were improving. Conversely, among those in good or intermediate outcome categories, at 5 years, just over one fifth relapsed and had a poor outcome at 20 years. This is a more serious prognostic error and is highlighted by the three deaths among the patients previously categorised as having a reasonably good outcome. The relapse rate after a period of apparent recovery can be compared with Theander's (1970) estimate of 12%. One lesson to be learned is the value of very long follow-up studies and the limitations of those of intermediate duration.
3. The comparison of the two follow-up studies also throws light on the validity of the outcome measures. It was found that the good and intermediate outcome groups were similar to each other. Accordingly, the categorisation of an intermediate outcome hardly means that this is less favourable than a good outcome but reflects instead our uncertainty in assessing the patient. When the patient's adjustment is considered over a long period of time it can be concluded that one third of the patients in the intermediate outcome group function well in spite of their relatively low weight. This observation, together with the low correlation between the patients' weights at 5 and 20 years, suggests

that body weight is a less reliable indicator of long-term compared with intermediate outcome.

Do Treatments Influence the Long-Term Outcome of Anorexia Nervosa?

Until recently it was generally supposed that the long-term course of anorexia nervosa was more dependent on the natural outcome of the patient's illness rather than any treatment she may have received (Russell 1977). In fact, the majority of follow-up studies appear to discount any possible influence of the treatments previously administered. Yet certain treatments are strikingly effective in weight restoration in the short term. Among these must be counted the intensive nursing care of patients admitted to a unit specialising in the treatment of this illness (Morgan and Russell 1975; Russell 1983; Andersen 1985). Family therapy has also been demonstrated to induce a general improvement in younger patients, a change which is clearly evident at the end of 1 year of therapy (Russell et al. 1987). It has hitherto been assumed that the longer the follow-up, the less is the likelihood of any contamination from treatments given during the earlier phase of the follow-up.

It is virtually impossible to assess the effectiveness of the treatment programmes available in different clinical units by comparing the rates of recovery reported by these units. If a specialised unit is established to treat patients with anorexia nervosa, the referrals to that unit will be from practitioners who have concluded that more developed therapeutic facilities are indeed required by their patients. They have made a judgement that their patients present a difficult therapeutic problem or that they may already have shown this by failing to respond to the endeavours of a previous psychiatrist or physician. Hence the specialist unit is certain to receive referrals of relatively resistant patients who as a group are like the tip of an iceberg, whose submerged mass stands for patients more likely to respond to simpler treatments or with illnesses so mild that they do not even seek treatment. The tip of the iceberg is unrepresentative of the complete population of anorexic patients. Each specialist unit will attract different patient populations: patients accepted with the minimum of delay from furthest afield will tend to include a high proportion who suffer from the most entrenched illnesses. Hence comparisons of therapeutic outcomes from different centres are more likely to reveal the features of the populations of patients referred to them rather than the relative effectiveness of their treatment programmes.

There is much evidence to support the overwhelming pull of specialist treatment services in attracting towards them the more resistant patients whose response to the more intensive treatments is surprisingly disappointing. For example, patients treated in a general medical setting tend to progress better than patients treated in psychiatric units (Bassoe and

Eskeland 1982; Askevold 1983). When selected patients from a medical setting are deemed to require psychiatric treatment, they tend to have a poorer outcome, for example when they are passed on to a psychiatric clinic (Bassoe and Eskeland 1982) or are referred for long-term psychotherapy (Askevold 1983). Within a given psychiatric series, patients admitted for treatment tend to fare less well than those treated as outpatients (Hsu et al. 1979), demonstrating again the operation of selection factors. The outcome of patients referred to specialized centres accepting distant referrals is less favourable than in psychiatric centres with a more local commitment (Szmukler and Russell 1986). A paradoxical law may thus be formulated: treatment-seeking behaviour which leads to more intensive psychiatric treatment is accompanied by a worse outcome. This is not to say, of course, that a patient who is well motivated will fail to respond favourably to a well-designed treatment programme. The paradoxical findings described above merely highlight the importance of selection factors.

The time has come to question whether treatments that are effective in the short term (say 6 months) are devoid of any benefits in the long term (e.g. after 3–5 years). A randomised controlled trial of family therapy has demonstrated that the therapeutic outcome at the end of 1 year was significantly superior in younger patients to that obtained with a control treatment (individual supportive therapy; Russell et al. 1987). The authors kept an open mind about the long-term effects of the family therapy. They postulated that the natural outcome of the illness, likely to be more favourable in younger patients, might overtake and neutralize the initial differential response to the two treatments. In the event, a preliminary analysis of the therapeutic outcome at the end of 3–5 years reveals that the patients who had previously received family therapy continue to progress more favourably. A detailed analysis of the full findings will be published as soon as possible.

It may be concluded that the long-term benefits of the more powerful treatments in anorexia nervosa can no longer be ignored. This probably signifies that the age of the retrospective follow-up study is past. In future, only prospective follow-up studies will carry much value and then only if homogeneous populations of patients have been randomly allocated to the different treatments capable of evaluation. Such an approach is imperative because of the ethical and practical objections to having a control group subjected to a purposive and complete avoidance of treatment.

Follow-up Studies in Bulimia Nevosa

It would be superfluous to go over the ground already well covered by Manfred Fichter in his chapter on the course of bulimia nervosa. It will suffice to draw together his observations with the principles already put forward in this chapter for anorexia nervosa.

A comparison can be made between the results of the German Longitudinal Bulimia Nervosa Study (Fichter et al., this volume) and the Harvard, Boston study by D.B. Herzog and his colleagues (Herzog et al. 1988; Keller et al. 1989). The 207 patients in the German Longitudinal Study had all received vigorous treatment, including admission to hospital with a broad spectrum behaviour therapy programme. At the end of a 2-year follow-up 50% no longer suffered from a major eating disorder (bulimia nervosa, anorexia nervosa or obesity). So far the Harvard study has been confined to 30 patients with bulimia nervosa who received a wide range of treatments, including drugs, but in no way controlled by the investigators. By the end of 18 months' follow-up some 57% of the patients were deemed to have recovered at some point. The resemblance between the two studies ceases here because the patients in the Harvard study were frequently found to relapse after an apparent recovery – 9 out of 17 in the course of the follow-up period.

In the German Longitudinal Study the outcome was found to be less favourable in patients whose weight was low on admission or who satisfied DSM-III-R criteria for one or other of the personality disorders or the affective disorders. Patients in the Harvard Study also showed a high rate of affective disorder (57% at intake), but the presence of an affective disorder had apparently little effect on recovery from bulimia nervosa. In some patients the course of the two disorders ran independently of each other.

In the German Longitudinal Study the more favourable outcome results in the "comparison" group illustrates well the primary role of selection factors in determining outcome. Fichter is careful to point out that the comparison population of subjects responding to advertisements did not constitute a true control group even though they satisfied DSM-III-R criteria for bulimia nervosa. They had been defined as individuals who had received no treatment or relatively little treatment. In contrast the index group of patients with bulimia nervosa had been "negatively selected" in consequence of their seeking active treatment. In accordance with the paradoxical law formulated for anorexia nervosa it could be predicted that their treatment-seeking behaviour would be associated with a less favourable outcome.

There is, however, a striking difference between follow-up studies in bulimia nervosa and anorexia nervosa. The former have been of relatively short duration so far, and as a result it is virtually impossible to disentangle the effects of treatment from those of natural outcome. It is encouraging that Fichter proposes to extend the German Longitudinal Study for up to 7 years.

Conclusions: Clinical Prognosis in Anorexia Nervosa

In the preceding account the limitations of follow-up studies have been stressed. Even with long periods of follow-up these results are limited in their general application because they are largely dependent on selection factors which may be intangible. Where their findings coincide with clinical experience they may be applied with caution. Thus the predictors of prognosis which are consistently found by different investigators may prove useful to the clinician called upon to make a prognosis. In particular, if the illness is of relatively short duration (e.g. under 18 months) and begins at a young age (especially below 20 years) the outcome is likely to be favourable so long as the patient previously had a stable personality and was able to enjoy warm and affectionate relationships within the family. It will also be in the patient's favour if her weight loss is not extreme and she displays no bulimic behaviours such as vomiting or laxative abuse, or if these are only transient. With such a patient the clinician will recommend family therapy and make every endeavour to secure the continued co-operation of the patient and her family. With such an approach admission to hospital may not be required so long as there is early and sustained progress with steady weight gain. If progress is delayed or if weight loss is life-threatening, admission to hospital will be essential, especially if the illness had a pubertal or prepubertal onset (Russell 1985). The course and outcome in male patients is remarkably similar to that in females as is the identification of predictors of outcome (Burns and Crisp 1984, 1990). Accordingly the same principles should guide the prognosis for male patients.

The overall outcome of anorexia nervosa influences greatly the menstrual status, psychosexual adjustment and psychosocial functioning achieved by the patient. Thus if a patient maintains a weight above 85% average body weight, she will resume menstruation in due course. These patients can establish satisfactory sexual relationships, marry and have children. Indeed some pregnancies can occur when the mothers are still amenorrhoeic. Patients with a good outcome can also achieve a good psychosocial adjustment, but even in this group there may persist a strong dependence on their family of origin, a restriction of social activities and a reluctance to seek employment (even in a single woman). Symptoms which persist most doggedly are difficulties with eating and preoccupations with thoughts about body weight and shape.

The clinician will be cautious with his prognosis in patients whose illness is influenced by prognostic predictors which are the converse of those described above. If the illness has persisted several years and has defied previous therapeutic attempts, the outlook will probably be more serious. This view is in keeping with that of other clinicians (Andersen 1985; Vandereycken and Pierloot 1983). A later age of onset may also indicate a worse prognosis but this is an area of uncertainty. Certainly the "anorexia tardive" (Dally et al. 1979) beginning after 30 or even 40 years is particularly resistant to treatment, and this may also be a feature of patients whose illness commences after marriage. Even so, it is important for the clinician to activate a vigorous treatment programme for those patients who at first sight look unpromising. For them admission to hospital leading to the resumption of a healthy weight may bring about a most gratifying improvement which may be sustained if admission is followed by outpatient psychotherapy. Therapeutic optimism is justified by occasional recoveries even after 15 years of illness (Ratnasuriya et al. 1991). All prognoses must be tempered by the likelihood that the course of the illness is often prolonged (2–3 years) even in favourable instances; there is a risk of death especially in very chronic illnesses; and recoveries may be followed by relapses or transitions into bulimia nervosa.

References

American Psychiatric Association (1987) Diagnostic and statistical manual of mental disorders, 3rd edn: revised (DSM-III-R). American Psychiatric Association, Washington DC

Andersen AE (1985) Practical comprehensive treatment of anorexia nervosa and bulimia. Johns Hopkins University Press, Baltimore, pp 183–184

Askevold F (1983) What are the helpful factors in psychotherapy for anorexia nervosa? Int J Eating Dis 2(4): 193–197

Bassoe HH, Eskeland I (1982) A prospective study of 133 patients with anorexia nervosa: treatment and outcome. Acta Psychiatr Scand 65:127–133

Bryant-Waugh R, Knibbs J, Fosson A, Kaminski Z, Lask B (1988) Long term follow up of patients with early onset anorexia nervosa. Arch Dis Child 63:5–9

Burns T, Crisp AH (1984) Outcome of anorexia nervosa in males. Br J Psychiatry 145:319–325

Burns T, Crisp AH (1990) Outcome of anorexia nervosa in males. In: Andersen AE (ed) Males with eating disorders. Brunner/Mazel, New York, pp 163–186

Dally P (1969) Anorexia nervosa. Heinemann, London

Dally P, Gomez J, Isaacs AJ (1979) Anorexia nervosa. Heinemann Medical Books, London

Eckart E, Halmi K (1990) Course and outcome predictors in eating disorders. Paper presented at the 4th international conference on eating disorders, 27–29 April 1990, New York

Hall A, Slim E, Hawker F, Salmond C (1984) Anorexia nervosa: long-term outcome in 50 female patients. Br J Psychiatry 145:407–413

Herzog DB, Keller MB, Lavori PW, Ott IL (1988) Short-term prospective study of recovery in bulimia nervosa. Psychiatry Res 23:45–55

Hsu LKG (1980) Outcome of anorexia nervosa: a review of the literature (1954–1978). Arch Gen Psychiatry 37:1041–1046

Hsu LKG (1988) The outcome of anorexia nervosa: a reappraisal. Psychol Med 18:2000–2006

Hsu LKG, Crisp AH, Harding B (1979) Outcome of anorexia nervosa. Lancet 1:61–65

Isager T, Brinch M, Kreiner S, Tolstrup K (1985) Death and relapse in anorexia nervosa: survival analysis of 151 cases. J Psychiatr Res 19:515–521

Kalbfleisch JD, Prentice RL (1980) The statistical analysis of failure time data. Wiley, New York

Keller MB, Herzog DB, Lavori PW, Ott IL, Bradburn IS, Mahoney EM (1989) High rates of chronicity and rapidity of relapse in patients with bulimia nervosa and depression. Arch Gen Psychiatry 46:480–481

King MB (1986) Eating disorders in general practice. Br Med J 293:1412–1414

Morgan HG, Hayward AE (1988) Clinical assessment of anorexia nervosa: the Morgan-Russell outcome assessment schedule. Br J Psychiatry 152:367–371
Morgan HG, Russell GFM (1975) Value of family background and clinical features as predictors of long-term outcome in anorexia nervosa: four-year follow-up study of 41 patients. Psychol Med 5:355–371
Morgan HG, Purgold J, Wellbourne J (1983) Management and outcome in anorexia nervosa: a standardised prognostic study. Br J Psychiatry 143:282–287
Ratnasuriya RH, Eisler I, Szmukler GI, Russell GFM (1991) Anorexia nervosa: outcome and prognostic factors after 20 years. Br J Psychiatry 158:495–502
Russell GFM (1970) Anorexia nervosa: its identity as an illness and its treatment. In: Price JH (ed) Modern trends in psychological medicine, vol 2. Butterworths, London, pp 131–164
Russell GFM (1977) General management of anorexia nervosa and difficulties in assessing the efficacy of treatment. In: Vigersky R (ed) Anorexia nervosa. New York, pp 270–289
Russell GFM (1983) Anorexia nervosa and bulimia nervosa. In: Russell GFM, Hersov LA (eds) The neuroses and personality disorders. Cambridge University Press, Cambridge, pp 285–298
Russell GFM (1985) Pre-menarchal anorexia nervosa and its sequelae. J Psychiatr Res 19:363–369
Russell GFM (1991) Anorexia nervosa of early onset and its impact on puberty. In: Cooper PJ, Stein A (eds) Monographs in clinical pediatrics: feeding problems and eating disorders
Russell GFM, Szmukler GI, Dare C, Eisler I (1987) An evaluation of family therapy in anorexia nervosa and bulimia nervosa. Arch Gen Psychiatry 44:1047–1056
Schwartz DH, Thompson MG (1981) Anorexia nervosa: a quantitative approach to follow-up. J Am Acad Child Psychiatry 20:167–183
Society of Actuaries (1959) Build and blood pressure study, vol 1. Chicago. Reprinted in: Documenta Geigy (1970) Scientific tables, 7th edn. Geigy, Basle
Steinhausen H-C, Glanville K (1983) Follow-up studies of anorexia nervosa: a review of research findings. Psychol Med 13:239–249
Steinhausen H-C, Rauss-Mason C, Seidel R (1991) Follow-up studies of anorexia nervosa. A review of four decades of outcome research. Psychol Med, 21:447–454
Swift WJ (1982) The long-term outcome of early onset anorexia nervosa. A critical review. J Am Acad Child Psychiatry 21:38–46
Szmukler GI, Russell GFM (1986) Outcome and prognosis of anorexia nervosa. In: Brownwell KD, Foreyt JP (eds) Handbook of eating disorders. Basic Books, New York, pp 283–300
Theander S (1970) Anorexia nervosa: a psychiatric investigation of 94 female patients. Acta Psychiatr Scand [Suppl] 214:1–194
Theander S (1985) Outcome and prognosis in anorexia nervosa and bulimia: some results of previous investigations, compared with those of a Swedish long-term study. J Psychiatr Res 19:493–508
Vandereycken W, Pierloot R (1983) Long-term outcome research in anorexia nervosa. Int J Eating Dis 2(4):237–242
World Health Organization (1987) International classification of diseases, 10th revision (ICD-10). Chapter V (f): Mental behavioural and developmental disorders. Clinical descriptions and diagnostic guidelines. World Health Organization, Division of Mental Health, Geneva

Chronicity in Anorexia Nervosa: Results from the Swedish Long-Term Study

S. Theander

The Concept of Chronicity

If a patient's illness is defined as "chronic", it should only mean that the illness has been afflicting the patient for a long period of time. The word chronic, however, has come to suggest that the illness is incurable. Lay people, but also people in the medical professions, now habitually use it when the illness is regarded as inveterate and permanent and when restitution to full health is no longer held to be possible. Examples from somatic medicine are chronic rheumatoid arthritis and chronic renal failure, and from psychiatry chronic brain disorder and chronic schizophrenia.

It is well known that a protracted course is a common occurrence in anorexia nervosa. In the present author's long-term follow-up study, more than one third of those patients who have recovered had to endure their illness for 6 years or more (Theander 1985). Clinical experience has repeatedly shown that patients with anorexia nervosa can recover even after a period of illness amounting to more than 10 years, and that a patient will seldom develop serious sequelae even after long-standing anorexia nervosa.

In order to avoid misconceptions and an unnecessarily pessimistic outlook, the author recommends that the term chronic anorexia nervosa be reserved for those patients whose illness is of very long standing, say a continuous illness of more than 15 years. Patients who have been ill for a long time, but not for as long as 15 years, should be referred to as suffering from long-standing or protracted anorexia nervosa.

The Course of Illness in Anorexia Nervosa

Anorexia nervosa is characterized by great variability in the course of illness and in the final outcome (Russell 1977). In illnesses where such variability is a characteristic feature, knowledge of factors which can be used as predictors is especially important. In the case of anorexia nervosa, knowledge of prognostic factors is still insufficient, and in clinical practice it is difficult to make a reliable prognosis in individual patients – especially in the early stages of the illness.

The variability in the natural course of anorexia nervosa makes the study of treatment effects difficult. Favourable changes which appear to be results of the treatment may in fact mainly be due to the natural course of the illness and its inherent tendency to self-healing. It has, in fact, been claimed that most of the variation in the outcome of eating disorders is due to the natural course of the illness: "Anorexia nervosa and bulimia nervosa will usually remit spontaneously; it is uncommon to see the full syndrome of either disorder after the age of 40" (Pope and Hudson 1989). This view is gaining support by results from outcome studies, according to which more patients successively recover as the observation period is prolonged. (Vandereycken and Pierloot 1983; Theander 1985).

Three main results are possible in the further course and outcome of anorexia nervosa: (1) untimely death as a result of suicide, medical complications or starvation; (2) the patient still suffers from severe or less severe anorexia nervosa or bulimia nervosa; and (3) recovery from the eating disorder, with or without other psychiatric problems. When a series of anorexia patients are being followed in repeated investigations, changes will continuously take place in patient distribution among the outcome groups. These changes will mainly take a favourable direction: more and more patients will get well, or less severely ill. However, changes towards a more serious state of health may also take place: patients may die as a consequence of the illness, relapses may occur, patients with a less severe illness may suffer exacerbations, and a syndrome shift may occur, e.g., from a restricting to a bulimic subtype of anorexia nervosa.

Relapses are quite often seen immediately or fairly soon after treatment aiming at weight restitution. However, relapses may also take place much later, sometimes after a remission of 10–15 years. In the present author's follow-up study (Theander 1970), relapses were seen in 12% of the cases. Fairly often, a relapse would occur after an obvious life crisis (see cases 54 and 69 below). During remission, the patient is apparently totally recovered from her eating disorder. She has no eating disturbance of clinical importance, her menstruation is normal and she may also give birth to one or several children. The fact that very late relapses can happen should be borne in mind whenever the onset of anorectic symptoms appears to occur at a comparatively late stage of a patient's life. Careful study of her history may reveal that she in fact suffered from anorexia nervosa at some point during her teens (Ziolko 1978; Dally 1984).

Chronic Anorexia Nervosa

In many clinical descriptions of anorexia nervosa, reports are given on patients with an extremely protracted course of illness. These patients are emaciated and have very disturbed eating habits. Many are affected by complications such as osteoporosis or renal damage (Fenley et al. 1990).

These chronic patients lead an isolated and restrictive life. Their anorectic attitude and behaviour have become ingrained, and the sufferer seems to identify completely with the role of being an anorexia patient. The following quotations supply a good description of the patient's situation: "The peculiar attitudes of such patients towards food virtually become features of the personality" (Kay and Schapira 1965). "Anorexia nervosa has become a mode of existence. The desire to be thin and the dread of becoming fat have taken on the leading and controlling role and imprison the patient in her body" (Casper 1987). "... The individual anorectic's lifestyle ... may become incorporated with the individual's coping mechanisms and way of life. Sadly, for some this may be the only adaptation that they can make to life" (Stonehill 1988).

These chronic patients have very poor insight into their situation; they rigidly deny their illness, and they seldom seek help spontaneously. They may enter hospital as emergency cases because of medical complications, but as a rule they discharge themselves as soon as the immediate crisis has been overcome. Many clinicians have expressed the opinion that only limited treatment goals should be proposed to these patients. Panic or depression and an acute risk of suicide may result if the patient is confronted with demands that she make great changes in her life style, especially in regard to weight and eating habits. In point of fact, experienced clinicians have claimed that a chronic patient may be better off if she remains anorectic. Treatment should, in many people's view, be limited to stabilizing the illness on a less dangerous level and alleviating the suffering of the patient herself and of her family (Crisp 1980; Hall 1982; Casper 1987).

It could be argued, however, that chronic anorexia nervosa according to the proposed definition – continuous illness for more than 15 years – is seldom met with and that therefore even therapists with a special interest in the treatment of eating disorders will only see a small number of these patients. Clinical experience and clinical research concerning chronic anorexia nervosa are clearly unsatisfactory. It is by no means proven that recovery is impossible in chronic anorexia nervosa; there are in fact anecdotal reports of patients who have recovered after an illness of more than 20 years' duration (Dally 1969). More treatment resources should be directed towards the long-standing and chronic patients, especially as a major proportion of the patients with a recent onset can be expected to recover spontaneously. Conceivably, chronic anorexia nervosa patients may become more motivated to seek help if special treatment programmes can be developed for this category.

Mortality

In reviews of outcome studies in anorexia nervosa, the great variation in mortality between various studies has been emphasized (Hsu 1980;

Steinhausen and Glanville 1983). Generally speaking, two factors are responsible for this variation: the selection of cases, and the length of the observation period (Vandereycken and Pierloot 1983). The selection of cases is of great importance. Factors such as long duration of the illness, many previous hospitalizations and late age of onset are related to poor outcome. Special units designed for the treatment of eating disorders will receive a selection of more severe cases. Consequently, an increased mortality among these patients is to be expected, especially when compared with the results from an outpatient clinic which mainly sees young patients with a recent onset. The other factor which exerts a decisive influence on the mortality rate is the length of the observation period. It has repeatedly been shown (see Theander 1985 for review) that patients who die of anorexia nervosa have often been ill for many years. A short observation period will therefore only cover a minor proportion of the expected mortality.

The Relationship Between Chronicity and Mortality

Not only the rate of recovery and improvement is increased when the observation period is extended; there is also an increase in the number of patients who die as a consequence of their eating disorder. In other words, there is a clear relationship between long duration of illness and an increased risk of death from starvation, medical complications or suicide. After a short- or medium-term follow-up period, mortality is low, but a large percentage of the patients are still anorectic. The findings derived from a long-term follow-up study would be expected to take the opposite direction: the mortality figure would be increased, while the number of patients who remain anorectic would be substantially diminished. A comparison between three British follow-up studies (Morgan and Russell 1975; Hsu et al. 1979; Morgan et al. 1983), on the one hand, and the present author's long-term study, on the other, resulted in the following findings: in the British studies, after a mean observation period of 5–6 years, the mortality rate was 3%, while 23% of the patients still suffered from anorexia nervosa; in the Swedish study, after a mean observation period of 33 years, the figures were reversed: 18% had died and a mere 6% remained anorectic (Theander 1985).

Data from a second follow-up study of the Morgan-Russell series, after a mean observation period of 20 years, have recently been presented (Ratnasuriya et al. 1991). This study corroborates the prediction offered above. Mortality has increased to 17% whereas 20% of the patients still have anorexia nervosa. (It should be observed that the original Maudsley series from 1959–1966 is a selection of severely ill patients.)

In a discussion of their findings from a long-term follow-up study, mainly the increase in the rate of recovery and the concurrent increase in mortality, Meyer and co-workers (1986) coined the expression that the time factor is "a polarizing element" in the outcome of anorexia nervosa.

The Swedish Long-Term Follow-up Study of 94 Female Patients

A clinical study of anorexia nervosa was started in 1962. In order to find a sufficiently large series of anorexia nervosa patients, the diagnosis registers in both psychiatric and somatic clinics were systematically searched. The study covered a defined region of southern Sweden during the years 1931–1960 (a regrettable fact is that no search was performed for male cases). The investigation resulted in a series of 94 female patients who fulfilled the diagnostic criteria used by the author (Theander 1970). A successive increase in incidence was noted in that 11, 25 and 58 patients, respectively, were found during the three decades. The mean incidence was 0.24 cases per 100000 inhabitants per year. During the 1950s, the incidence was 0.45. It should be emphasized that all patients had been hospitalized, which in itself means a selection tending towards seriously ill patients. However, some of the patients who had been admitted to a somatic clinic for a diagnostic checkup had a rather slight variety of the illness.

Two follow-up investigations have been performed, mainly by way of personal investigations; the first followed the patients up to 1966, the next was performed in 1982–1984. The patients are still continuously monitored, mainly regarding mortality, but the author has also stayed in contact with many patients by means of letters, telephone calls and personal visits.

Figure 1 supplies a sketch of the three decades and examples of patients entering the series at various points in time; the time limits for the first and second follow-up are also presented. In the first outcome study, the observation period after the index hospital stay was at least 6 years. By now, fairly reliable data are available for almost every patient for at least 25 years after the index hospital stay. In many cases, though, the observation period is much longer.

The total results of the long-term follow-up study are presented in Fig. 2. It should be noted that the results are shown at various intervals counted from the onset of the disorder, and not from the time of the index hospitalization. The patients' state of health is summarized at four points in time: at 3 years, 6 years, 12 years and 24 years after onset. From the diagram, the two tendencies in the course of the illness – increased recovery rate and increased mortality – are easily observed. The mortality columns contain deaths from anorexia nervosa and suicide, but not deaths from accidents or diseases unrelated to the eating disorder.

From a comprehensive diagram showing the mortality over time (Fig. 3), the relationship between chronicity and mortality is clearly seen. Many

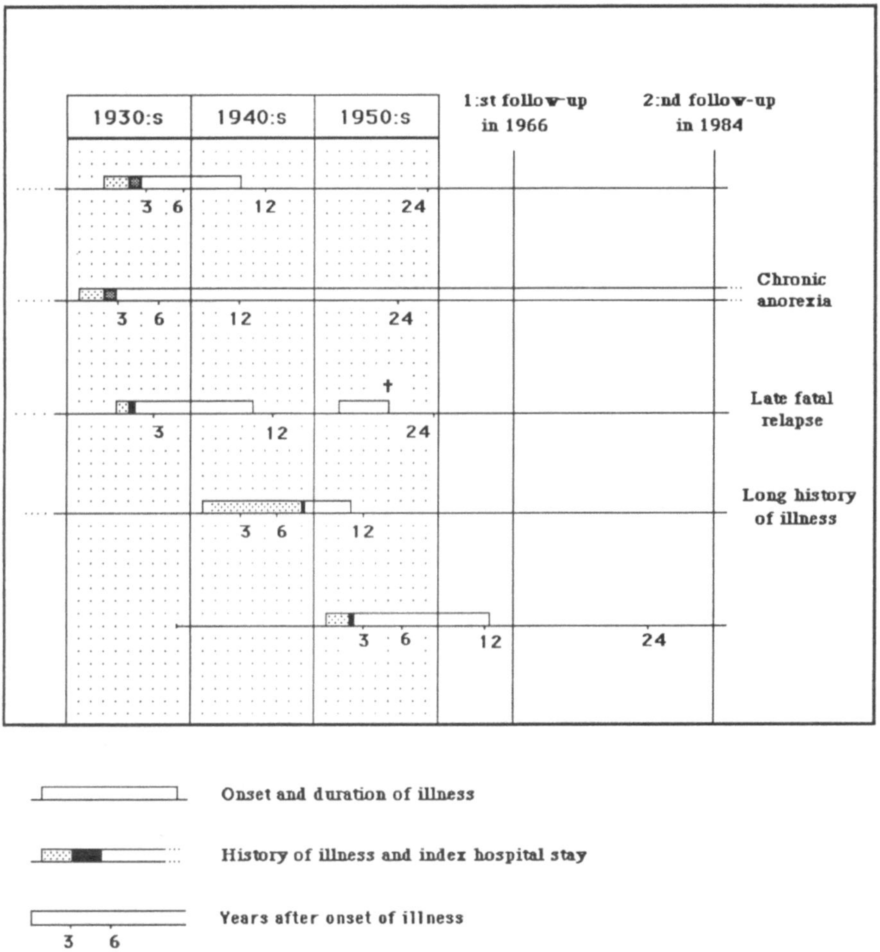

Fig. 1. The Swedish long-term follow-up study. Examples are given of patients from three decades entering the series by way of their index hospital stay. Variable lengths of history and duration of illness is illustrated as is the state of health of patients at various points of time counted from the onset of the illness

patients die during a late stage of the illness, i.e., during chronic anorexia nervosa. It is also noteworthy that in the author's study, quite a few patients died at an early point during their illness. This pattern is not often seen nowadays, a fact which explains the low mortality rate (0%–4%) in recent follow-up studies. In accordance with the discussion above, though, an increased mortality would probably still be found if the patients in these studies were to be reviewed after 10 or 20 years, as some patients with long-standing or chronic anorexia nervosa will eventually succumb to their illness.

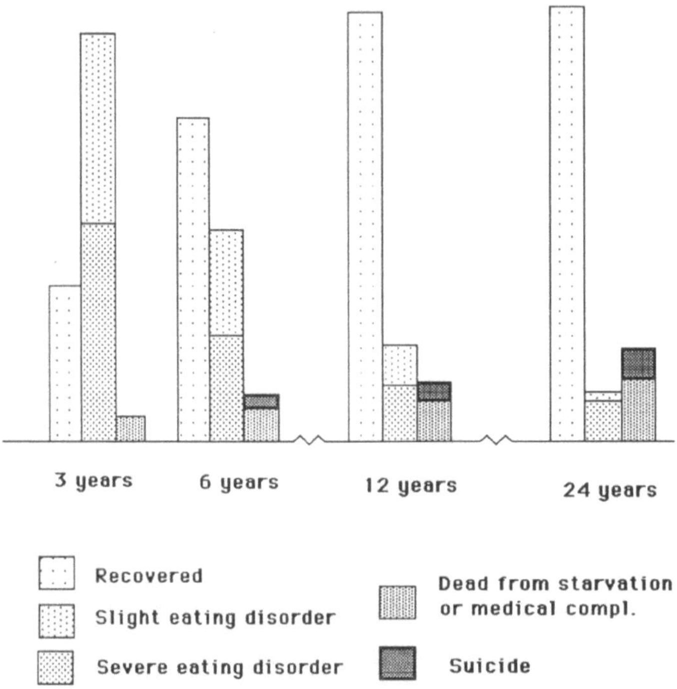

Fig. 2. Summary of the 94 patients' state of health at various points of time counted from the onset of the illness

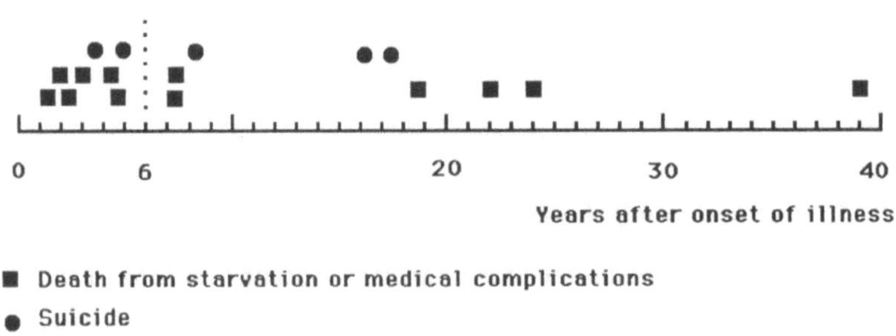

Fig. 3. Mortality. The time and cause of death is presented in relation to the onset of the illness. (Some patients had a remission followed by a relapse; and in these cases the duration of the illness was shorter than the span of time given in the diagram.)

The Fate of Patients with Long-Standing Anorexia Nervosa

In order to analyze the problem of chronicity, a study has been performed on the further course and final outcome of those patients from the author's original series who had to endure a long period of continuous illness. A time limit of 6 years was arbitrarily chosen. There were altogether 37

Table 1. Outcome in 37 patients with long-standing anorexia nervosa

Long-term outcome (*n*)	Comments
Good (19)	Recovery after 6–9 years (12 patients)
	Recovery after 10–11 years (7 patients)
Intermediate (6)	Amelioration of the eating disorder
Poor (12)	Death from anorexia nervosa (6 patients)
	Suicide (3 patients)
	Chronic anorexia nervosa (3 patients)

(39%) out of the original 94 cases who had been continuously ill for 6 years. A recurrent case was included only if one of the episodes had a duration of at least 6 years.

It should be observed that eight patients had died from anorexia nervosa or suicide before they had reached the 6-year limit (Fig. 3). It is not possible to know what would have happened to these patients in the long run if they had survived that crisis. Some of them might have recovered; others might have developed a long-standing or chronic illness.

According to the long-term results, the 37 patients were divided into three outcome groups: good outcome (19), intermediate outcome (6) and poor outcome (12). In the present study, this subdivision was made according to broad clinical criteria; a closer analysis of the follow-up results, using the Morgan and Russell (1975) and other outcome measures, will be performed in the near future. In the following pages, some data on the patients in these outcome groups will be presented, as well as a clinical description of some patients from each group. The results are summarized in Table 1.

The Good Outcome Group

In 19 cases (51%) out of the 37 patients who had been ill for 6 years or more, the outcome was good. Twelve of them got well within another 3 years; their illnesses thus had a total duration of 6–9 years. Six of them have given birth to children. Another seven patients were restored after having been continuously ill for 10–11 years; four of them have given birth to children.

In sum: the 19 patients in the good outcome group got well after a period of illness lasting for 6–11 years; 10 of the 19 have given birth to children. Two illustrative case histories from the good outcome group are presented. In order to preserve anonymity, names and some irrelevant data in the histories have been changed.

Case 4 (Ingrid). Ingrid had a very typical eating disorder. The illness started as a restricting anorexia nervosa at the age of 14. Her body weight dropped from 63 kg

(Body Mass Index, BMI 23.7) to 33 kg (BMI 12.4). After a couple of years she started binge eating and vomiting, but Ingrid was still thin and amenorrhoic, which means that she developed a bulimic type of anorexia nervosa. Finally, for 2 years, she had a normal-weight bulimia nervosa. Ingrid recovered at 22 years of age. The total duration of the illness was 8 years.

Case 54 (Mildred). The history of Mildred is an example of a late relapse. The first episode was a typical teenage anorexia of the restricting type. Mildred recovered after 4 years when she was 19 years of age. Her eating behaviour, weight level and menstruation became normal. She started a course of education with a view to becoming a teacher. She met a man during her academic years and married at 23 years of age. She had great problems in her marriage due to her husband's extreme jealousy. She became pregnant several times, but each time she had a spontaneous abortion. She divorced at 26, but a few years later she and her former husband again tried to live together. The man had at that time begun to consume alcohol and Mildred also got used to liberal alcohol consumption. After some rather unhappy years, when Mildred was 31 years of age, she again started to lose weight. Thus, the second episode of the eating disorder occurred after a remission of 12 years. She lost about 15 kg and her menstruation stopped. During the second period of her eating disorder, she also developed a very severe bulimia. Her alcohol problem got worse. She had periods of heavy drinking, in the course of which she did not, as a rule, have any bulimic episodes. During periods when she was sober, however, she was bingeing and vomiting all the time. She had also got into the habit of using an old-fashioned water enema every day. Mildred has reported that she has regained normal body weight, that her menstruation has returned (at about 40 years of age) and that she has entirely stopped bingeing and vomiting. She has a new boyfriend and says that she is very happy.

The Intermediate Outcome Group

There are only six patients in this group. They have all suffered from a long and complicated eating disorder, but their present state indicates that great improvement has taken place. There are good reasons to expect that these patients will ultimately belong to the good outcome group. With regard to the "polarizing" effect of the time factor (see above), it would be logical for patients with anorexia nervosa – after a very long observation period – to split up into two groups only: a good outcome group, which will include those patients who sooner or later have overcome their illness and a poor outcome group, including those patients who died and those whose disorder is truly chronic. Of the intermediate group, one patient is presented. Like Mildred, she had a late relapse of a bulimic type.

Case 69 (Irene). (Fig. 4) This patient was presented under the same case number and name in the author's monograph (Theander 1970). Irene had a period of restricting anorexia nervosa at the typical age – in her late teens. In the monograph, this case was characterized as a recurrent and atypical case of the depressive type. Her teenage anorexia occurred in two spells, with a short remission in between. Irene was moderately overweight before the onset of her illness (70 kg, BMI 24.5). During her first anorectic episode, her weight dropped to 49 kg (BMI 17). Irene recovered at the age of 21; she worked in a bank, and later she became a teacher. She married and had two children. When Irene was 36 years of age, her husband had a severe heart attack. When he returned from hospital, Irene's chief ambition was to keep the children quiet and the

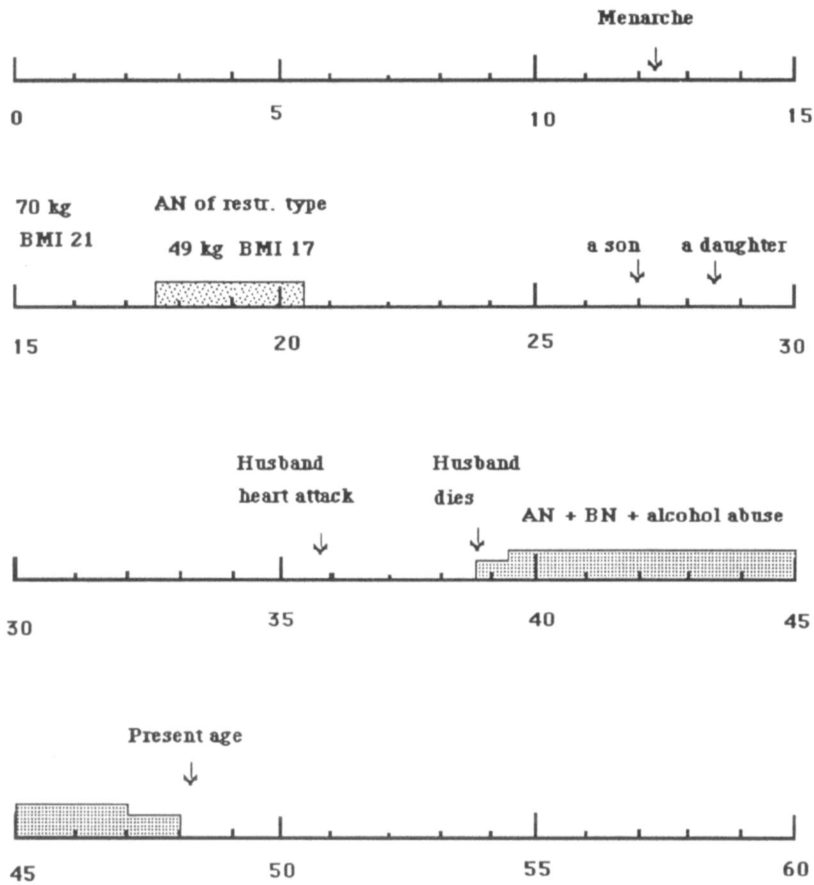

Fig. 4. Case 69: Irene, year of birth 1942. Example of a patient in the intermediate outcome group. The diagram gives the life and illness history of this patient who had a typical teenage anorexia nervosa but also a late relapse with anorexia and bulimia with onset at 39 years of age after the death of her husband

home in good order. During this stressful period, Irene became used to benzodiazepines in high doses. Her husband suddenly died 3 years later after another heart attack. Irene could not allow herself to feel sorrow and to show weakness; instead, she wanted to be strong and able to manage her job and her duties towards her children. Within a short time she again started to lose weight, and her menstruation stopped. Now it was not only a restricting type of anorexia. Irene dieted during the day, but in the evenings she had typical binges. She also had a tendency to abuse both alcohol and sedatives, but all the time she was able to work full time and to take care of her two adolescent children. By now Irene has had a continuous eating disorder during the recurrent episode for about 9 years. During the last year, however, she has been in a much better state; her weight level is almost normal and her bingeing less pronounced. Irene is a person who has a great sense of responsibility. As her children will soon be grown up, it is realistic to think that her duties, and the stress she is under, will be diminished in the near future. Hopefully Irene will then be able to overcome her eating disorder.

The Poor Outcome Group

Out of the twelve patients in the poor outcome group, nine have died – six of starvation and medical complications, three of suicide. The remaining three patients have chronic anorexia nervosa.

Patients Who Died

As nine patients have died out of 37 who had been continuously ill for 6 years, the crude mortality among patients with a long-standing anorexia nervosa amounts to 24%. As was mentioned above, another eight patients died before reaching the 6-year limit, which means that the total crude mortality in the long-term follow-up study is 17 deaths from anorexia nervosa and suicide among 94 patients, or 18%.

Three patients died rather soon after the 6-year limit, one of suicide and two of starvation. The history of one patient is presented below.

Case 74 (Ulrika). This patient was 34 years of age at the start of her illness. Her age of onset was in fact the highest in the whole group. Ulrika had a congenital malformation of the large vessels (coarctation of aorta). Both before and after an operation at 22 years of age, she seems to have been a weak and asthenic woman. She was advised by a physician never to have any children. She married at 30 years of age, and a few years later she started to vomit after every meal. She became extremely emaciated and died, weighing only 28 kg, after a total period of illness amounting to $6\frac{1}{2}$ years.

Another six patients in the poor outcome group had a chronic eating disorder, which eventually caused their death. In the majority of cases death was the result of extreme starvation, but in two cases the cause of death was suicide. These six patients died after a period of illness lasting for 18 years, 19 years, 20 years, 23 years, 25 years and 40 years. Two examples are presented:

Case 89 (Johanna). Johanna died at the age of 31 after having had anorexia nervosa since the age of 11. One striking fact is that this patient weighed only 25 kg when she died. The explanation of this incredibly low weight for a woman in her thirties is her extremely low height, only 153 cm (BMI 10.7). The case of Johanna is an example of the result produced by a very early onset of anorexia nervosa, combined with a long duration of illness. The patient's growth certainly stopped as a consequence of her anorexia. Johanna became ill before her menarche, and she never had any menstruations. Complications of prepuberal anorexia nervosa in the form of retarded growth and inhibited breast development have been described by Russell (1985).

Case 93 (Rakel). There is insufficient information available on this patient. However, she probably had anorexia nervosa continuously for about 40 years until she died, extremely emaciated, at 58 years of age. Possibly she had a remission at about 30 years of age, when she had irregular menstruations for a short period. Rakel's anorexia started at the age of 18, and she had menstruated for 3 years before the onset. She was short (157 cm), and she is said to have weighed 24 kg only when she died. At the first follow-up investigation in 1963, she was visited in her home. She was an extremely emaciated woman who lived in the country with her brother in very poor conditions. She looked some 20 years older than her real age and was not very willing to give

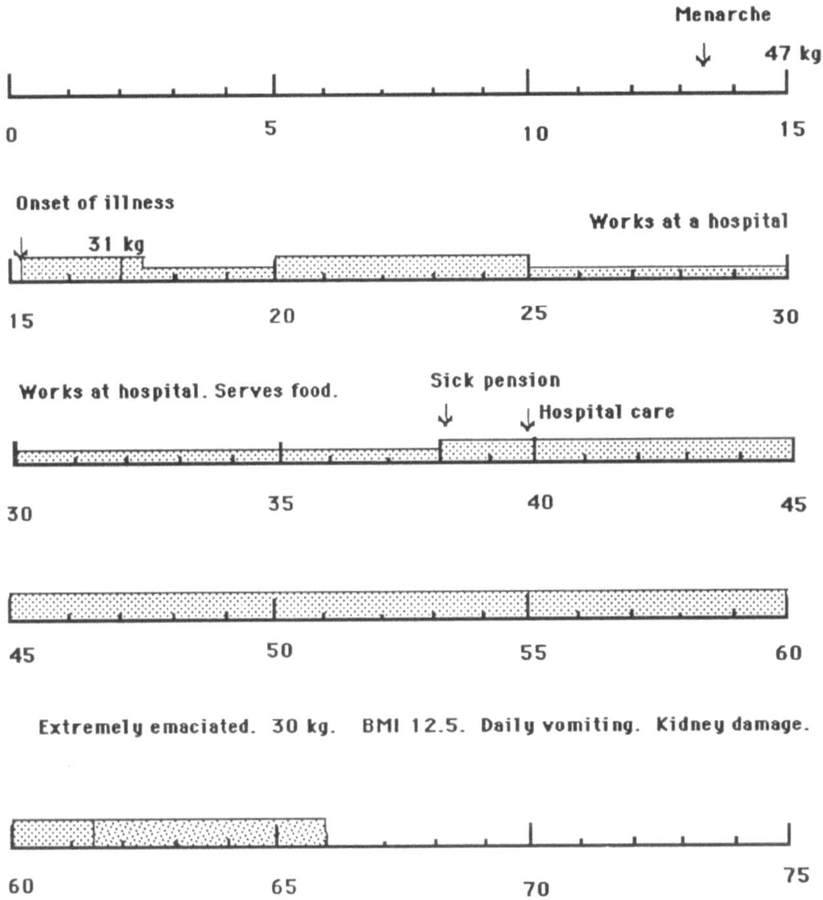

Fig. 5. Case 60: Ellen, year of birth 1924. A patient with a chronic anorexia nervosa

information about her life or her state of health. The impression was that she led a very impoverished and restricted life.

Patients with Chronic Anorexia Nervosa

Three patients suffering from chronic anorexia nervosa are still alive. As was mentioned previously, another six patients had a chronic illness which resulted in untimely death. Thus, out of the original series, nine patients (10%) – of whom two thirds have died – developed chronic anorexia nervosa according to the proposed definition.

All three surviving patients with chronic anorexia nervosa are extremely thin, having weighed about 30 kg for years. Two of these patients will be presented:

Case 60 (Ellen). Ellen has been continuously ill with anorexia nervosa for about 50 years by now (onset at the age of 15, present age 66; Fig. 5). She was hospitalized several times during the first 3–4 years of the illness. At the age of 40, she was again admitted several times to a psychiatric hospital. As she did not agree to stay for the recommended treatment programme, she was compulsorily admitted for rather a long period of treatment. In spite of all efforts, she relapsed almost immediately after her discharge. Ellen is extremely emaciated. She vomits after every meal, but she has no typical binges. She has multiple somatic complications of the anorexia, but in spite of all her problems, her vitality is remarkable.

Case 76 (Isabella). This patient fell ill at the age of 17. She was married between 24 and 30 years of age, still having severe anorexia nervosa. As she could not become pregnant, the couple adopted a child. At the divorce when she was 30, her body weight was extremely low (27 kg; BMI 10.2) and she was treated in hospital for 6 months, but without much success. In spite of her serious illness, Isabella managed to take care of her adopted child after the divorce and to work full time as a qualified secretary. Since the age of 40 she has been working part time, having a partial disability pension. She is now 55 years of age and still working part time. It is astonishing that a patient who has had very severe anorexia nervosa for almost 40 years can still have some working capacity.

Summary and Conclusions

The concept of chronicity in anorexia nervosa is discussed, and it is argued that the term "chronic anorexia nervosa" should be reserved for those patients who have been continuously ill for 15 years or more. It is emphasized that there is a great need for better knowledge of anorexia patients with a long-standing illness, and that special programmes should be developed in order to motivate and treat these patients.

An analysis is presented on the long-term outcome of those 37 patients in a Swedish follow-up study comprising 94 female patients who had been continuously ill for 6 years or more. The results are as follows: Nineteen patients (51%) recovered after a total period of illness amounting to 6–11 years. Nine patients died (24%). Six of them had chronic anorexia nervosa, and they died after having been ill for 18–40 years. Out of the total of 94 patients, 9 (10%) developed chronic anorexia nervosa. Six of them have died, but three are alive, having been seriously ill for 40–50 years.

References

Casper RC (1987) Psychotherapy in anorexia nervosa. In: Beumont PJV, Burrows GD, Casper RC (eds) Handbook of eating disorders, part 1. Elsevier, Amsterdam

Crisp AH (1980) Anorexia nervosa: let me be. Academic, London

Dally P (1969) Anorexia nervosa. Heinemann Medical Books, London

Dally P (1984) Anorexia tardive – late onset marital anorexia nervosa. J Psychosom Res 28:423–428

Fenley J, Powers PS, Miller J, Rowland M (1990) Untreated anorexia nervosa: a case study of the medical consequences. Gen Hosp Psychiatry 12:264–270

Hall A (1982) Deciding to stay an anorectic. Postgrad Med J 58:641–647
Hsu LKG (1980) Outcome of anorexia nervosa: a review of the literature (1954–1978). Arch Gen Psychiatry 37:1041–1046
Hsu LKG, Crisp AH, Harding B (1979) Outcome of anorexia nervosa. Lancet 1:61–65
Kay DWK, Schapira K (1965) The prognosis in anorexia nervosa. In: Meyer J-E, Feldmann H (eds) Anorexia nervosa. Thieme, Stuttgart
Meyer A-E, Von Holtzapfel B, Deffner G, Engel K, Klick M (1986) Psychoendocrinology of remenorrhea in the late outcome of anorexia nervosa. Psychother Psychosom 45:174–185
Morgan HG, Russell GFM (1975) Value of family background and clinical features as predictors of long-term outcome in anorexia nervosa: four-year follow-up study of 41 patients. Psychol Med 5:355–371
Morgan HG, Purgold J, Welbourne J (1983) Management and outcome in anorexia nervosa; a standardized prognostic study. Br J Psychiatry 143:282–287
Pope HG Jr, Hudson JI (1989). Eating disorders. In: Kaplan HI, Sadock BJ (eds) Comprehensive textbook of psychiatry, 5th edn. Williams and Wilkins, Baltimore
Ratnasuriya RH, Eisler I, Szmukler GI, Russell GFM (1991) Anorexia nervosa: outcome and prognostic factors after 20 years. Br J Psychiatry 158:495–502
Russell GFM (1977) General management of anorexia nervosa and difficulties in assessing the efficacy of treatment. In: Vigersky R (ed) Anorexia nervosa. Raven, New York
Russell GFM (1985) Premenarchal anorexia nervosa and its sequelae. J Psychiatr Res 19:363–369
Steinhausen H-C, Glanville K (1983) Follow-up studies of anorexia nervosa: a review of research findings. Psychol Med 13:239–249
Stonehill E (1988) Anorexia nervosa – a psychobiopathological defence: what happens in the long term? In: Hardoff D, Chigier E (eds) Eating disorders in adolescents and young adults. Freund, London
Theander S (1970) Anorexia nervosa: a psychiatric study of 94 female patients. Acta Psychiatr Scand [Suppl] 214
Theander S (1985) Outcome and prognosis in anorexia nervosa and bulimia: some results of previous investigations, compared with those of a Swedish long-term study. J Psychiatr Res 19:493–508
Vandereycken W, Pierloot R (1983) Long-term outcome research in anorexia nervosa. Int J Eating Dis 2:237–242
Ziolko HU (1978) Zur Katamnese der Pubertätsmagersucht. Arch Psychiatr Nervenkr 225:117–125

What Can We Learn from Long-Term Outcome of Anorexia and Bulimia Nervosa?

K. Tolstrup

Introduction

The reflections on long-term course proposed in this chapter are based on 40 consecutive years of clinical experience in the intensive treatment of very ill patients with anorexia/bulimia nervosa (Tolstrup 1990) and on being involved in a systematic follow-up study of patients from different departments at a university hospital (Tolstrup et al. 1985). This combined experience may perhaps increase the advantages and moderate the drawbacks of the two different ways of obtaining knowledge of a disease course. The longitudinal, personal and individual contact gives more profound insight into the nuances of the course in the individual patient; however, this obviously entails the risk of generalization on the basis of casuistic anecdotes, both in the positive and negative sense. In contrast, a cross-sectional study of a great number of patients, with statistical analysis of well-defined individual data, serves to eliminate personal bias. It implies at the same time, though, the risk of losing essential data on individual courses that cannot be recorded and treated statistically, with the result that individual differences are blurred.

Anorexia/bulimia nervosa (AN/BN) is defined here as a continuum of severe eating disturbances, ranging from the classic syndrome with restrictive food intake (AN), over bulimic variants to bulimia nervosa (BN) as defined by Russell (1979). Behind the uniform outward appearance inflicted by the disease, these patients differ individually. They have the following characteristics in common: large weight loss, fear of obesity, distorted body image, and abnormal eating habits. There is no known etiology, but the coincidence with puberty and adolescence in the majority of patients suggests that the characteristics of this developmental period are of central etiologic importance. Furthermore, physical, mental, social, cultural, and hereditary causal factors are of relevance, too. AN/BN is a disease in the proper sense of the word, in that it is potentially fatal. Even if its course is nonfatal in the vast majority of patients, AN/BN must still be considered a disease, because it requires treatment owing to the many complaints and problems involved for the patients and the people in their surroundings. In many cases, though, the pecularities as to eating and body image overlap those seen within the limits of normal puberty.

Types of Course

Classification

AN/BN is a disease which involves all the essential physical and mental functions: nutrition, weight, growth, menstruation and other sexual functions, social contact, mood, sleep, and physical activity – with the consequent social complications. The ideal approach would be to include all these factors in the evaluation of the course, but obviously this is not possible in practice. One must try to utilize those single factors which best contribute to the total picture. The issue of the disease is an arrest of development, rightly compared with hibernation, in an especially vital period of growth, puberty. The issue is also one of whether to survive or perish. With the present, insufficient, knowledge, the course can best be classified under the broad headlines: *cure*, chronic or *intermittent course*, *death*.

Within this rough division, a few, easily determinable characteristics have been chosen, all telling something essential about the course. The selection is marked by my particular experience with young patients (mean age 19 years, range 8–43 years) and by the areas which The Copenhagen Follow-up Study (Tolstrup et al. 1985) dealt with specifically: the patients' later pregnancies, deliveries, and relations to their children; growth; relapse; and death. Furthermore, the influence on the course of age and sex will be discussed.

Cure

Among other things, cure means that the vital basic functions have returned to normal. They include reproduction, mothering, and growth.

Reproduction and Mothering

Amenorrhea, primary or secondary, is a cardinal symptom in AN, either conditioned by undernutrition or psychogenic in nature. In an ethological perspective, AN has been regarded as "a reflection of the female ability to alter maturation rates and reproductive function in response to environmental conditions." (Surbey 1987). According to this view, it is *appropriate* that during disease, undernutrition, and mental strain, a young woman is *not able* to bear children.

The Copenhagen Follow-up Study (follow-up period 4–22 years, mean 12.5) included a subproject concerning the probands' pregnancies,

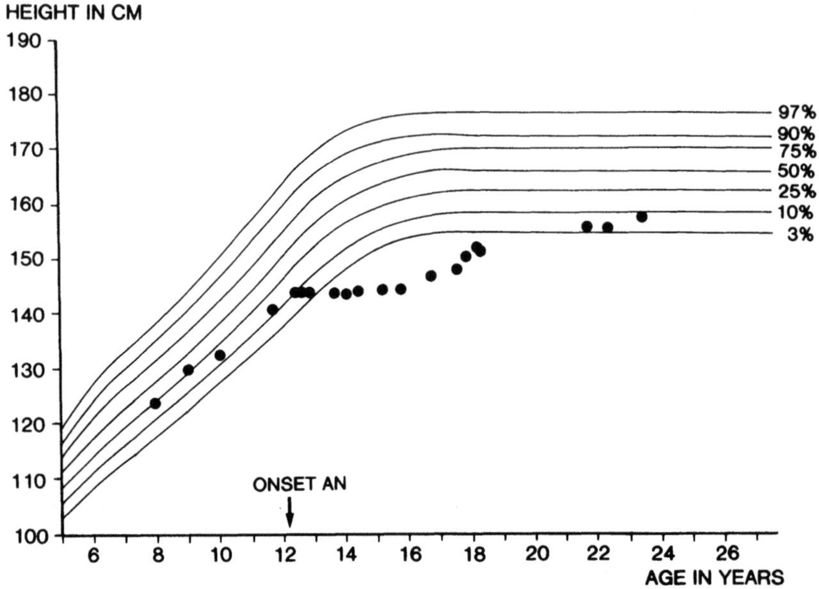

Fig. 1. Growth in relation to the normal values in Danish school girls (Andersen et al. 1974). At onset of anorexia nervosa (*AN*) at the age of 12, growth stops. From the age of 16, growth is resumed and has not yet stopped when the patient is 23 years old

deliveries, and parental function (Brinch et al. 1988). Four years later, on the average 16.5 years after the primary contact, a further follow-up was made on perinatal data.

The latter study (unpublished) showed that 60 of the original 140 female probands (43%) had delivered 110 children in the period 1943–1985. This corresponds to between one half and one third of the deliveries in the reference population in the same period. The high infertility was voluntary. Involuntary infertility was not more frequent than among other Danish women. Thus, by and large, previous AN/BN patients had the children they wanted, even if cure was preceded by very protracted disease. As a casuistic curiosity, the patient whose growth is graphically illustrated in Fig. 1 had a normal wished-for child in her late twenties, when she had recovered completely from disease of many years' duration.

Furthermore, the study by Brinch et al. (1988) showed that, compared with the normal population, the probands had more spontaneous abortions, their children a lower birth weight, there were more premature children, and a higher perinatal mortality. The significance of these findings is not yet clarified, but the results may be considered to support the ethological theory of the reduced ability of these children to survive (Surbey 1987).

The primary hypothesis of the study was that pregnancy and delivery often provokes relapse of AN/BN, because of increased weight during pregnancy and motherhood itself intensifying the conflicts concerning female identity, which are considered central psychogenic factors in the development of the disease. The frequency of coincidence of pregnancy and relapse of AN/BN, however, was low.

The early nursing and upbringing of the children, whose mean age at the follow-up was 7 years (range 0–38), involved no more problems than were seen in the general population, not regarding eating either. However, one child died from AN at the age of 13 years.

Childbearing itself is not a sufficient indication of cure. This can also be seen from the fact that 9 of the 60 women conceived while suffering from AN/BN, as judged by presence of oligomenorrhea/amenorrhea, low body weight, etc. (unpublished data). Only one conceived following hormonal treatment. However, according to the study by Brinch et al. (1988), completion of a normal pregnancy and a satisfactory parental function are strongly suggestive of cure. Consistently, a global clinical evaluation described significantly more mothers than nonmothers as functioning well. In addition, only one of nine deaths occurred among mothers.

Growth

Especially in the youngest, not yet full-grown patients, growth is a good criterion of cure. The causal relationship between height and AN/BN, however, is complicated. The Copenhagen Follow-up Study (unpublished data) thus showed that the five women whose children died in the perinatal period were relatively short. We do not know whether the low height and the deaths of the children had a common causal factor, or what the role of nutrition and AN/BN was, or whether there a causal relationship at all.

It has been asserted that generally low height predisposes to AN/BN (Rollins and Piazza 1981). Truly, there exist casuistic examples of low height possibly being a contributing etiological factor, either genetically or psychologically conditioned (Brinch and Manthorpe 1987). However, premorbid data in another Danish study (Nielsen 1985) on the growth in 66 AN/BN patients do not suggest such a general trend. The annual measurements by the school health services, performed at least 2 years before onset of AN/BN, showed that the premorbid growth of the girls did not differ from that in the normal population of Danish girls, whereas the boys were taller than those of the same age. We can offer no explanation of the latter phenomenon.

Secondarily, growth may be delayed in patients with onset of disease during or before puberty, most markedly in the case of onset of AN before the prepubertal growth spurt. The most obvious reason is of course undernutrition, but the immediate and direct role of emotional factors should not be underestimated – refer to the diagnostic concept "psychosocial dwarfism" (Green et al. 1984). As already pointed out by Janet (1908), specifically in AN an important element of the disease is the patient's intensive wish to remain a child, concretely expressed as repugnance to growing. One patient who had not grown during the 5 years of AN reacted strongly when she had to get new shoes in a larger size – a marked sign of return of growth.

Return of growth in patients with early onset AN can be a good indication of cure being within reach. Measurement of height is accurate and easy – and therapeutically encouraging – since, in contrast to weight, height does not undergo the capricious relapses that are so frustrating for the therapist. Permanent improvement can be hoped for when a standstill of several years' duration is replaced by resumed growth, seen in the height curve as a plateau followed by a rising curve (see the clinical example in Fig. 1). On the other hand, the psychotherapeutic problem may arise of whether the patient *accepts* the increase in growth and thereby the body image of normal puberty.

Course with Relapse

The Copenhagen Follow-up Study is a historical-prospective catch-up study (see Rutter 1981). This means that the original material at the primary contact contains a number of data which, without being specifically established in view of follow-up, are so well defined that they can be related to the same or other well-defined data at follow-up. The Copenhagen Follow-up Study is a cross-sectional study which was performed over a 2-year period (Tolstrup et al. 1985). Even if the follow-up is intensive, this method may render it difficult to obtain applicable data for the "bridging period" between the primary contact and the follow-up. Variables for the bridging period sufficiently well defined are, however, relapse and death. These variables formed the basis of a substudy (Isager et al. 1985).

In the latter study, relapse was defined as a weight loss of 15% or more within 1 year. At the end of the primary treatment, 31 of 151 probands were not in remission, i.e., they were still ill. The condition in the remaining 120 probands was so satisfactory that it was meaningful to speak about a possible later relapse. At follow-up after 4–22 years, relapses totaled 31, with 17 occurring during the 1st year and 4 during the 2nd, whereafter the annual number was strongly declining; the last relapse occurred after 14 years. On the basis of a survival analysis, the risk per year of relapse was calculated to be 14% for the 1st year, 4% for the 2nd, and declining thereafter. For the total period, the risk per year of relapse averaged 3%.

It can be concluded that there is a marked trend for relapses to accumulate during the 1st years of the catamnesis period.

Fatal Course

Death, unfortunately, is a most well-defined criterion of course. All extensive follow-up studies with a follow-up period of at least 4 years show fatal courses. Since death, among other things, is a product of time, mortality should always be related to the duration of the follow-up period. The annual mortality, i.e., the estimated risk per year rate on the basis of survival analysis, in the Copenhagen Follow-up Study thus was 0.5% compared with 0.07% in all Danish women in the 20–40 year age group. The deaths were evenly distributed over the first 13 years, whereas no deaths occurred during the last 9 years of the 22-year study period (Isager et al. 1985).

The reported surveys often distinguish between death caused directly by AN/BN and death from other causes. In my opinion, a realistic picture of the extra mortality can be obtained only by using the total numbers, irrespective of the immediate cause of death. For instance, it is farfetched to consider suicide a form of death not related to AN. In the Copenhagen Follow-up Study thus 6 of 9 deaths were suicides (Tolstrup et al. 1985). However, other causes of death should be included in the total mortality. For instance, who ventures to preclude that a fatal myocardial infarction is precipitated by debility owing to previous undernutrition?

Our knowledge about predictors of premature death in AN/BN patients is incomplete. Our substudy (Isager et al. 1985) shows that previous hospitalization in psychiatric departments is connected with increased mortality. The 20 probands hospitalized at least once prior to primary contact with the participating departments had an risk per year rate of 3% versus 0.2% in the 122 probands not previously hospitalized in a psychiatric department; this difference is highly significant ($p < 0.001$). A possible explanation is that marked psychopathology is an expression of severe disease, with its consequent poor prognosis. On the other hand, any experienced therapist can tell about patients who have recovered completely, despite long-term disease and several stays in psychiatric departments. This typically illustrates the caution with which the individual course should be predicted on the basis of statistical mass analyses.

In most cases of fatal AN/BN, plausible somatic explanations can be offered for the immediate cause of death. On the other hand, it remains a mystery why, by continuing to refuse to eat and possibly also to vomit with the consequent severe undernutrition, certain patients seem to head inevitably, almost determinedly towards death. It is tempting to use terms such as death urge or inability to live.

In my opinion the severely ill AN patients display two extremely different types of reaction: the "rebellious" and the "resigned".

The "rebellious" patient is incessantly combating her surroundings, is being ceaselessly provocative by her demonstrative refusal to eat, and her disease seems to be a message – even if this message may be difficult to interpret for the people in their surroundings and possibly is not realized by herself either. Raimbault and Eliacheff (1989) use the term "les indomptables" for this type of personality and exemplify her untameable attitude to life by Antigone, the figure in the ancient Greek tragedy.[1] Treatment of this type of patient is like an exhausting, but meaningful fight, because she is a good fighter, loaded with aggressions, and "savagely" furious. The therapist feels optimistic with regard to her long-term prognosis.

On the other hand, it is difficult to have any optimism when treating the other type, the "resigned" AN patient. They demand special attention. The patient is defeatist and self-effacing, and she conceals her illness as long as possible; she does not provoke openly, but rather systematically, and without speaking about it, she reduces her weight. Her relations to those in her surroundings are evasive and unengaged; the question of conflicts cannot be raised, nor her problems discussed. Nurses experienced in long-term treatment of AN may say that "she blends into the background"; she disappears, so to speak, in the daily inpatient therapy if you are not specifically attentive to her. This type of patient may seem to head for death. She might be termed the Ismene type, after the compliant and subdued sister of the rebellious Antigone.

In some cases the risk of death is strengthened by an apparent family interaction. AN is an extremely straining disease for the relatives, but the negative family reactions are more often secondary than primary elements in the development of the disease. Some families, though, give up more easily than others. The parents may refuse hospitalization of the girl at a time when her somatic condition obviously is potentially fatal, with the words: "Let her die in peace now." A mother may state that she would rather see her daughter die than be hospitalized for another, strenuous treatment. The drawings by an AN girl may express her phantasies on how her mother in her own deep depression wishes her daughter to be dead. When such phantasies are strong enough, and uncorrectable through therapy, they may contribute to break down an already weak will to live.

Do Sex and Age Influence the Long-Term Course?

In a 1965 study on 28 AN patients, 24 females and 4 males, with a catamnesis period of from 6 months to 12 years, I asserted that there were no essential differences in symptoms between the youngest and the oldest patient, who at onset of disease were 7 and 24 years old, respectively; nor were there any differences as to course (Tolstrup 1965). I still believe that sex and age do not decisively determine the prognosis, even though more extensive and better supported studies during the intermittent 25 years

---

[1] Petzold (this volume, p. 367) has used the Antigone metaphor in a different sense.

require that modifications and nuancing of this categoric assertion be made (Jacobs and Isaacs 1986).

Sex

The male prevalence is so low that even large total samples do not allow general conclusions to be made as to the influence of sex on the course of disease. It should be noted, however, that in the Copenhagen Follow-up Study and a substudy 4 years later (unpublished data), none of the 11 males died during the 8–26 years of follow-up. Nor had any of them become fathers when their mean age was 33 years. In comparison, 60 of the 140 women in the same sample had given birth to 110 children. The follow-up data do not suggest any sexual abnormalities in these men.

Age

It was recognized early on that late onset of disease, after the age of 20, was associated with a relatively poor prognosis (Theander 1970). In contrast, it has been almost a dogma that early onset implied a good prognosis. However, recent studies suggest that a very early onset, before the age of 11, *possibly* is predictive of a poor course (Bryant-Waugh et al. 1988).

The controversy as to the significance of age is due to the many other age-related elements which condition and influence the disease. Thus, two opposite prognostic tendencies pertain to the young patients. The *positive* tendency is connected with the improved therapeutic possibilities. In the very young patient, the family seeks medical attention on her behalf, and the family generally is motivated for treatment, including formal family therapy, which has been shown to be especially effective in young patients (Russell et al. 1987). Also, in children and very young subjects with a life-threatening disease, it is easier to accomplish the rather drastic therapeutic procedures such as, possibly compulsory, tube feeding. On the other hand, the *negative* prognostic tendency should be seen in the light of the general child psychiatric experience that early severe psychopathology and disturbances in family function are indicative of a long-term and poor course.

It would be interesting in a large sample with a wide age distribution to test the hypothesis that both very *late* and very *early* onset of disease imply poor prognosis – with respect to onset in these extreme age groups outside the "natural" period of maturation, namely, puberty.

Can Outcome Be Influenceds?

It is difficult to evaluate the influence of treatment on the long-term course and the ultimate outcome, since there exist no single or combined therapeutic methods that are both rational, well tested, and generally accepted (Tolstrup et al. 1984). The choice is between somatic methods, numerous psychotherapeutic methods, cognitive therapy, and behavior therapy. Furthermore, these methods are often used in a combination that is insufficiently described.

One of the purposes of the Copenhagen Follow-up Study (Tolstrup et al. 1985) was to compare outcome in the three different participating departments in the same university hospital: a child psychiatric (64 probands), a psychiatric (51 probands), and a medical department (36 probands). The three subsamples were comparable as to extent of weight loss. The mean age was lowest (15.2 years) and the mean duration of disease shortest (1.4 years) in the child psychiatric subsample, highest (24.2 years) and longest 3.2 years, respectively, in the psychiatric one. Despite these differences, the patients, who came from all over Denmark, had so many similarities that the well-characterized *differences* in treatments could be claimed to have some influence on the course of disease. The child psychiatrists employed the longest and the psychotherapeutically most intensive method, the treatment by the internists was of the shortest duration and concentrated on the undernutrition, and the psychiatrists employed a regimen that was closer to that of the child psychiatrists, though of less psychotherapeutic intensity.

Our hypothesis was that the intensive child psychiatric treatment would clearly prove to be more effective, with a markedly better outcome than the treatments in the other two departments. The results of the follow-up, however, showed no statistically significant differences in the total outcome, evaluated on the basis of the criteria mortality, general somatic outcome, socioeconomic status, global clinical evaluation, and psychiatric diagnosis (Tolstrup et al. 1985). The results in the psychiatric subsample, however, tended to be somewhat poorer.

There are three immediate explanations of the lack of significant differences in outcome ascribable to treatment.

First, it is possible that AN/BN is refractory to the regimens employed and that the disease, consequently, unchallenged has taken its "natural" long-term course.

Second, there may exist a common and effective basic element in the otherwise different regimens of the three departments. If so, this common element could be the uncompromising attitude to the demand of a relatively rapid and massive weight gain. This basic therapeutic philosophy may also explain the fact that no deaths occurred in any of the departments during the primary treatment of this selected and very affected patient sample.

A third possible explanation is that the significance of a number of individual factors should be analyzed before the final conclusion on the effectiveness of the treatments can be drawn. Thus, in the above-mentioned substudy (Isager et al. 1985) a survival analysis showed a higher rate of risk of death for those probands who had been hospitalized in a psychiatric department prior to primary contact, which exactly was a characteristic of the adult psychiatric subsample. Another analysis showed an inverse correlation between duration of therapeutic contact and relapse rate. This finding, obviously, should be interpreted with caution, or as the authors express it: "We cannot say whether longer treatment led to a better outcome, or whether better motivated patients were likely to continue with treatment".

Concluding Remarks

Our present incomplete knowledge of AN/BN obviously makes any classification into types a general one and more or less random and individual. Thus, the present description does not consider certain, presumably important prognostic factors, such as presence of bulimia, and – of special interest to a psychiatrist – personality characteristics (Sohlberg 1987); nor does it take environmental background and family structure into consideration.

One thing is certain: there is no single deciding factor of the course, which in the individual patient is unpredictable. At the start of any treatment there exist both the chance of cure and the risk of death or chronicity. A large number of patients are cured within a few years, but the therapist's pride should also contain a portion of professional humbleness in the realization that he or she has presumably been instrumental in promoting a spontaneous recovery. Those patients who do *not* recover within a few years need long-term support, with the modified primary aim of inhibiting the tendency of the disease to become chronic or fatal, or at any rate, help in living an acceptable life with the disease. On the other hand, it is part of the unpredictability of the disease that even after many years, a marked improvement may occur (Theander 1985). Time, patience, and consistency probably are the most effective elements in the treatment of this enigmatic disease.

References

Andersen E, Hutchings B, Jansen J, Nyholm M (1974) Heights and weights in Danish school children 1971–72. Ugeskr Laeger 136:2799–2802

Brinch M, Manthorpe T (1987) Short stature as a possible etiological factor in anorexia nervosa. Acta Psychiatr Scand 76:328–332

Brinch M, Isager T, Tolstrup K (1988) Anorexia nervosa and motherhood: reproduction pattern and mothering behaviour of 50 women. Acta Psychiatr Scand 77:611–617

Bryant-Waugh R, Knibbs J, Fosson A, Kaminski Z, Lask B (1988) Long-term follow-up of patients with early onset anorexia nervosa. Arch Dis Child 63:5–9

Green WH, Campbell M, David R (1984) Psychosocial dwarfism. A critical review of the evidence. J Am Acad Child Adolesc Psychiatry 23(1):39–48

Isager T, Brinch M, Kreiner S, Tolstrup K (1985) Death and relapse in anorexia nervosa. Survival analysis of 151 cases. J Psychiatr Res 19:515–521

Jacobs BW, Isaacs S (1986) Pre-pubertal anorexia nervosa: a retrospective controlled study. J Child Psychol Psychiatry 27:237–250

Janet P (1908) Les obsessions et la psychasthénie, 2nd edn, vol 1. Alcan, Paris, pp 33–41

Nielsen S (1985) Evaluation of growth in anorexia nervosa from serial measurements. J Psychiatr Res 19:227–230

Raimbault G, Eliacheff C (1989) Les indomptables. Figures de l'anorexie. Odile Jacob, Paris

Rollins N, Piazza E (1981) Anorexia nervosa – a quantitative approach to follow-up. J Am Acad Child Psychiatry 20:169–183

Russell GFM (1979) Bulimia nervosa: an ominous variant of anorexia nervosa. Psychol Med 9:429–448

Russell GFM, Szmukler GJ, Dare C, Eisler J (1987) An evaluation of family therapy in anorexia nervosa and bulimia nervosa. Arch Gen Psychiatry 44:1047–1056

Rutter M (1981) Longitudinal studies: a psychiatric perspective. In: Mednick SA, Baert AE (eds) Prospective longitudinal research. Oxford University Press, Oxford, pp 326–336

Sohlberg S (1987) Anorexia nervosa and bulimia nervosa. Diagnosis, treatment and the role of impulsivity for long-term outcome. Dissertation, University of Uppsala.

Surbey MK (1987) Anorexia nervosa, amenorrhoea and adaptation. Ethol Sociobiol 8:47–61

Theander S (1970) Anorexia nervosa. A psychiatric investigation of 94 female patients. Acta Psychiatr Scand. [Suppl] 214

Theander S (1985) Outcome and prognosis in anorexia nervosa and bulimia: some results of previous investigations, compared with those of a Swedish long-term study. J Psychiatr Res 19:493–508

Tolstrup K (1965) Die Charakteristika der jüngeren Fälle von Anorexia nervosa. In: Meyer J-E, Feldman H (eds) Anorexia nervosa. Thieme, Stuttgart, pp 51–58

Tolstrup K (1990) Treatment of anorexia nervosa: current status. In: Simeon J, Ferguson HB (eds) Treatment strategies in child and adolescent psychiatry. Plenum, New York, pp 115–132

Tolstrup K, Brinch M, Isager T (1984) Therapie und Verlauf der Anorexia nervosa. In: Remschmidt H (ed) Psychotherapie mit Kindern, Jugendlichen and Familien, vol 2. Enke, Stuttgart, pp 45–53

Tolstrup K, Brinch M, Isager T, Nielsen S, Nystrup J, Olesen NS, Severin B (1985) Long-term outcome of 151 cases of anorexia nervosa. The Copenhagen anorexia nervosa follow-up study. Acta Psychiatr Scand 71:380–387

Part III
Special Issues

Genetic Determinants in Anorexia Nervosa: Results of Studies in Twins*

H. Schepank

Twins, and especially identical twins, arouse peculiar feelings in a person. It is not only the curiosity of it all, but such a meeting also touches on the problems of one's own identity development, individuality, and narcissism. In the study of human genetics, classical research in twins, that is, comparing monozygotic and dizygotic pairs of twins as to the degree of correspondence/rate of concordance, is one of the most exact and reliable methods of researching problems of heredity versus environmental influence in medicine and psychology. Nevertheless, this controversy sparks strong emotional resistance among researchers, with the remarkable consequence that in the past decades studies in twins – especially in Germany – have scarcely been carried out and existing results hardly taken into consideration.

In the same way, the horrifying appearance of anorexia nervosa (AN) patients and the development of the condition fascinates the general public and researchers alike. If one follows the scientific controversies of the past three centuries (the disease was described for the first time already in 1689 by Morton!) one will notice here, too, an ideological, deep-seated conflict which has caused considerable tension between those who consider it a physical condition and those who consider it a psychological matter. Indeed, the problem can be reduced to the century-old and still unresolved issue of body and soul.

Twins are not an everyday occurrence: one can count on running into 1 person born as a twin of 50 who are asked. Neither is AN a commonplace condition compared to other functional psychogenic disorders, and, hence, the coming together of both in one individual a real rarity. Considering the few existing and very clear-cut results from studies in twins as to AN in these individuals (Crisp et al. 1985; Schepank 1980, 1981, 1983) and from collective case studies compiled methodically and carefully (Fichter 1985), the prevailing ignorance in the field still always comes as a surprise.

---

* This work was supported in part by the *Deutsche Forschungsgemeinschaft* (DFG) and performed at the SFB 32 (Clinical Genetics) Department of the University of Heidelberg. I am also grateful to the Central Institute of Mental Health, Mannheim, who provided financial support for travel and examination costs.

A description of AN and its defining clinical features is given in other chapters of this book, as is a delimitation of AN from bulimia only, and so this will not be expounded upon in this chapter – also because in our study only individuals with more marked conditions of AN were included. The index twins were found among the patients of medical clinics, and so unequivocal diagnosis had already been made.

Methods of Research in Twins

Because of some basic misunderstandings and frequent misinterpretation of individual publications of observations in twins, first a few comments:

1. The *techniques* used to clarify scientific questions using twins are very different from one another, as can be the aim of the research – whether a question of genetics or environmental influence or only evaluation of the influence of specific environmental factors, etc. We are acquainted with prospective and even *genuine* experimental designs (for example, in developmental psychological or learning experiments). Most investigations are retrospective in nature, but there are some prospective genealogical test designs, for example, the brilliant "Trick" developed by Margaret Fisher in which the offspring of discordant monozygotic twins are investigated. Quite different is the discordance analysis of monozygotic twins: This test aims specifically at identifying pathogenetically relevant environmental factors (!). For example, if one, of course genetically identical, twin is mentally deficient, then the manifestation of the condition must be a result of environmental influence in this instance (!), e.g., as a result of brain trauma.

2. Disregarding individual case studies, *classical research in twins* usually has large numbers of twins to work with, whereby monozygotic twins are compared with dizygotic twins (which are just as important and necessary to research!) with respect to concordance within the pair of the trait being examined. It would not make sense just to study monozygotic twins with this method.[1] Rate of concordance is the deciding factor when comparing the two kinds of twins. A significant difference in the rate of concordance in terms of a greater similarity within the monozygotic than within the dizygotic twins indicates that genetic factors play a part, since the basic

[1] Less well informed psychologists often require that only monozygotic twins who have grown up together be compared with those who have grown up separately, turning a design around that is basically methodologically permissible. In such cases the independent variable genetic factor is kept constant while the ua variable environment changes. This, however, is in practical terms (due to such constellations occurring all too seldom) hardly possible to carry out and also for basic methodological reasons, particularly in psychogenic disease, the wrong procedure (see Schepank 1980/1981).

genetic makeup of monozygotic twins is almost the same, whereas only half of the genetic potential of dizygotic twins is identical. This assumes that the twins being examined have grown up in the same faimly, and that the environmental variables for twin siblings are just about the same. This method is theoretically sound and through twins a natural form of experimentation.

3. *Individual case studies* alone prove basically nothing – except discordant manifestation of a trait in one pair of twins. Here (but only in this individual case!), one can conclude that there *must* have been outside influence. Any further generalizations with respect to other test subjects or to an entire clinical picture, however, would be an impermissible and false conclusion. But just such a mistake is often made with regard to rare diseases! The aforementioned example merely proves that there do exist conditions of mental deficiency caused by environmental intervention but not that mental deficiency – or to express this in even simpler terms, intelligence – is basically an environmental condition and could not be hereditary.

4. It seems obvious that the *determination of zygosity* and unequivocal case definition and case identification must be ascertained for the investigated trait. But this banality is often disregarded.

5. Just as important as technique is *sampling*, i.e., the means by which the subjects are selected and where the sample is taken from. Naively collecting published accounts of individual case studies presents a great conceptual pitfall. One must always remember that the fact that scientific results have been published means that a preselection has already been made, which may also be the expression of a particular scientific trend: There were times when studies in mainly concordant monozygotic twins were published with the false conclusion that a particular disease is hereditary. In these studies the fact was neglected that of the four possibilities (concordance regarding the investigated trait or disease in monozygotic twins, concordance in dizygotic twins, discordance in monozygotic twins, and discordance in dizygotic twins) only discordance within monozygotes permits limited conclusions – and then only for the individual case! The other three constellations are – taken alone – scientifically insignificant. In the past few decades, though, numerous studies with reports of discordance within individual pairs of monozygotic twins have been published, presenting an interesting selection of results which apparently lend themselves well to the trend of psychologizing. If the results of such rarities are compiled, the incorrect conclusion of a disease having been proved to be nonhereditary emerges, based on the (published!) fact that a trait is most often discordant in monozygotes, that is to say, in only one of a pair of monozygotic twins. The fact is overlooked, though, that the other three possibilities – if they occurred in one pair – were not studied or the results

maybe not even considered worthy of publication, not least because such data might support an old-fashioned hypothesis of heredity.

6. Finally, in sampling one must take into account that studies in twins can lead to varying results as to the degree of hereditary influence, depending on whether the subjects have been selected from index twins from institutions where serious diseases are treated (for example, state psychiatric hospitals) or from those where less severe conditions are treated (outpatient services), or even from twins who have not used hospital facilities or have been found by random testing in the general population from birth registries, as is the case in some of the Scandinavian countries.

7. Another serious and seldom considered methodological error can be made in the interpretation of single findings or collected case studies. Publications often appear very quickly on discordant pairs of children or adolescents although a *period of risk* is not even near to being over. The probability that concordance will increase (but not discordance!) increases over time.

The most important methodological requirement for putting together a sample of twins to elucidate determinants of genetic versus environmental influence in a clinical picture is always that there should be no *(deliberate or not)* preselection *according to concordance or discordance or to whether twins are identical or fraternal.*

Finally, it should be added that research in twins and the interpretation of the results of such studies work on a phenomenological level and cannot be expected to provide knowledge of the special way of inheritance, of chromosomal mechanisms, or of factors localized in the genetic composition or possible biochemical mediators and, hence, of pathogenetic details.

For further information on the subject of methods in psychiatric disease, see Becker 1980, 1981; Meyer 1990; Propping 1989; Schepank 1974, 1980, 1981; Vogel 1961; Zerbin-Rüdin 1985; and many others.

The Literature

Twins research which is methodologically sound, specifically oriented, and proceeds from a hypothesis is scarcely being done in Germany anymore (to my knowledge not in any other field of medicine either, but especially no in psychiatry). The only empirical research in psychopathology using twins has taken place in our group in the field of psychosomatics and in psychogenic diseases (Schepank 1974; Heigl-Evers and Schepank 1980/1981; Schepank and Muhs 1990, 1991) and with the human genetics studies of Propping (1989, pertaining to the origin of alcoholism).

In the definitive German monographs on AN (Feiereis 1989; Fichter 1985; Meermann 1981; Mester 1981) the authors present individual pairs of twins from their samples. There also exist a number of publications dealing with individual case studies, which, however, are highly biased and which are cited over and over again or taken together as collected cases.

A general, critical observation is in order here, however: Due to the high degree of selection in the published literature – mostly only discordance in monozygotic twins – collecting case studies often leads to the kind of false conclusions already mentioned (Askevold and Heiberg 1979; Bruch 1969; Crisp 1966; Gifford and Pilot 1969; Halmi and Brodland 1973; Hill 1976; Hofmann and Kind 1980; Klotz et al. 1961; Nowlin 1983; Yamaoka et al. 1979). Dizygotic twins, which are necessary for comparison, are almost completely missing in the literature before 1980 and are still underrepresented. And this deficiency in methods is not appreciated well enough! On the contrary, from the USA to Europe to Japan the same references to publications of individual cases are cited over and over, and usually misinterpreted to refute the hypothesis of heredity. We will not offer critique to individual work because of the obvious, basic deficiencies of such selection.

We do not underestimate the considerable difficulties in conducting a hypothesis-based investigation: twins are a seldom occurrence, and so are people suffering from AN. Such an investigation requires competent information from various fields: genetic diagnosis of zygosity and competent knowledge of the psychopathology of AN. The Privacy Acts hinder all additional field surveys, and AN patients are known to have an aversion to physicians – to mention only a few of the problems involved.

Next to us, and independently of our studies, the research group of Crisp, Holland and coworkers in London is the only other one in Europe who are carrying out substantial twin research in AN patients, to our delight with almost the exact results as ours. Fichter (1985) is the only author who has compiled (along with some personal empirical observations) an extensive collection of case studies which are, insofar as possible, methodologically correct, and interpreted them, albeit with reserve.

Present Methods

In the 1960s in collaboration with A. Heigl-Evers we extracted and examined the twins from two extensive groups of patients with psychogenic disease who had been treated between 1950 and 1970 with respect to the question of heredity and environment. The subjects were taken from an outpatient psychotherapeutic institution, the Institute for Psychogenic Diseases in West Berlin (n = 26799), and from another well-known, large inpatient psychotherapeutic facility, the State Hospital of Lower Saxony in Tiefenbrunn bei Göttingen (n = 25202 psychotherapeutically treated patients). (Monographic summaries and earlier publications cited therein can be found in Schepank 1974 and Heigl-Evers and Schepank 1980/1981.) Among the 109 pairs of twins there were 3 with AN: 1 pair of monozygotic twins concordant and 2 dizygotic twin pairs discordant for this condition. Despite the fact that this distribution obviously fits the hereditary hypothesis (for case study, see Schepank 1974, pp 111–112), we did not venture any further-reaching conclusions at first evaluation in 1969/1970. It was not until 1971 that I began to systematically study AN twins when I came across another pair of monozygotic twins concordant for AN who were undergoing psychotherapy as inpatients in the Psychosomatic Clinic of the University of Heidelberg. I then began to inquire nationwide and consistently of colleagues, especially those researchers who had studied the treated larger numbers of AN patients, whether there were twins in their samples. In this endeavor I adhered strictly to the rule of examining *all* twins whose names were given, irregardless of whether they were monozygotic or dizygotic and did not choose any of the subjects on the basis of reported concordance or discordance in the manifestation of a particular clinical feature – two fundamental essentials of sampling if scientifically sound results are to be achieved! After initial disappointment, I became acquainted with AN twins sporadically over and over during the course of the next 2 decades whom I systematically invited for an appointment or visited. I learned of other patients in my responsibility as expert (for the assessment of applications for payment for psychotherapy by the public health insurance companies). Also, an AN patient who was a twin brought me into contact with yet another pair of twins she knew from her eating disorders self-help group. Data from cooperating colleagues I know well personally for four more sets of twins are also included in the statistics given below (one set each in publications by A. Dührssen 1950 and J.E. Meyer 1961 and another set each through personal communication by J.E. Meyer and H. Stierlin); it was possible in these instances to make specific further inquiries of the examiner or therapist. Here, too, I was very careful that no preselection was made according to concordance or discordance, i.e., in the direction of hereditary or environmental factors; I was motivated only by the possibility of being able to make detailed inquiries and of clarifying unknown findings (especially as regards the determination of zygosity) so that the most reliable results possible could be reached.

Results

The results are presented in the form of four-cell tables as seen in Table 1. If the numbers in cells a and d are significantly higher than those in cells b and c, i.e., if there is a significantly higher rate of concordance (X) of the monozygotic twins than of the dizygotic twins (Y), then this indicates a (more or less high) role of hereditary factors in the manifestation of the

Table 1. Example of four-cell table

	Concordant pairs	Discordant pairs	Concordance rate (%)
Monozygotic pairs	a	b	X
Dizygotic pairs	c	d	Y

n = Number of pairs of twins.

Table 2. Core group of 14 pairs of AN twins

	Concordant pairs	Discordant pairs	Concordance rate (%)
Monozygotic pairs	4	3	57
Dizygotic pairs	0	7	0

n = 14 pairs of twins; Fisher's Exact Test, $p < 0.035$.

Table 3. Total group of 25 pairs of AN twins

	Concordant pairs	Discordant pairs	Concordance rate (%)
Monozygotic pairs	8	6	57.1
Dizygotic pairs	1	10	9.1

n = 25 pairs of twins.[4]

[4] The method of calculation for such a constellation has not been totally agreed upon amomg geneticists since it is a seldom occurrence. After discussions and advising from Prof. P. Propping I decided to proceed as follows, whereby the distribution in this set of triplets spoke against the hereditary hypothesis: The particular triplets were dizygotic females. The disease had become manifest in only one of the monozygotic triplets (hence monozygotic discordance) and in the dizygotic girl, which was included once as dizygotic concordance in the data and again as dizygotic discordance. Thus the data was entered into the table three times, whereby two entries disproved and one entry supported the hereditary hypothesis.

condition. In our sample, except for four pairs of adolescent twins between the ages of 15 and 17 years all had passed the usual age of risk.

The numbers in Tables 2 and 3 differ; the data in Table 3 reflect an increase in the uncertainty especially of zygosity in the twins, because zygosity could not be determined. My experience with twins and being present during the genetic examinations of 50 pairs of twins which were performed in the 1960s by P.E. Becker, however, do provide some guarantee of reliability to my assessment of zygosity in cases where genetic determination (serologic examination, polysymptomatic similarity comparison, finger and hand printing, etc.) was not possible: sometimes, for example, each twin had to be examined separately. Essentially, AN was diagnosed according to the Feighner criteria. In some of the simpler cases or if diagnosis of AN could only be reconstructed from the patient history, I had to rely on my 30 years of clinical experience with patients with psychogenic disease – and among them a good percentage of AN patients.

In Table 2 the results are presented for the core group of patients in whom genetic determination of zygosity and AN had been reliably diagnosed. There were four instances of concordance and three instances of discordance in monozygotic twins, i.e., a concordance rate of 57%, compared to seven instances of discordance in dizygotic twins, i.e., a concordance rate of 0%. The results are significant at a rate of 0.035%.[2]

The results presented in Table 3 include data from pairs of twins who could not be completely examined (some had died), data from other authors, and also data from triplets. Here the concordance rate for the manifestation of AN is 57.1% in monozygotic twins and 9.1% in dizygotic twins. Three values were entered for the dizygotic triplets.[3]

Two other pairs of AN twins that I know of (nos. 23 X1 and 24 VVss in Table 4) are probably dizygotic twins discordant for the condition. This data has not been included in Tables 2 and 3, though, because of the uncertainty of the findings. Two other pairs of twins come from W. Herzog's sample. In one pair the zygosity is uncertain and they presented with a very mild form of disease; in the other pair one of the sisters had died at an early age and so no exact information is available.

Table 4 provides an overview of the necessary information, including possible inconclusive diagnoses.

We were not able to obtain new information as to correlation of AN manifestation with order of birth or psychosocial environmental influences in early childhood, or as to etiopathogenesis with respect to the event triggering the symptoms in this series of patients.

Discussion

As far as the literature on AN twins is concerned, the existing individual case reports on, almost exclusively monozygotic, twins with the condition are practically worthless since the selection is biased and there was no comparison made with dizygotic twins. An evaluation of such studies, if the results are inappropriately and uncritically summarized from collected

---

[2] Because there were less than five in the cell, data were calculated using Fisher's Exact Test and one-tailed analysis. There is no option in such questions of genetics for two-tailed analysis and so this would be nonsensical.

[3] In fact, 22 sets of twins and 1 set of triplets, yielding three data entries according to pairs, are included.

Table 4. Overview of sample

Table, cell	Pair no	Age[a] (years)	Sex[b]	Zygosity	Reliability of zygosity determination[c]	Concordant/discordant for AN	Reliability of AN diagnosis in other twin[d]	Degree of severity and course in index twin	Degree of severity and course in other twin	Comments and reference
2,a	1 Bmm	43	♀♀	M	+++	Concordant	+++	Severe, died at age 43	Severe, died at age 43	Course observed for over 20 years. Both died within 6 months of the other of AN. Index twin suffered from temporary exogenous psychosis (Schepank 1974, p. 111 ff; 1980/1981, p. 709)
2,a	2 HMgI	19	♀♀	M	+++	Concordant	+++	Severe, died at age 19	Severe (35 kg at height 1.71 m)	Index patient died of AN (Schepank 1980/1981, p. 709 ff)
2,a	3 AMrr	20	♀♀	M	+++	Concordant	++	Remission	Remission of AN. Presently suffering from compulsion neurosis	Mild course in both twins (Schepank 1980/1981, p. 710). Fichter (1985 p. 206 ff) claims diagnostic criteria for AN not completely certain
2,a	4 BBpkd	34	♀♀	M	+++	Concordant	+++	Severe, chronic	Remission	Brother commited suicide after becoming Schizophrenic. Questionable anorectic phase during psychotic episode
2,b	5 BKkk	28	♂♂	M	+++	Discordant	Definitely no AN	Moderate, but definite (42 kg at height 1.76 m). Remission after analysis and professional success	No AN	Information after 28 years, benign course

Table 4 (*continued*)

Table, cell	Pair no	Age[a] (years)	Sex[b]	Zygosity	Reliability of zygosity determination[c]	Concordant/ discordant for AN	Reliability of AN diagnosis in other twin[d]	Degree of severity and course in index twin	Degree of severity and course in other twin	Comments and reference
2,b	6 HHbh	47	F,F	M	+++	Discordant	Definitely no AN	Mild	No AN	Recovered after Frahm therapy
2,b	7 FFgII	17	♀♀	M	+++	Discordant	Definitely no AN	Mild, remission	Lost 3 kg, but stopped	Mild course in adolescents (Schepank 1980/1981, p. 710 ff)
2,c	–	–	–	–	–	–	–	–	–	–
2,d	8 BBnn	57	♀♀	D	+++	Discordant	Definitely no AN	Moderate, chronic relapse	No AN	Index twin professionally successful and stable. Marriage late in life. Suicide at the age of 59 after death of spouse. Drug abuse and alcohol problem. Other twin single and professionally stable (Schepank 1974 p. 205, no. 31; 1980/1981, p. 709)
2,d	9 MMdd	24	♀♂	P	+++	Discordant	Definitely no AN	Moderate	No AN	(Schepank 1980/1981, p. 710)
2,d	10 HHbb	33	♀♂	P	+++	Discordant	Definitely no AN	Favorable course	No AN	Residuum: single and hyposexual
2,d	11 HHkk	26	♀♀	D	++	Discordant	Probably no AN	Severe, remission	No AN	Severe depression in one sister
2,d	12 BBgg	17	♀♀	D	++	Discordant	To date no AN	Mild (but definite), remission	No AN	Mild course, spontaneous remission
2,d	13 BBrr	20	F,F	D	+++	Discordant	Definitely no AN	Mild (but definite)	No AN	Mild course

2,d	14 GGnn	22	♀♀	D	+++	Discordant	Definitely no AN	Severe	No AN	From parallel series of twins Tiefenbrunn (Heigl-Evers 1980/1981, p. 712; Schepank 1980/1981, p. 764)
3,a	15 BBdd	15	♀♀	M	++	Concordant	+++	Severe	Severe	Further course unknown (Dührssen 1950; Schepank 1980/1981, p. 712 ff)
3,a	16 MÜme	27	♀♀	M	++	Concordant	+++	Severe, remission	Severe, development into Schizophrenic psychosis	(J.E. Meyo 1961)
3,a	17 ERbr	34	♀♀	M	+	Concordant	+	Moderate, course unknown	Moderate, course unknown	Correspondence with therapist. Comprehensive description as part of report of psychotherapy (Schepank 1980/1981, p. 713 ff)
3,a	18 MKss	22	♀♀	M	+++	Concordant	+++	Severe (36 kg at height of 1.71 m). At age 16 development into paranoid psychosis; temporary catatonic schizophrenia	At age 14 definite AN (36 kg at height of 1.61 m), amenorrhea	Psychosis began at age 16 in index twin, well after onset of AN. The former psychotic index patient claimed she was a fraternal twin although genetic determination of monozygosity certain. See also Fichter (1985 p. 242 ff), case no. 3
3,b	19 HDss	35	♀♀	M	+	Discordant	Probably no AN	Severe (32 kg), benign course	No AN	Competent somatic diagnosis at psychosomatic clinic. At follow-up after 14 years patient had recovered spontaneously (W. Herzog's Sample)

Table 4 (*continued*)

Table, cell	Pair no	Age[a] (years)	Sex[b]	Zygosity	Reliability of zygosity determination[c]	Concordant/ discordant for AN	Reliability of AN diagnosis in other twin[d]	Degree of severity and course in index twin	Degree of severity and course in other twin	Comments and reference
3,b	20 EEbIII	19	♀♀	M	?	Discordant (?)	Probably no AN	Severe	Sister stopped dieting after losing 5 kg	Records were read but patient did not come to examination
3,b	Triplets HHll	25	♀♀♀	M	+	Discordant	Probably no AN	Marked	No AN	See footnote 4 to table 3
3,c	Triplets HHll	25	♀♀♀	D	+	Concordant	++	Moderate (?)	Moderate (?)	See footnote 4 to table 3
3,d	Triplets HHll	25	♀♀♀	D	+	Discordant	++	Moderate	No AN	See footnote 4 table 3
3,d	21 MHss	34	♀♀	D	+	Discordant	No indication of AN	Moderate	No AN	(Schepank 1980/1981, p. 712)
3,d	22 HDs	17	♀♀	D	+	Discordant	Probably no AN	Moderate	No AN	Report from colleague. At 1.59 m height weight reduced to 44 kg for past 4 years
3,d	23 MXl	Mid-20s	♀♀	D	+	Discordant	++	No AN	Moderate	Healthy individual, professonally competent to report AN, indicated definite diagnosis of AN in her fraternal twin sister. Twin could have been added to cell d, discordant dizygote. I did not meet both personally

3,d	24 VVss	25	♀♀	D (?) M (?)	???	Discordant	No diagnosis	Severe (35.5 kg at height of 1.66 m), amenorrhea, remission	Dieted at beginning, but stopped	Comprehensive, reliable reports from hospitalization and out-patient treatment in index patient with definite diagnosis. Also plausible reports of healthy sister. Father and twins claim definite dizygosity so they could be included in cell d. However, the twins are confused so often that I do not consider the zygosity determination (from physician who attended birth or public health officer) reliable enough, even though it fits the hereditary hypothesis

M, monozygotic; D, dizygotic; P, ♂-♀ pair; AN, anorexia nervosa.
[a] Age indicated is the one given at last examination or catamnesis information.
[b] The sex of the index twin is given first.
[c] +++, Definite Zygosity determination by competent human geneticist; ++, assessed by me to be definite by inspection, thorough history, confusions, status at birth, pictures, description, etc.; +, according to description probable; ?, questionable but definitely stated; ??, so contradictory or unclear so as to be not evaluable.
[d] +++, Definite, reliable diagnosis; ++, highly probable diagnosis, however, no substantiated findings; +, probable, based on anamnesis and data.

cases and offered in support of or as (apparent!) proof of the environmental hypothesis, may even be wrong.

Only very few extensive, methodologically sound twin studies have been done which inclued both monozygotic and dizygotic twins in whom zygosity had been determined and AN diagnosed. These (especially the 1985 study of Crisp et al., the overview by Fichter in 1985, and the results presented in this paper) all definitely indicate a great discrepancy between the relatively high rate of concordance in monozygotic pairs of twins (in each about 50%) and a very low rate of concordance in dizygotic twins (about 3%–9%). This suggests a strong genetic component. These rates are higher than those for psychogenic disease in general (37.7% in monozygotic pairs and 16.7% in dizygotic pairs; Schepank 1974).

Even so, the still obvious discordance in many cases of monozygosity of course justifies psychotherapeutic intervention and effort and should not lead to feelings of resignation. Therapeutic nihilism should not be the consequence of research. (Mental deficiency in phenylketonuria is called to mind here, a condition which is definitely genetically determined and which can be prevented by dietary intervention at a certain time in life!). Twin research results can, though, console clinicians and therapists when again and again they are faced with their own limitations in efforts to treat this dramatic and upsetting illness if they view their success self-critically and for a long enough time.

Supplementary Information

1. Among the women in this series of patients there is a high rate of mortality. After long-term follow-up, 4 of the 36 total patients with definite AN (index twins as well as concordant twin siblings) had died by the age of 41. Another twin had committed suicide at the age of 58 (no. 8 BBmn). Other twins who were dead could not be included in our results because of uncertain diagnosis of AN or unknown zygosity.
2. It also seems important to recall the often forgotten relationship between AN and the psychoses, a connection which L. Binswanger had already made in 1944 (Case study Ellen West) and J. Zutt in 1948. In the long term, AN develops into a real psychosis more often than does schizophrenia with its life-long risk in the general population of less than 1% (Häfner . . . 1985). (See also twins nos. 1, 16, and 18.)
3. We know of no twins samples that use cases of bulimia as an index (Feiereis in 1989 reported on a monozygotic pair with bulimia, p. 88.) Our sample of AN twins indicates a hereditary connection between AN and bulimia only. Cases of bulimia with no emaciation are, then, only a special form of the disease entity with a primarily psychic disturbance of wanting to lose weight.

Acknowledgements. I am very grateful to those colleagues all over the country who provided me with data from twins or made it possible for me to examine the twins in person, especially Dr. Bodenstein, Berlin; Dr. Broser, Erlangen; Prof. Dührssen, Berlin; Prof. Engel, Bochum; Dr. Fahrig, Heidelberg; Prof. Heigl-Evers, Düsseldorf; Dr. W. Herzog, Heidelberg; Dr. Mais, Berlin; Prof. J.-E. Meyer, Göttingen; Dr. Schneller-Reindell, Weinheim; Prof. Stierlin, Heidelberg; and Dr. Wunderlich, Berlin, I am also indebted to Dr. Steffens of the Human Genetics Institute of the University of Heidelberg who carried out the necessary examinations and assisted me in zygosity determination of the twins. The genetic zygosity determinations in the research project carried out in 1960s were performed mainly by Prof. P.E. Becker, the geneticist in the research group in Göttingen. The Institute for Documentation and Statistics of the German Cancer Research Center, Prof. G. Wagner, was responsible for statistical evaluation and methodological advising. Our special thanks go to these two emeritus professors and sponsors of our works as well as to Prof. A. Dührssen, with whom we have worked since 1960, and who in conjunction with Prof. W. Schwidder initiated the twins project.

References

Askevold F, Heiberg, A (1979) Anorexia nervosa – two cases in discordant MZ twins. Psychother Psychosom 32:223–228

Binswanger L (1944/45) Der Fall Ellen West. Eine anthropologisch-klinische Studie. Schweiz Arch Neur 63:255; 64, 69 u. 330; 65:16

Bruch H (1969) The insignificant difference: discordant incidence of anorexia nervosa in monozygotic twins. Am J Psychiatry 126:85–90

Camerer JW, Schleicher R (1936) Beitrag zur Frage der konstitutionellen Fett- und Magersucht an Hand von Beobachtungen an eineiigen Zwillingen. Z Menschl Vererb Konstit Lehre 19:32–39

Crisp AH (1966) Anorexia nervosa in an identical twin. Postgrad Med J 42:86–92

Crisp AH, Hall A, Holland AJ (1984) Nature and nurture in anorexia nervosa: a study of 34 pairs of twins, one pair of triplets, and an adoptive family. Br J Psychiatry 145:414–419

Dührssen A (1950) Zum Problem der psychogenen Ess-Störung. Psyche 4:56–72

Feiereis H (1989) Diagnostik und Therapie der Magersucht und Bulimie. Marseille-Verlag, Munich

Fichter MM (1985) Magersucht und Bulimie. Springer, Berlin Heidelberg New York

Gifford S, Pilot ML (1969) Anorexia nervosa in one of identical twins (zit. nach H Bruch)

Häfner H (1985) Sind psychische Krankheiten häufiger geworden? Nervenarzt 56:120–133

Halmi K, Brodland G (1973) Monozygotic twins concordant and discordant for anorexia nervosa. Psychol Med 3:521–524

Hill OW (1976) Anorexia nervosa. In: Hill OW (ed) Trends in psychosomatic medicine. Butterworth, London

Hofmann F, Kind H (1980) Ein eineiiges Zwillingspaar diskordant für Anorexia nervosa. Praxis Kinderpsychol Kinderpsychatr 29:293–298
Klotz HP, Balier C, Javal J (1961) Un cas d'anorexie mentale chez une jumelle homozygote. (Etude psychosomatique d'un couple). Rev Med Psychosom 3:37–44
Köhle K, Simons C (1979) Anorexia nervosa. In: Uexküll T von (ed) Lehrbuch der Psychosomatischen Medizin. Urban and Schwarzenberg, Munich
Meermann R (ed) (1981) Anorexia nervosa. Enke, Stuttgart
Mester H (1981) Die Anorexia nervosa. Springer, Berlin Heidelberg New York (Monographien aus dem Gesamtgebiete der Psychiatrie, vol 26)
Meyer JE (1961) Das Syndrom der Anorexia nervosa. Arch Psychiat Nevenkr 202:31–59
Nowlin NS (1983) Anorexia nervosa in twins: case report and review. J Clin Psychiatry 44.3:101–105
Propping P (1989) Psychiatrische Genetik. Springer, Berlin Heidelberg New York
Schepank H (1974) Erb- und Umweltfaktoren bei Neurosen. Springer, Berlin Heidelberg (Monographien aus dem Gesamtgebiete der Psychiatrie, vol 11)
Schepank H (1980, 1981) Anorexia nervosa. In: Heigl-Evers A, Schepank H (ed) Ursprünge seelisch bedingter Krankheiten, vol 2. Vandenhoeck and Ruprecht, Göttingen, pp 705–719
Schepank H (1983) Anorexia nervosa in twins: is the etiology psychotic or psychogenic? In: Krakowski AJ, Kimball CP (eds) Psychosomatic medicine. Plenum, New York
Vandereycken W, Pierloot R (1981) Anorexia nervosa in twins. Psychother Psychosom 35:55–63
Yamaoka Y, Kusakabe Y, Kacasaw H (1979) Case of anorexia nerovsa observed in one of monozygotic twins (in Japanese). Clinical Psychiatry 21:659–665
Zerbin-Rüdin E (1985) Vererbung und Umwelt bei der Entstehung psychischer Störungen, 2nd edn. Wissenschaftliche Buchgesellschaft, Darmstadt
Ziolko HU (1966) Aotrexia nervosa. Fortschr Neurol Psychiatr 34:353–396
Zutt J (1948) Das psychiatrische Krankheitsbild der Pubertäts-Magersucht. Arch Psychiatr Z Neurol 180:776–849

Menstrual Disorder and Other Hormonal Disturbances

P.J.V. Beumont

Introduction

Endocrine dysfunction is intrinsic to the concept of anorexia nervosa. A disturbance of menstruation, which is its most obvious manifestation, was noted by Gull (1873) in his classical description of the illness. It is cited in all the major definitions of anorexia nervosa, often as a necessary criterion (Table 1), and similar references are proposed for both ICD-10 and DSM-IV. Indeed it is true to say that the occurrence of regular menstrual periods would cause most clinicians to doubt the diagnosis.

Although hormonal disturbance is not as prominent in the newly described syndrome of bulimia nervosa, and although there is no reference to it in current definitions of this latter condition (Russell 1979, 1983a; American Psychiatric Association 1980, 1987; Fairburn and Garner 1986), nevertheless it is an important feature of this illness as well, albeit in a more subtle form. Once again, a disorder of menstrual function is the abnormality which has received most attention. Thus, Russell (1979) found 11 patients with amenorrhoea among the 28 women who were the subjects of his seminal 1979 report. Fairburn and Cooper (1984), using DSM-III criteria to diagnose 35 bulimic patients, reported that 46% were on oral contraceptives (and hence of uncertain menstrual status), 21% amenorrhoeic, 37% had very irregular cycles and only one patient appeared to have a normal menstrual pattern. Of the patients decribed by Pirke et al. (1985) 50% reported amenorrhoea as did more than 20% of those of Leon et al. (1986).

Furthermore, bulimic patients' self-reports of their menstrual status is likely to underestimate the prevalence of disordered hormonal functioning. Pirke (1990) studied 15 bulimic women of whom 4 claimed to have normal menses, 7 oligomenorrhoea and only 3 amenorrhoea. He reported that over the period of an average cycle, plasma oestradiol remained consistently under 120 pg/ml in 7 patients, indicating that follicular development had not occurred. In a further 7 patients, the luteal phase rise of progesterone was either absent or deficient despite an acceptable oestradiol peak. In only one patient was the menstrual cycle truly normal.

Table 1. References to menstrual disorder in various definitions of anorexia nervosa

Study	Definition
Dally (1969)	Amenorrhoea of at least 3 months duration when menstruation has previously been regular – or if menstruation had been irregular with gaps of 2 or 3 months, a period of amenorrhoea of at least 6 months
Russell (1970)	Evidence of an endocrine disorder – amenorrhoea in the female and loss of sexual interest and potency in the male
Feighner et al. (1972)	At least two of the following manifestations: amenorrhoea, lanugo hair and bradycardia
Garrow et al. (1975)	A secondary endocrine disorder of the hypothalamic-anterior pituitary-gonadal axis, manifest in the female as amenorrhoea and in the male by diminution of sexual interest and activity
ICD-9 (World Health Organization 1978)	Amenorrhoea is usual
DSM-III-R (American Psychiatric Association 1987)	In females, absence of at least three consecutive menstrual cycles when otherwise expected to occur (primary of secondary amenorrhoea). (A woman is considered to have amenorrhoea if her periods occur only following hormone, e.g. oestrogen administration)

Menstrual Status of Patients at Follow-up

Because of their importance in the clinical presentation, most authors who have undertaken the follow-up of eating disorder patients have been at pains to include reference to endocrine factors, at least as they are reflected in menstrual function, in their assessment of outcome. Theander (1985) comments: "The most important criteria for a favourable outcome in an eating disorder are *spontaneous and regular menstruation*, stable body weight within 15% of average . . . normal eating habits . . . and a reasonably normal conception of the body" [italics added]. Morgan and Russell (1975), whose criteria have received wide acceptance, recognize three categories of outcome: good with body weight maintained within 15% of average for height, age and sex and with *regular cyclical menstruation*; intermediate for patients whose weight has risen only intermittently to within 15% of average and/or in whom there is *continuing menstrual disturbance*; and poor for those whose body weight has not approached minus 15% of average and whose *menstruation has been absent or sporadic* [italics added].

Menstrual function is thus a key aspect of outcome in most methodologically sound follow-up studies of anorexia patients. However, it is not always possible to extract precise information about menstrual status from the published reports.

Table 2. Menstrual status of anorexia nervosa patients at follow-up

Study	Duration of study	Number of female patients	Surviving patients	
			Regular menses (%)	Amenorrhoeic (%)
Theander (1970)	8 years	94	66	13
Morgan and Russell (1975)	4–10 years	38	47	39
Hsu et al. (1979)	4–8 years	102	51	28
Morgan et al. (1983)	4–8 years	73	58	19
Hall et al. (1984)	4–12 years	49	50	32

Data from a number of key studies from which such data could be obtained is given in Table 2. It is apparent that about 50% of patients had resumed regular menstruation while about 25% remained amenorrhoeic.

Unfortunately some of the published reports of excellent long-term studies which have been undertaken in recent years do not refer directly to menstrual functioning, although the date must be available (Isager et al. 1985; Theander 1985). However, Ratnasuriya et al. (1991), in a 20-year follow-up of the patients originally described by Morgan and Russell in 1975, did refer to the current menstrual functioning of 31 patients. Nineteen reported regular periods, three irregular cycles and nine had been amenorrhoeic since the time of their admission.

Because body weight and menstrual status are presented as separate aspects of outcome in the influential study of Morgan and Russell (1975), the impression may be gained that they are somehow independent of each other. Of course, this is not really true. Thus Ratnasuriya and his colleagues (1990) commented that there was a clear relationship between body weight and menstrual function at 20-year follow-up in their patients. All patients whose weights were below 75% of average were amenorrhoeic, while all but one whose weights were above this level were menstruating. Similarly, in a much shorter term study, Halmi and Falk (1983) reported on 40 female anorexia nervosa patients 1 year after completion of treatment. The 15 patients who were menstruating were much closer to their ideal body weights than were the 25 patients who had remained amenorrhoeic. The latter were also noted to have maintained typically anorexic attitudes and to persist in typically anorexic behaviours.

In respect to all such studies of menstrual function at follow-up, two issues warrant special consideration. First, it is likely that the percentage of patients whose menstruation has resumed is considerably higher than the proportions whose ovulatory functioning has actually been restored to normal. Just as in bulimia nervosa (see above, Pirke 1990), luteal phase insufficiency is common in anorexia nervosa patients even when their menstrual periods have resumed (Beumont et al. 1973). The return of menstrual bleeds does not necessarily imply the restoration of normal hormonal functioning. The matter will be referred to again in a later section of this chapter.

The second issue which now needs to be addressed is this. Because endocrine dysfunction, particularly menstrual disturbance, is so intrinsic a part of the clinical presentations of anorexia and bulimia nervosa, is it meaningful to consider its outcomes separately from the outcome of the illness themselves? In order to discuss the question, it is necessary to review briefly the pathogenesis of hormonal abnormalities in these conditions.

The Pathogenesis of Endocrine Dysfunctions in the Dieting Disorders

Despite previous views to the contrary, there can now be little doubt that the endocrine symptoms of anorexia and bulimia nervosa are epiphenomena. They are not the cause of illness but its effects (Beumont 1984a). Nevertheless their presentation is strikingly stereotyped and they constitute a characteristic syndrome which is perhaps best understood as a series of adaptations to an inadequate energy intake. Perhaps it would be appropriate to term anorexia and bulimia nervosa dieting disorders rather than eating disorders. Both are characterized by behaviours directed at rigidly controlling or losing body weight. Always there is at least some element of restriction, e.g. dieting, food refusal and strenuous exercise. Often there are also purging behaviours such as self-induced vomiting and laxative abuse. It is important to stress this point in the present context because the weight losing behaviours as well as the undernutrition which results are the direct causes of the endocrine dysfunctions associated with these illnesses.

The evidence for this assertion in respect to anorexia nervosa is extensive (see reviews by Beumont and Russell 1982, and Russell and Beumont 1987). It is perhaps paradoxical that so much of our current knowledge of the endocrinology of undernutrition derives from the study of anorexic patients. Unlike the victims of famine, starvation or protein calorie malnutrition, they are easily accessible to research workers with sophisticated laboratory facilities.

As the menstrual disturbance is the best known of the endocrine complications of anorexia nervosa, it has received the most attention. The disorder is basically one of anovulation. Undoubtedly undernutrition appears to be the major factor in its causation.

At low body weights, circulating values of luteinizing hormone (LH) are depressed. The depression of LH levels relates to the severity of emaciation and, as weight is regained during refeeding, circulating LH is restored to normal tonic values. The critical weight for the restoration of normal levels is about 80% of standard body weight, i.e. the average weight of individuals of the same age, sex, and height.

The cause of the low LH values is not to be found in the pituitary. If the gland is stimulated by a course of several days' administration of LH releasing hormone (LHRH), normal values of LH are restored without any necessary weight gain. This finding suggests that the primary fault may be inadequate stimulation of the pituitary gonadotrophins by LHRH, and hence that the site of disturbance may be the hypothalamic level.

The dynamics of the dysfunction are illustrated by the use of various provocative tests of the hypothalamic-pituitary-ovarian axis. For instance, the inhibiting effect of ethinyl oestradiol on LH can not be discerned at very low body weights. Clomiphene, an antioestrogen which blocks the inhibiting feedback of endogenous oestrogen, causes by this means a rise in LH values. In emaciated anorexia nervosa patients the effect is minimal or nonexistent. Further, the stimulatory effect of a bolus or short-term infusion of exogenous LHRH is markedly depressed in these subjects.

The simplest explanation of these findings is that the pituitary gonadotrophs undergo disuse atrophy as a result of inadequate LHRH stimulation during emaciation.

As the patient is refed and regains weight, normal tonic levels of LH are restored. Ethynyl oestradiol once more exerts its inhibitory effect, clomiphene produces a short-term rise in LH, and LHRH may even evoke an exaggerated LH response reminiscent of that seen during puberty.

Plasma follicle-stimulating hormone (FSH) levels tend to follow a similar pattern to those of LH. They are often depressed in patients at very low weight, but values are restored to normal by increases in weight smaller than those associated with the restoration of LH levels. Similarly, the FSH response to LHRH becomes exaggerated at an earlier phase than the LH response, a phenomenon which accounts for the finding of an "inverted" gonadotrophin response – high FSH, low LH – that has been reported in some studies of anorexia nervosa patients.

However, the association between undernutrition and decreased circulating levels of gonadotrophins is not sufficient to account for all aspects of the menstrual disorder of anorexia nervosa. Other factors must be involved to explain at least some of the observed abnormalities such as the early onset of menstrual disturbance. Amenorrhoea often antedates significant weight loss, sometimes by many months. Rather than so-called psychogenic influences, the importance of which remains unproven, it is more likely that it is the weight losing behaviours themselves, such as variations in the composition of the diet and increased physical activity and exercise, that are responsible. Presumably similar factors are involved in the menstrual disturbance in bulimia nervosa patients.

The relation of the nonreproduction endocrine abnormalities to the severity of emaciation is less obvious. In most instances, they appear to be a direct response to the patient's inadequate energy intake. For instance, basal growth hormone (GH) levels are elevated in about 50% of patients. Most authors have been unable to demonstrate an association between GH levels and the extent of emaciation. Rather, GH appears to be raised in patients whose current diet is deficient in calories, and the abnormality is corrected promptly by the institution of a high calorie diet. The inverse relationship between GH and certain amino acids, particularly alanine, which has been described in protein depleted patients with kwashiokor, does not appear a significant factor in anorexia nervosa. Rather the elevated GH is probably an adaptation to the low energy intake, and it is necessary to combat the accompanying hypoglycemia by mobilizing fat tissue.

A euthyroid sick syndrome with low T3 levels is also usually found. It, too, may be a compensatory mechanism helping to conserve energy in the presence of undernutrition. It is one of the factors contributing to the patients' hypothermia.

Sheehan and Summers in 1948 pointed out that one of the distinguishing factors between anorexia nervosa and hypopituitarism was that adrenal function was preserved in the former. In fact, circulating levels of cortisol are elevated. One reason is that the half-life of cortisol is prolonged

Table 3. Some endocrine abnormalities found in anorexia nervosa, bulimia nervosa and experimental starvation. (From Fichter and Pirke 1990)

Hormonal disturbance	Anorexia nervosa literature	Bulimia nervosa literature	Munich fasting experiments
Gonadotrophins			
Decreased plasma levels	+	+	+
Reduced response to LHRH	+	?	+
Regression of 24-h LH secretory pattern	+	+	+
Prolactin			
Decreased nocturnal secretion	+	+	+
Reduced response to TRF	+	+	+
Adrenal			
Increased plasma cortisol	+	−	+
Abnormal 24-h cortisol secretion	+	+	?
Impaired suppression by dexamethasone	+	+	+
Reduced catabolism of cortisol	+	−	+
Increased production of cortisol	+	−	+
Growth hormone			
Increased tonic plasma levels	+	+	+
Thyroid			
Decreased T3	+	+	+
Reduced response to TRF	+	+	+

LHRH, luteinizing hormone releasing hormone; LH, luteinizing hormone; TRF, thyrotropin releasing factor.

and the tetrahydrocortisol to cortisol ratio increased, presumably due to reduced activity of the relevant enzyme. Another is that there is an increased hypothalamic drive on cortisol production so that the respose to insulin-induced hypoglycemia is deficient, the diurnal variation of plasma cortisol is eliminated and there is early escape from dexamethasone suppression. The latter is quickly reversed as soon as the patient regains weight.

In recent years there has been an increased recognition of the endocrine abnormalities associated with bulimia nervosa. Not only are the observed abnormalities similar to those found in anorexia patients, but they can be mimicked by fasting experiments in normal subjects. This has been an area of particular interest to a group of Munich workers (see Pirke et al. 1985; Fichter et al. 1988; Schweiger et al. 1988; Fichter and Pirke 1990) whose findings are summarized in Table 3.

There is a further aspect of the endocrine changes that needs to be considered. It has been proposed that they contribute to a vicious circle effect whereby the behavioural disorder is perpetuated (Ploog and Pirke 1987). This suggests that they may be similar in nature although different in form to the kinds of changes which accompany other spontaneous hypophagias, such as those with rutting, hibernation and incubation in lower animals (Beumont 1984b).

The Prognosis of the Hormonal Disturbance

What then is the long-term prognosis of these endocrine abnormalities? To a major extent, it depends on whether or not the behavioural disturbance has been corrected. As long as anorexic behaviours such as food restriction and purgation continue, as long as a state of continuous or intermittent semi-starvation persists, endocrine abnormalities will occur as a direct result of the illness. Once normal eating and nutrition are restored, they may be expected to abate. Thus, the "prognosis" of the endocrine syndrome is simply the "prognosis" of the illness itself. Of course, this interpretation is too simplistic as it confuses outcome with persistence of illness. To put the question in a more meaningful way, it is necessary to define precisely what is meant by prognosis or outcome.

Theander (1985) states: "The outcome of a disease may be defined as the long-term result of a pathological process. At its final stage, some patients are found to have *recovered completely*. In others the disease process has come to an end but has resulted in *various defects* which hamper bodily or mental functioning. In some patients the disease is *still active* in a severe or less severe form. Some patients may have *died* from the disease or from health problems casually connected with the disease. The disease may also *predispose to other illness*, those occurring more frequently among patients who survives" [italics added]. This definition is useful in the present context as it leads to a number of questions which are relevant to important clinical issues.

1. In those patients who appear to have recovered completely, has endocrine function been fully restored to what it had been prior to illness? Or are there patients in whom there is full remission of all psychological and behavioural abnormalities, but a persistence of endocrine dysfunction? If so, what is the nature of this defect? If recovered patients later relapse, does endocrine dysfunction recur, i.e. do the hormonal features "breed true?"
2. Is the picture of endocrine dysfunction different in patients who are chronically ill as opposed to those in the acute phase of illness? Are these endocrine abnormalities dependent on whether the chronic form of illness is of a severe or less severe type? Also, can the endocrine dysfunction remit despite the presence of a chronic form of the illness?
3. Do endocrine factors contribute to the high mortality rate that has been reported in anorexia nervosa patients? Do similar factors pose a risk to life for bulimia nervosa patients?
4. What other illnesses arise as complications of dieting disorders? Are endocrine factors important in their pathogenesis? As a corollary, to what extent does the endocrine dysfunction contribute to the psychological disturbance and to the suffering and discomfort caused by the illness?

None of these questions have been addressed directly in the various follow-up studies that have been reported to date. At best, some tentative answers can be derived from the literature.

Is Endocrine Function Fully Restored in Patients Who Have Recovered from Anorexia Nervosa?

As a general rule, the persistence of hormonal abnormalities suggests the persistence of anorectic, weight losing behaviour. However, as with all generalities, there are exceptions. Confusion is most likely to arise in the patient with anorexia who has regained weight to her premorbid level. While it is to be expected that some months may elapse before regular menstrual cycles resume, clinical experience teaches that the delay is often much longer than this. A tardy return of menses is as common as is the early onset of amenorrhoea, about which comment has already been made.

Various explanations have been proposed. Frisch (1977), on the basis of her work concerning the role of body fat in determining the onset of menarche, sugggests that the proportion of fat in body composition has not attained a critical value in those patients whose amenorrhoea persists. Lean tissue rather than fat has accounted for their weight gain. As oestrogens are metabolized in fat tissue, the net effect is a decrease in the biological activity of the sex hormones and a less effective feedback to the hypothalamus. Observations that female athletes and ballet dancers commonly show anovulation in the presence of a disproportionately low proportion of fat in body composition offer indirect support for Frisch's suggestion. More direct support would be provided by a demonstration of low body fat in a weight restored anorexia patient, but the literature in this area remains controversial (Mitchell and Truswell 1987).

Another possibility is that the patients' propensity for strenuous exercise is the responsible factor. In ballet dancers, a relationship has been demonstrated between intense physical activity and amenorrhoea. If training is interrupted in these subjects, menses return without alterations in body weight or the lean to fat ration (Abraham et al. 1982). Similarly, it has been reported that almost 50% of women with amenorrhoea associated with acute exercise and training are not excessively lean (Cumming and Rebar 1985). Exercise has been shown to have a direct effect on the pulsatile release of LH in female athletes, even in the presence of menstrual cycles (Cumming et al. 1985).

On the other hand, the delay may be due simply to a persistent insensitivity of the relevant hypothalamic pituitary mechanisms. During the normal menstrual cycle, oestrogen exerts two opposing effects on circulating gonadotrophins. Below a critical value it depresses LH levels, but once this threshold is exceeded it actually increases pituitary capacity and induces a rise in LH. Similar effects can be noted following the administration of exogenous ethinyl oestradiol and of clomiphene. In many weight restored anorexia patients, the negative feedback effects of these drugs can be evoked, but not their positive effects (Wakeling et al. 1977; Beumont et al. 1973).

Further evidence concerning the nature of this insensitivity has emerged from studies using a 4-h infusion of LHRH which normally evokes a biophasic response. This indicates that there are two pools of LH in the pituitary. The first is dependent entirely on priming by LHRH, the second by the positive feedback effects of oestrogen. In emaciated anorexia patients, LHRH infusion evokes no response. After weight restoration, the first LH peak only is evoked. LHRH stimulation has resumed and hence the first pool has been reconstituted. However, a full biphasic response is apparent only after further rehabilitation. Presumably the delay is due to defective oestrogen feedback effects (Beumont and Abraham 1981).

Another abnormality of endocrine function which persists after weight restoration in anorectic patients is that of arginine vasopressin (AVP). Plasma and cerebrospinal fluid (CSF) levels of AVP may be simply deficient or, more commonly, they may show a pattern of erratic and osmotically uncontrolled release (Ebert et al. 1984). Perhaps this abnormality is responsible for the difficulty some anorectic patients have in conserving water. Because AVP regulates brain water permeability (Raichle et al. 1980), it may also be responsible for the structural brain alterations that are frequently found on cranial computed tomography (Lauer et al. 1990).

These alterations are characterized by enlarged CSF spaces and dilated ventricles, occur in up to 80% of anorectic and 50% of bulimic patients, and often persist long after apparent remission of illness (Ploog and Pirke 1987). However, their association with AVP as yet is merely speculative. Because enlarged CSF spaces also are found in Cushing's disease, it may well be that hypercortisolism is more relevant to their occurrence (Krieg et al. 1986).

In relation to all such observations of hormonal disturbance persisting after apparent remission, the nature of the remission has been poorly documented. A crucial question is whether subtle endocrine abnormalities persist in patients who have a full recovery from all psychological and behavioural abnormalities, and not just a restoration of body weight to within the normal range and a reversal of the more obvious stigmata of undernutrition. The need to clearly ascertain that "recovered" patients are satisfactorily readjusted in the emotional and social spheres has been recognized (Schwartz and Thompson 1981), but as yet there have been no reports of endocrine assessment during apparent remission. As a minimum requirement, such patients should not only fulfill the criteria of recovery proposed by Morgan and Russell (1975), but they should also be shown to be eating an adequate diet with regular meals, not to be overactive and not to induce vomiting or purgation.

Is the Picture of Endocrine Dysfunction Different in Patients with Chronic than in Those with Acute Conditions?

I am not aware of any studies which have been directed to this question, and yet it would appear to be important. Hsu (1987) reviewed the major outcome studies of patients with anorexia and found that almost 50% of subjects were reported either to have remained severely undernourished or be in a suboptimal nutritional state (the poor and intermediate outcome categories of Morgan and Russell 1975). The proportion of bulimia nervosa patients whose illness becomes chronic is unknown, but possibly is of a similar magnitude. The management of patients with chronic anorexia and bulimia nervosa is a major concern of responsible clinicians working in this area, and a knowledge of their particular endocrine problems might well be relevant to their optimal care. Does a gradual adaptation occur to the undernutrition? The clinical observation that some patients with chronic illness resume menstruation while still at a low body weight suggests that this may be so.

Do Endocrine Factors Contribute to the Mortality Rate?

An unacceptably high proportion of patients with anorexia die prematurely. In their analysis of 12 major outcome studies over the preceding 15 years,

Schwartz and Thompson (1981) calculated a mean mortality rate of 8%. In recent longer-term studies, Theander (1985) found a mortality rate of 18% and Ratnasuriya et al. (1991) of 15% after 20 years. Isager et al. (1985) using "survival methods" of analysis, reported 9 deaths among 151 patients followed for between 4 and 22 years. In all of these studies, relatively few deaths occurred in the first few years of follow-up, most patients dying only after long periods of chronic illness.

While two thirds of the deaths in the series of Isager et al. (1985) were the result of suicide, Schwartz and Thompsons (1981) found a relatively low suicide rate (1%). By far the majority of deaths in the series they reviewed were directly related to self-starvation (6% of patients, ranging from 0% to 21.5% between different studies). In Theander's (1985) study, 5 deaths were by suicide and 12 directly as a result of the illness, while in the Maudsley series, 3 were due to suicide, 3 to the illness itself and 1 to an unrelated cause (Ratnasuriya et al. 1990). Although there is considerable variation between these figures, it is clear that the high mortality rate in anorexia nervosa is due mainly to the illness itself rather than to an increased suicide rate (see chapter Herzog et al., this volume).

The precise causes of death are not easy to retrieve from the published reports. Some are due to infection in a setting of gradually increasing disability. Ratnasuriya et al. (1991) refer to "electrolyte imbalance", and this certainly is often an important factor leading to cardiac arrhythmia. Serious disturbances of blood chemistry in anorexia patients are usually related to vomiting and purging behaviours (Hall et al. 1989), and endocrine factors are not likely to be directly relevant.

Much less is known about mortality rates in bulimic patients. Because vomiting and purgation are prominent features of this illness as well, it may be assumed that similar factors operate as in the vomiting and purging category of anorexia patients. However, Hall et al. (1989) reported fewer *serious* electrolyte disturbances in bulimic than in purging anorexia patients, and the risk of death is probably considerably lower.

Anorexia nervosa has profound effects on carbohydrate metabolism and insulin (see review by Russell and Beumont 1987) and potentially fatal hypoglycemia may occur (Ratcliffe and Bevan 1985). However, in addition to these direct effects, there is an ominous association between anorexia and bulimia nervosa and insulin-dependent diabetes mellitus (IDDM). During the 1980s several authors reported patients in whom the management of IDDM was compromised by the coexistence of an eating disorder. Such patients sometimes admit that they use their diabetes in a direct and purposeful way to regulate their weight, e.g. "to lose calories through urine", or that they do not compensate for bulimic episodes by increasing their insulin dosage (Szmukler and Russell 1983).

Rodin et al. (1985) in a systematic study of adolescent girls with IDDM, found a six fold increase in anorexia nervosa and a two fold increase in bulimia as compared to nondiabetic peers. The presence of an eating disorder was associated with an increased incidence of major medical complications and hence with an increased risk of premature death.

What Other Illnesses Arise from the Endocrine Dysfunction in Dieting Disorders?

Anorexia nervosa in a prepubertal child usually lends to a delay in growth and in sexual maturation. Early studies of young patients suggested that these delays were merely temporary. Thus Warren (1968) concluded that menarche occurred on recovery from the illness, albeit some years later than normal, and that there was no interference with growth. Similarly, Silverman (1974) reported that most young girls with anorexia nervosa progressed well physically, and that persistent morbidity was likely to be psychological rather than physical.

Lacey et al. (1979) reported a rather worse prognosis for the young anorexia patients. Skeletal maturation was delayed during the phase of undernutrition, but development resumed when adequate nutrition was restored. However, there was no phase of "catch-up" growth, and hence an overall stunting effect. This finding is consistent with the observation that patients in whom anorexia nervosa begins in early adolescence tend to be shorter in stature than their peers (Crisp 1969).

Russell (1983b) reported a series of nine girls who presented with anorexia nervosa at an unusually young age (12–14), prior to the onset of menarche. The breasts were undeveloped in six, and only three patients had reached the 50th percentile in height for their age group. In most of those who were successfully refed, normal pubertal changes occurred, but two patients required hormonal treatment in order to achieve breast growth. One patient had a persistent disturbance of menstruation despite weight restoration and only three patients grew significantly in stature after weight gain. The patients with a poorer outcome were those in whom the illness had been of longer duration, and Russell concluded that when undernutrition persists for more than 3 or 4 years, there is significant risk that subsequent physical development will be incomplete.

Osteoporosis is a common and important complication of chronic anorexia nervosa in older patients. Ayers et al. (1984) reported serious osteopenia in seven patients all of whom had been amenorrhoeic for at least 12 months. A retarded bone age was demonstrated in five patients and, again, appeared to relate to the duration of amenorrhoea. Rigotti et al. (1984) studied 18 patients using direct photon absorptiometry, and reported a significantly reduced bone density as compared to controls, with vertebral compression fractures in two patients. However, these authors found no relationship between the bone changes and the duration of amenorrhoea. It seems likely that a combination of factors is responsible.

Low dietary intakes of calcium, fat and vitamin D with low oestradiol and high cortisol may all play a role, while increased physical activity is perhaps protective. Osteoporosis in anorexia nervosa patients has important clinical implications, and painful and incapacitating pathological fractures may occur (Brotman and Stern 1985). Clinicians are well advised to investigate complaints of bone and back pain carefully (see chapter Herzog et al., this volume).

Anovulation is characteristic of anorexia nervosa and many instances of infertility are undoubtedly due to undiagnosed dieting disorders (Fries 1974). Although many patients are able to conceive following weight restoration, pregnancy after anorexia nervosa is reported relatively infrequently (Hsu et al. 1979). However, pregnancy may be induced even during the active stage of emaciation by the use of pulsatile LHRH treatment. Such intervention is injudicious. The presence of an active eating disorder whether it be bulimia or anorexia nervosa correlates highly with low birth weight, low Apgar scores, a higher incidence of congenital malformations, and a higher than expected incidence of prematurely and perinatal mortality and subsequent poor parenting (see review by Woodside and Shekter-Wolfson 1990). However, these matters are really the concern of K. Tolstrup and have been discussed in another chapter in this volume.

References

Abraham SF, Beumont PJV, Fraser LS, Llewelyn-Jones D (1982) Body weight, exercise and menstrual status among ballet dancers in training. Br J Obstet Gynaecol 89:507–510

American Psychiatric Association (1980) Diagnostic and statistical manual of mental disorders, 3rd edn. Washington DC

American Psychiatric Association (1987) Diagnostic and statistical manual of mental disorders 3rd edn: revised. Washington DC

Ayers JWT, Gidwai GP, Schmidt IMV, Gross M (1984) Osteopenia in hypoestrogenic young women with anorexia nervosa. Fertil Steril 41:224–228

Beumont PJV (1984a) Endocrine function in magersucht disorders. In: Pirke KM, Ploog DW (eds) The psychobiology of anorexia nervosa. Springer, Berlin Heidelberg New York

Beumont PJV (1984b) A clinician looks at animal models of anorexia nervosa. In: Bond NW (ed) Animal models of psychopathology. Academic, Sydney

Beumont PJV, Abraham SF (1981) Continuous infusion of LHRH in patients with anorexia nervosa. Psychol Med 11:477–484

Beumont PJV, Russell J (1982) Anorexia nervosa. In: Beumont PJV, Burrows, G (eds) Handbook of psychiatry and endocrinology. Elsevier, Amsterdam

Beumont PJV, Carr PJ, Gelder MG (1973) Plasma levels of luteinizing hormone and of immunoreactive oestrogens (oestradiol) in anorexia nervosa: response to clomiphene citrate. Psychol Med 3:495–501

Brotman AW, Stern TA (1985) Osteoporosis and pathological fractures in anorexia nervosa. Am J Psychiatry 142:495–496

Crisp AH (1969) Some skeletal measurements in patients with primary anorexia nervosa. J Psychosom Res 13:125–142

Cumming DC, Rebar RW (1985) Hormonal changes with acute exercise and with training in women. Sem Reprod Endocrinol 3:55–64
Cumming DC, Vickovic MM, Wall SR, Fluker MR, Belcastro AN (1985) The effect of acute exercise on pulsatile release of luteinizing hormone in women runners. Am J Obstet Gynecol 141:482–485
Dally P (1969) Anorexia nervosa. Grune and Stratton, New York
Ebert MH, Kaye WK, Gold PW (1984) Neurotransmitter metabolism in anorexia nervosa. In: Pirke KM, Ploog DW (eds) The psychobiology of anorexia nervosa. Springer, Berlin Heidelberg New York
Fairburn CG, Cooper PJ (1984) The clinical features of bulimia nervosa. Br J Psychiatry 144:268–269
Fairburn CG, Garner DM (1986) The diagnosis of bulimia nervosa. Int J Eating Dis 5:403–420
Feighner JP, Robins E, Guze SB, Woodruff RA, Winokur G, Munoz R (1972) Diagnostic criteria for use in psychiatric research. Arch Gen Psychiatry 26:57–63
Fichter MM, Pirke KM (1990) Endocrine dysfunctions in bulimia (nervosa). In: Fichter MM (ed) Bulimia nervosa. Wiley, Chichester.
Fichter MM, Pirke KM, Poellinger J, Wolfram G, Brunner E (1988) Restricted caloric intake causes neuroendocrine disturbances in bulimia. In: Pirke KM, Vandereycken W, Ploog D (eds) The psychobiology of bulimia nervosa. Springer, Berlin Heidelberg New York
Fries H (1974) On functional amenorrhoea. Acta Universitatis Upsaliensis 207 (abstracts)
Frisch RE (1977) Food intake, fatness and reproductive ability. In: Vigersky R (ed) Anorexia nervosa. Raven, New York
Garrow JS, Crisp AH, Jordan HA, Meyer JE, Russell GFM, Silverstone T, Stunkard AJ, Van Itallie TB (1975) Pathology of eating. Group report. In: Silverstone T (ed) Dahlem Konferenzen. Life Sciences Research, Berlin, report 2
Gull WW (1873) Anorexia nervosa (apepsia hysterica). Br Med J 2:527–528
Hall A, Slim E, Hawker F, Salmond C (1984) Anorexia nervosa: long-term outcome in 50 female patients. Br J Psychiatry 145:407–413
Hall RCW, Hoffman RS, Beresford TP, Wooley B, Hall AK, Kubasak L (1989) Physical illness encountered in patients with eating disorders. Psychosomatics 30:174–191
Halmi KA, Falk JR (1983) Behavioural and dietary discriminators of menstrual function in anorexia nervosa. In: Darby PL, Garfinkel PE, Garner DM, Coscina DV (eds) Anorexia nervosa: recent developments in research. Liss, New York
Hsu LKG (1987) Outcome and treatment effects. In: Beumont PJV, Burrows GD, Casper RC (eds) Anorexia and bulimia nervosa. Elsevier, Amsterdam, pp 371–378 (Handbook of eating disorders, part 1)
Hsu LKG, Crisp AH, Harding B (1979) Outcome of anorexia nervosa. Lancet 1:61–65
Isager T, Brinch M, Kreiner S, Tolstrup K (1985) Death and relapse in anorexia nervosa: survival analysis of 151 cases. J Psychiatr Res 19:515–521
Krieg JC, Backmund H, Pirke KM (1986) Endocrine, metabolic and brain morphological abnormalities in patients with anorexia nervosa. Int J Eating Dis 5:990–1005
Lacey JH, Crisp AH, Kart G, Kirkwood BA (1979) Weight and skeletal maturation – a study of radiological and chronological age in an anorexia nervosa population. Postgrad Med J 55:381–385
Lauer CJ, Laessle RF, Fichter MM, Pirke KM, Krieg JC (1990) Structural brain alterations and bingeing and vomiting behaviour in eating disorder patients. Int J Eating Dis 9:161–168
Leon GR, White Whelan P, Kelly JT, Patten SR (1986) The symptoms of bulimia and the menstrual cycle. Psychosom Med 48:415–422

Mitchell PB, Truswell AS (1987) Body composition in anorexia nervosa and starvation. In: Beumont PJV, Burrows GD, Casper RC (eds) Anorexia and bulimia nervosa. Elsevier, Amsterdam, pp 45–78 (Handbook of eating disorders, part 1)
Morgan HG, Russell GFM (1975) Value of family background and clinical features as predictors of long-term outcome in anorexia nervosa. Psychol Med 5:355–371
Morgan HG, Purgold J, Welbourne J (1983) Management and outcome in anorexia nervosa: a standardized prognosis study. Br J Psychiatry 143:282–287
Pirke KM (1989) Menstruation and neuroendokrine Störungen der Gonadenachse bei Bulimia. In: Fichter MM (ed) Bulimia Nervosa. Enke, Stuttgart
Pirke KM, Pahl J, Schweiger U, Warnoff M (1985) Metabolic and endocrine indices of starvation in bulimia: a comparison with anorexia nervosa. Psychiatry Res 15:33–37
Ploog DW, Pirke KM (1987) Psychobiology of anorexia nervosa. Psychol Med 17:843–859
Raichle ME, Grubb RL, Eichling LO (1980) Central neuroendocrine regulation of brain water permeability. Brain Res Rev 1:219–235
Ratcliffe PJ, Bevan JS (1985) Severe hypoglycemia and sudden death in anorexia nervosa. Psychol Med 15:679–681
Ratnasuriya RH, Eisler I, Szmukler GI, Russell GFM (1991) Anorexia nervosa: outcome and prognostic factors after 20 years. Br J Psychiatry 158:495–502
Rigotti NA, Nussbaum SR, Herzog DB, Neer RM (1984) Osteoporosis in women with anorexia nervosa. N Engl J Med 31:1601–1606
Rodin GM, Daneman D, Johnson LE, Kenshole A, Garfinkel P (1985) Anorexia nervosa and bulimia in female adolescents with insulin dependent diabetes mellitus: a systematic study. J Psychiatr Res 19:381–384
Russell GFM (1970) Anorexia nervosa: its identity as an illness and its treatment. In: Hill OW (ed) Modern trends in psychological medicine, vol 2. Buttaworths, London
Russell GFM (1979) Bulimia nervosa: an ominous variant of anorexia nervosa. Psychol Med 9:429–448
Russell GFM (1983a) Anorexia nervosa and bulimia nervosa. In: Russell GFM, Herso L (eds) The neurosis and personality disorders. Cambridge University Press, Cambridge, pp 285–298 (Handbook of psychiatry, vol 4)
Russell GFM (1983b) Delayed puberty due to anorexia nervosa of early onset. In: Darby PL, Garfinkel PE, Garner DM, Coscina DV (eds) Anorexia nervosa: recent developments in research. Liss, New York
Russell J, Beumont PJV (1987) The endocrinology of anorexia nervosa. In: Beumont PJV, Burrows GD, Casper RC (eds) Anorexia nervosa. Elsevier, Amsterdam, pp 201–234 (Handbook of eating disorders, part 1)
Schwartz DM, Thompson MG (1981) Do anorectics get well? Current research and future needs. Am J Psychiatry 138:319–323
Schweiger U, Laessle RG, Fichter MM, Pirke KM (1988) Consequences of dieting at normal weight: implications for the understanding and treatment of bulimia. In: Pirke KM, Vandereycken W, Ploog D (ed) The psychobiology of bulimia nervosa. Springer, Berlin Heidelberg New York
Sheehan HL, Summers VK (1948) The syndrome of hypopituitarism. Q J Med 18:319–378
Silverman JA (1974) Anorexia nervosa: clinical observations in a successful treatment plan. J Pediatr 84:68–73
Szmukler GI, Russell GFM (1983) Diabetes mellitus, anorexia nervosa and bulimia. Br J Psychiatry 142:305–308
Theander S (1970) Anorexia nervosa: a psychiatric investigation of 94 female patients. Acta Psychiatr Scand [Suppl] 214:1–194
Theander S (1985) Outcome and prognosis in anorexia nervosa and bulimia: some results of previous investigations compared with those of a Swedish long-term study. J Psychiatr Res 19:493–508

Wakeling A, de Souza VA, Beardwood CJ (1977) Assessment of the negative and positive feedback effects of administered oestrogen on gonadotrophin release in patients with anorexia nervosa. Psychol Med 7:397–405
Warren W (1968) A study of anorexia nervosa in young girls. J Child Psychol Psychiatry 9:27–40
Woodside DB, Shekter-Wolfson LF (1990) Parenting by patients with anorexia nervosa and bulimia nervosa. Int J Eating Dis 9:303–309
World Health Organization (1978) International classification of mental disorder, revision 9. Health Organization, Geneva

Aspects of the Natural History of Normal and Disordered Eating and Some Methodological Considerations

G. Rathner

Introduction

Research into anorexia nervosa (AN) and bulimia nervosa (BN) has increased exponentially in recent years (Fichter 1989). However, the large number of publications on demographic and clinical descriptions, aetiological theories, diagnostic investigations and treatment studies stand in stark contrast to the neglected area of epidemiological surveys. Even more noticeable amongst epidemiological work is the dearth of longitudinal studies. Screening studies are available in general populations, clinical and other selected groups. Yet fewer than 20 reports have considered the natural longitudinal course of eating behaviour, attitudes or disorders. It is imperative that we become aware of the natural course of untreated, disordered eating symptoms. Without this knowledge we have no grounding upon which to base our assessment of other important aspects of eating disorders (EDs), such as the chronicity of the problems, the efficacy of treatment and the stability of diagnoses over time.

Despite the need for categorical diagnosis like DSM and ICD, which has fostered scientific progress in the past, the continuum hypothesis (Vandereycken and Meermann 1984) is probably more pertinent to eating behaviour. Eating behaviour can range from no concern to a 'normal' concern with weight and shape, through a short episode of ED which spontaneously remits and is never seen by any health professional to the serious chronic course, or, even the worst outcome, death. There is no doubt that a continuum of severity exists in the general population. What is less certain is whether an individual person will move along this continuum in the course of their eating history, i.e. is the continuum dynamic, or is each individual naturally static upon it (saving treatment, intervention).

At present the natural history of even clinical EDs is generally unknown. It has been suggested that the long-term outcome of AN is more dependent upon the natural history of the patient's illness than any treatment she may have received (Russell 1977) – that the illness usually breeds true. But why does AN in one person evolve to a chronic state or even death and in another to a mild and limited episode in adolescence only?

Like many other psychiatric disorders, EDs show a considerable variation of outcome and in the individual case, the prognosis is unpredictable.

What do we know about *spontaneous recoveries*? How many girls show an anorectic episode, a mild or transient form of the disorder, which might be interrupted by a caring family physician, by school health personnel, teachers or a friend. These nonexperts may know more than us about spontaneous recoveries, because these subjects never will be seen at a hospital. Sours (1980) remarked that many mild or subclinical cases of ED are not included in epidemiological studies or evaluated by specialists. How many such cases have been overlooked? A recent report by Fenley et al. (1990) of the devastating physical sequelae in a woman following 30 years of untreated AN shows us not only an alarming natural course – a chronic one – but should also make us pessimistic about the expertise of medical doctors, who made several false diagnoses despite numerous admissions of this patient.

Every clinician is conscious of the issue of spontaneous recovery and the need to disentangle this in order to evaluate the true efficacy of treatment (Yager et al. 1987). Without this knowledge we can not really know what part of our work is effective – which interventions are helpful to the patient. More simply, we cannot know what works. Apart from a few fundamental guidelines (e.g. Bruch 1988; Crisp 1980; Lacey 1985; Russell 1977; Vandereycken and Meermann 1984) we still have no specific answers! Compared to the treatment of AN, even less is known about the longitudinal course of treated BN; only recently have careful outcome studies of BN become available (see chapters in this book).

The course of ED in the general population and of patients who never arrive at a clinic may be quite different from the natural history of our patient samples, and we still do not know for sure the typical course of the clinical disorder. Which factors influence the help-seeking behaviours in ED, and what is the difference between those who seek help and those who do not?

We need to know: (1) the interrelationship between clinical AN (and its subgroups of restricters and bulimics) and BN; (2) the relationship of clinical EDs to subclinical or partial forms; (3) what determines the movement between diagnostic categories; and (4) how subclinical forms evolve. Up to now all we know is that severity can wax and wane, that individuals move back and forth between diagnostic categories, and that a noticeable proportion of AN patients later develop BN following treatment (Hsu 1988).

Other chapters of this book describe long-term follow-up studies in treated clinical samples. It has been shown that differences in outcome are more related to original selection criteria than to treatment approach, and that – in the course of time – the impact of different treatments diminishes (and becomes comparable; see Herzog et al., this volume), so these studies may assess the natural outcome independent of therapy. Complementing this chapter (Herzog et al., this volume) the present one reviews all available longitudinal studies in samples drawn from the general population, and studies of high-risk groups such as ballet students. Some results of the author's ongoing longitudinal epidemiological study are presented. Lastly, some methodological recommendations for future studies are suggested.

Most studies on the natural course of eating attitudes and ED have been published only very recently and as yet no comprehensive review is available. These studies are of two types: the methodologically weaker screening studies, and the state of the art two-stage surveys. Only the data relevant for the natural course are considered, as treatment studies have been reviewed elsewhere (Herzog et al., this volume).

Screening Studies

Screening studies involve assessment using a variety of screening instruments – usually questionnaires – at two time points. Most involve resampling of individuals from a group of subjects, although a method of cohort resampling (rather than specific individuals) has been employed. Because of its relative ease, screening studies are the most common. However, they share the methodological weaknesses of any epidemiological survey based upon self-report questionnaires (Rutter and Sandberg 1985; Pyle 1983–85; Szmukler et al. 1985; Rathner 1989). Often questionnaires are not standardized and their reliability and validity are unknown, but extrapolation to case identification and diagnosis is dubious, even with well-standardized instruments. Response bias is a major concern, particularly as large surveys often rely upon mailed questionnaires, which produce a notoriously poor response rate. With two time points the response biases are exacerbated. Finally, sample selection is often guided by ease of collection rather than by the wish to represent fully the general population.

Huenemann et al. (1966) was one of the first longitudinal studies to be published in the field, reporting gross body composition and teenagers' views on shape, weight and eating practices. The sample consisted of approximately 1000 teenagers at a public high school in California, followed from age 14.5 to age 17.3, with investigations made annually. Only 206 boys and 250 girls were investigated on all four occasions. Their methods and questionnaires might be questionable compared with present standards, but the results are interesting as one of the earliest investigations of normative dieting. Unfortunately, no exact data are available for this longitudinal sample, but the authors indicate that, by comparing the longitudinal to the cross-sectional sample, in no category did the percentages differ by more than 6%, and in all instances changes over time were equally apparent in each set of data. Figure consciousness (felt fat) in-

Table 1. Review of longitudinal screening studies: population and sampling characteristics

Authors	Location	Subjects	Sample finding procedure	Sample size (T_0) (n)	Age (years) mean (SD) (range)	Participation rate[a] (%)	Representativity
Huenemann et al. 1966	Urban USA	Male/female adolescents	Public high school	Female 500 Male 438	14.5	n.r.	n.r.
Yager et al. 1987, 1988	n.r. USA	Adult women	Magazine advertisement	1729	24.7 (7.3)	–	Sample of convenience
Drewnowski et al. 1988	Urban USA	Female students	College freshman class	931	n.r.	41.9	US (midwest) university; social class representative for entire freshman class
Attie and Brooks-Gunn 1989	Metropolitan USA	Female adolescents and mothers	Private middle school	193 girls and mothers	13.9 (1.1) (12–15)	Mothers >85	Bias to upper social classes (95%)
Striegel-Moore et al. 1989	Urban USA	Male/female college students	College freshman class	Female 450 Male 590	Female 17.7 (0.6) Male 17.9 (0.5)	Female 79.6 Male 73.8 (Mean 76.2)	Competitive US (northeastern) university; social class not assessed

Johnson et al. 1989	Urban USA	Female adolescents	High school	1268 T0 1085 T1	14–18	98.1 T_0 96.0 T_1	n.r.
Rosen et al. 1990	Urban USA	Female adolescents	Secondary boarding school (grade 9–12)	162	15.9 (1.1) (14–18)	65.3	Bias to upper social classes
B. Dolan 1990, personal communication	Metropolitan UK	Adult women	Family planning clinic	365	26.0 (7.4) (16–48)	99.2	Social class and ethnic distribution representative, but only sexually active women

n.r., not reported.
[a] Response rate concerning selected sample.

Table 2. Review of longitudinal screening studies: longitudinal sampling, methods and results

Authors	Longitudinal samples (*n*)	Follow-up duration	Participation rate follow-up[b] (%)	Methods[a]	Diagnostic criteria	Results
Huenemann et al. 1966	Female 250 Male 206	3 years (annual follow-up)	n. r.	Anthropometric, questionnaires	Dieting, feeling fat	Horizontal and longitudinal data virtually the same
Yager et al. 1987, 1988	628	Mean = 20 months (min 15 months)	36.3	EPQ, EDI, BSI, PDQ, FHQ	Simulated DSM-III (BAN, NWB, SDED)[c]	Minor improvements to subclinical level, psychosocial impact unchanged, EDI improved
Drewnowski et al. 1988	599	6 months	64.3 (27.0)	Questionnaire (clinically validated, data n.r.)	Simulated DSM-III-R (BN)	Partial remissions offset by new "cases"
Attie and Brooks-Gunn 1989	173 girls, approx. 145 mothers	2 years	Girls 89.6 (n.r.) Mothers 75 (n.r.)	Anthropometric, SIQYA, EAT, FES, YBP	Problem eating (defined by mean EAT, nine items version)	Maturation, body image and psychopathology predict problem eating; EAT stable

Striegel-Moore et al. 1989	Female 330 Male 342 Total 672	approx. 9 months	Female 73.3 (58.4) Male 58.0 (42.8) Total 64.6 (49.3)	Questionnaires and rating scales, EDI, WOFO, Perc. Stress Scale	Probable, suspected DSM-III-R BN	BN prevalence virtually unchanged, new cases more likely than remissions
Johnson et al. 1989	–	5 years (cross-sectional)	–	Questionnaire (Pyle), EDI	Simulated DSM-III Bulimia	Second sample showed 50% reduction of bulimic behaviour
Rosen et al. 1990	143	4 months	88.3 (57.7)	TEFQ, Adolescents Perceived Events Scale, BSI	None	Stability of self-reported behaviour and symptoms
B. Dolan 1990, personal communication	63	1.5 years	61.2 (17.1)	EAT, BSQ	None	EAT stable

AN, anorexia nervosa; BAN, bulimic AN; NWB, normal weight bulimia; SDED, subdiagnostic eating disorder; n.r., not reported.
[a] For abbreviations of methods see text.
[b] Numbers in parenthesis are response rates related to elected sample at T_0.

creased over 3 years (from 43% to 56%), although the proportion of obese girls remained the same (about 25%). Weight control effort, i.e. dieting, was mentioned in 65% of 9th grade girls and tapered off to 55% in 12th grade girls, but few succeeded in weight loss. These figures on eating attitudes are replicated, more or less, in the numerous screening studies done in the 1980s (e.g., Button and Whitehouse 1981; Mann et al. 1983).

After this study there is a long gap in the literature. Only recently has interest in the natural history of ED created stimulating prospective and longitudinal screening studies. Many of these use self-report questionnaires, so obviously a diagnosis can not be reliably made. Most of these studies report simulated, probable or suspected diagnosis for ED (see Tables 1, 2).

Yager et al. (1987, 1988, 1989) is the first of the only two longitudinal screening studies in adult samples as yet reported. An invitation in *Glamour* magazine for women with bulimic problems secured 1729 women who completed an extensive mailed questionnaire battery. At follow-up 628 women replied, resulting in the usual response rate of one third for postal surveys. The test battery included the Eating Problems Questionnaire (Johnson et al. 1982), Brief Symptom Inventory (Derogatis et al. 1974), Eating Disorder Inventory (EDI; Garner et al. 1983) and a newly developed help-seeking and satisfaction questionnaire. Out of these data a simulated diagnosis of DSM-III AN (with bulimic features), of DSM-III normal-weight bulimia (NWB) and of subdiagnostic eating disorder (SDED) was made. Accordingly, 34 women had bulimic AN, 392 had NWB and 202 had SDED. The majority of these 628 women (58.1%) were still in the same diagnostic category 20 months later; 28.3% reduced frequencies of symptoms so as to be classified as subdiagnostic and 10.7% increased symptoms to a significant clinical level for AN or NWB; 2.9% changed from AN to NWB or vice versa, developing mostly towards NWB. One could argue with the author's conclusion of a general tendency towards improvement in this self-selected sample, as the AN to NWB interchange is not an improvement and since only one quarter improved to a subdiagnostic level, but none showed total remission. Those respondents who sought professional help between initial survey and follow-up reported no more improvement than those who did not seek help. But the longitudinal sample represented only 36% of the original survey group, thus it is possible that those who sought help and were now asymptomatic did not reply to the second survey.

Considering the longitudinal development of test scores, the EDI scores tended to improve significantly from entry to follow-up, with few exceptions especially in the AN group. Severity scores for symptomatic behaviour showed a reduction at follow-up, but no change was reported for the stressful impact of these behaviours on various subjective and objective aspects of everyday life. So, the burden was still considerable and especially the AN group showed the worst prognosis and an increase of impulsive behaviours (alcohol and tobacco abuse).

Dolan et al. (1990) conducted a screening study on eating behaviour and attitudes to weight and shape in British women from three ethnic groups (n = 365 Caucasian, n = 71 Afro-Caribbean and n = 43 Asian women) with a battery of standardized tests, including, for example, the Eating Attitudes Test (EAT-26; Garner et al. 1982). The rescreening of an albeit small subsample is of interest, because it represents the only adult sample not self-defined as ill. Women attending a family planning clinic were asked to complete a second questionnaire some time later, which would be mailed. In all only 103 (28.2%) women (mean age 26.0 years) gave an address, but comparing those 103 who provided an address with those who did not showed no significant difference in age or EAT-26 score. Of these women 62.1% completed a further questionnaire approximately 18 months later, and a further 18.4% were untraceable. The mean EAT-26 score of this group on outset was 8.9 (SD 8.3), compared to a follow-up score of 9.9 (SD 10.7), revealing no significant differences (paired sample t-test: t = 0.88; p = .384). Despite this stability, one should not forget the high number of women who did not want to be bothered by follow-up questionnaires. The study was representative of the sexually active women in the general population of the local area (B. Dolan 1990, personal communication).

Drewnowski et al. (1988) reported the first longitudinal study on bulimia in female college students and applied a self-constructed questionnaire. Although response rates were poor (42%), some tendencies are shown towards a shift of suspected cases to cases in partial remission. The new cases at second assessment showed only subthreshold symptomatic behaviour on initial assessment. Of the 27 initial cases of BN, 18 were followed up for 6 months, 8 of them still being classified as bulimic, whereas the remaining 10 showed evidence of spontaneous, if only partial, remission, but kept on bingeing at least monthly and indicated the feeling of loss of control over eating. Interestingly, only a minority of suspected bulimics (20%) sought professional treatment.

Striegel-Moore et al. (1989) surveyed 962 male and female college students at the beginning and at the end of their college freshman class. They tried to elicit information about possible indicators of BN using a number of nonstandardized rating scales and self-report questionnaires containing weight, desired weight, feelings about weight, perceived attractiveness, and two subscales of the EDI (perfectionism, ineffectiveness), a newly developed Disordered Eating Symptoms Scale (DESS) with frequencies of disturbed behaviour and two more general tests (Work and Family Orientation Questionnaire, and Perceived Stress Scale). Inclusion as a suspected case of BN required the stringent criterion of binge eating three times a week, to err in the more conservative direction. At baseline, the estimated prevalence of BN was 3.8% in females and 0.17% in males (virtually the same at follow-up); the one male dropped out at follow-up.

There was a significant increase in dieting over the year with over three times as many women starting dieting as stopping, but 70.4% of the total female sample did not change (57.3% stable dieting, 13.1% stable non-dieting). Similar results were found for men. The net increase in binge eating was less impressive. But again, the majority of the longitudinal female sample remained in their category: 33.1% kept on bingeing, overall 43.5% were stable nonbingers. The self-rating is probably overinclusive, but the stability of subjects within categories and the exacerbation rather than amelioration of symptoms remains.

The 17 probable cases of BN showed the following course: 11 remained, 5 no longer met the diagnostic criteria for BN, but still reported symptomatic behaviour and 1 dropped out. Three new cases emerged at the follow-up who had not reported sufficient frequencies of symptomatic behaviour at first assessment. The worsening of disordered eating among females was associated with increasingly dysphoric feelings about weight, decreased ratings of attractiveness, high perceived stress, increased weight dissatisfaction and increased feelings of ineffectiveness.

Rosen et al. (1990) measured daily and major life stress, psychological symptoms, and dieting (Three Factor Eating Questionnaires, Adolescents Perceived Events Scale and Brief Symptom Inventory) in 143 adolescent school girls at the beginning and the end of a short, 4-month, prospective study. The longitudinal participation rate was quite high (88.3%), but their longitudinal sample could include only 57.7% of the total. Although this study adopted a longitudinal screening approach the time interval was rather short. At longitudinal evolution, all mean scores of these questionnaires remained virtually the same and stable. Multiple regressions and linear structural relationships (LISREL) analysis showed that stress and psychological symptoms do not contribute to the prediction of future dieting after the stability of dieting behaviour over time is taken into account. Future levels of psychological stress, but not symptoms, were predicted by weight-reducing behaviour, when baseline scores were controlled. As previous reports indicate, stress predicts psychological symptoms in adolescents, and dieting behaviour may eventually have an influence on the manifestation of psychological symptoms through the mediation of stress. The authors conclude that weight reducing in this age group may not be as innocuous as people first think when they refer to the nearly normative dieting efforts in adolescents. However, the stability of self-reported behaviour and symptoms (albeit over a very short time interval) is emphasized.

Attie and Brooks Gunn (1989) carried out the only screening study which also included the mothers of female adolescents in order to investigate a broader maturational, body image, developmental and family frame. The age range and follow-up duration was chosen to cover the supposed critical period of change from late puberty to early adolescence (from age 14 to age 16). A large test battery (EAT, Self-Image Questionnaire for Adolescents, Center for Epidemiological Studies – Depression Scale, Youth Behaviour Profile, EDI and Family Environment Scales) was used, but unfortunately most of these only at second assessment. In addition, the definition of the independent variable problem eating seems questionable, as it is based only on an abridged nine-item version of the EAT-26. This lends doubt regarding clinical relevance of the measure and the statistical procedures are complicated. However, girls who at first assessment felt most negatively about their bodies were more likely to develop problem eating; this negative body image was the only predictor of long-term change. Psychopathology and maturational factors, i.e. growing body fat, were associated with problem eating at initial assessment and not at follow-up. At follow-up body image of the girls, maternal body image and depression, but also maternal ratings of the family milieu (less cohesion, organization and expressiveness) were now associated with daughters' eating problems. But the strong predictive influence of body image should be emphasized as the main result.

Johnson et al. (1989) assessed the extent to which bulimic behaviours reach a stable morbidity in our culture in an interesting study. They adopted a longitudinal cross-sectional approach, i.e. conducted a repeated assessment with the same methodology and with the same cohort (but different subjects) 5 years later. They studied over 1000 female high school students with a fairly high participation rate (98.1%) and repeated it 5 years later with the same success (participation rate of 96%). The long interval was chosen to ensure that most of the youngest students would have graduated, thus enabling a longitudinal comparison between two demographically equivalent, but temporally different cohorts. Simulated DSM-III diagnosis of bulimia (including a frequency criterion of bingeing/purging of at least once a week to rule out exaggerated estimates of morbidity) showed a decline from 4.1% in 1981 to 2.0% in 1986. Associated bulimic behaviour also decreased significantly. A 10% decrease in the number of students ever on a diet (from 70% to 60%) and a 6% decrease in being on a diet now (from 36% to 30%) was reported, whereas the desired body weight and the Body Dissatisfaction Scale (EDI) remained unchanged. The later sample also showed significantly reduced (mean) Drive for Thinness scores (from 6.19 to 5.33), but this change of 0.86 does not seem to be so clinically meaningful.

The authors conclude that the shift appears to be purely one of attitude, as desired body weight and its percentage of normal weight were the same, and they relate this shift to changed cultural pressures. They suggest that whereas in earlier years publicity may have fostered the increase in bulimia, now mass media influence may work in the opposite direction, by replacing the drive to achieve thinness by the pursuit of strength, but they emphasize the bad news that the stable body-image dissatisfaction could be the breeding place for unhealthy manipulations of the body in these youngsters, be it dieting, bingeing and purging or exercise abuse. In contrast to Johnson et al.'s suggestions, an alternative explanation of the decrease could be considered. Although the findings are statistically significant, the absolute changes are small and may not represent a clinically meaningful reduction. It could also be that the increased publicity of bulimia resulted in a better knowledge of some words (especially binge) written in these questionnaires and therefore decreased the self-attribution of some symptoms, which are necessary for such simulated diagnoses.

A summary of the seven longitudinal screening studies is shown in Tables 1 and 2. The variety of diagnostic and screening measures applied makes comparison difficult. Duration of follow-up ranged from 4 months to 3 years (mean 22 months), and response rate varies from 36% to 90%. Despite these inconsistencies a general picture of results can be drawn. The overall tendency is for group mean questionnaire scores to remain stable over time; only in Yager's adult group did mean EDI score decrease, but that may represent a regression to the mean in such a selected sample. Rosen et al. (1990) found levels of eating behaviour and symptoms to be stable over 4 months in adolescents, but Huenemann et al. (1966) found attitudes still stable after 3 years with annual follow-up. Finally, some optimism can be taken from Johnson's cohort resampling study which suggests that over 5 years bulimic behaviour and dieting have become less common in adolescents.

Change in clinical status can also be considered. Yager et al. (1988) found only partial remission into subclinical EDs in some (28%) adult women, although 60% of his sample remained in the same category over 20 months. In student samples Drewnoski et al. (1988) and Striegel-Moore et al. (1989) both report that partial remissions of BN are offset by the incidence of new cases.

As a final caveat, it should be noted that all but one of these studies were conducted in the United States of America, only two studies surveyed adult women and all were carried out in urban or metropolitan populations. Until longitudinal surveys are available from general populations of all ages, in different countries and cultures, we should be wary of generalizing from the findings reported above.

Two-Stage Investigations

Psychiatric epidemiology has advanced in recent years in sophistication of sampling techniques, improvements in case identification, clear definition and specification of rules for diagnoses and, finally, in the development of standardized interview schedules. Screening studies have many disadvantages, as outlined above, the major problem being the unreliability of a clinical diagnosis based upon a questionnaire score. Ideally the minimum standard in epidemiological work should be a two-stage approach as introduced by Rutter in 1970 (see Rutter and Sandberg 1985). Unfortunately, until recently the two-stage approach has been infrequently used in epidemiological research into eating disorders. The two-stage approach involves an initial questionnaire screening of the chosen population, followed by direct interviewing of individuals in the sample who have a high risk for the target diagnosis as identified by the questionnaire scores. The assumption that questionnaire scores detect all cases at risk must also be tested by including a random sample of the low scoring group in the interview stage. This two-stage approach can be further improved by the use of third-party informants (family, friends) or by using several screening methods with different types of observation (e.g. self-report, objective measures). So far only eight studies of eating behaviour are available which have have adopted this two-stage approach.

Schleimer's (1983) outstanding piece of longitudinal research is a 10-year follow-up study, extending the seminal study of Nylander (1971). From a representative sample of 1129 boys and 1241 girls aged 14–21 years, who were screened by Nylander, the author drew a high-risk sample of 130 girls according to predefined risk criteria (dieting in connection with possible anorectic symptoms, weight loss of at least 10 kg and/or psychiatric referral). Some 111 subjects (patricipation rate 85.4%) could be interviewed personally (using a semi-structured format) and were clinically evaluated. Of these, 97 were re-evaluated personally 10 years later, giving an impressive 87.4% trace rate (74.6% of the selected sample). In addition, third-party information on some of the drop-outs, which also contained three girls with AN, is provided. AN was strictly diagnosed according to the criteria of Garrow et al. (1975) and anorectic behaviour was diagnosed according to Fries' criteria (1977).

From the high risk group of 130 15 girls were diagnosed as having AN, and 3 (20%) of these had never been in treatment in spite of severe symptoms. The cases diagnosed as AN had all had anorectic problems as early as at original screening time, and no new case of AN had occurred during the 10-year observation period. Anorectics show a similar course to treatment samples: some develop a bulimic AN (chronic or intermittent), some remain stable abstainers, and about 50% are healthy at the 10-year follow-up. Dieting was common, taking place in brief (<6 months) recurring periods over this 10-year interval.

Schleimer (1983) suggests that progression from ordinary teenage dieting is due to neurotic mechanisms. Neither the girls with no anorectic behaviour at the early stage nor those showing some anorectic behaviour progressed to AN at follow-up (despite protracted efforts to diet). Only dieting combined with low self-esteem, fears about interpersonal relationships and difficulties in interoceptive awareness seemed to lead to anorectic behaviour or AN. It is still possible that, with a longer time interval, further regression might occur because the subjects had not yet passed the high-risk age period. The three untreated anorectic women were still ill 10 years later. The overwhelming evidence from this study is that anorectic behavior and clinical AN are distinct categories, and subjects do not move between them over time. These results question the dynamic continuum hypothesis, as put forward, for instance, by Vandereycken and Meermann (1984).

Szmukler et al. (1985) used a two-stage procedure in a sample of 100 adolescent ballet dancers. There was a low participation rate of approximately 50%, which the authors attribute to the need for informed consent. The first stage included the EAT-40 and a demographic and eating behaviour questionnaire. Inclusion criteria for the interview, which was repeated at the 1 year follow-up, were high EAT score (>30), current secondary amenorrhoea (>4 months), or current primary amenorrhoea (for age 16 and above). A few low scorers were also interviewed, but no data are provided. In total 25 interviews were conducted. The authors discussed the difficulties of correct diagnosis within an environment, in which disturbances of weight and menstruation and preoccupation with shape were common, and therefore presented only numbers of possible cases of AN. Using the DSM-III criteria for AN, the first interview assessment revealed seven possible cases of AN, two borderline cases and one girl with previous AN. After 1 year they only reinterviewed seven girls from their diagnosed sample who were still at school and provided weight data for the remaining three girls or their case sample. The seven girls showed a mean weight gain from 80% to approximately 90% matched population mean weight (MPMW), one had resumed her regular periods and three girls had begun to menstruate irregularly. The authors particularly note that in this school, any marked weight loss was followed with pressure to gain weight, including sending home one girl for 2 weeks in order to gain weight (which proved successful, at least in terms of weight). The authors conclude that the longitudinal course in this specific environment showed a considerable improvement in all cases, in terms of weight gain without any medical intervention. Although these girls went on in their career, no hints are given as to how psychopathology or eating habits developed in the future. The low participation rate, the development of the two borderline cases and the previous AN case places doubt on their overall conclusion.

Garner et al. (1987) adopted a two-stage procedure in a sample of 55 female students, also ballet dancers, aged 11–14 years (mean 13.0 ± 0.6 years) from a highly competitive school in Canada. In the first screening stage, a test battery was applied (Eating Disorders Inventory, Family Assessment Measure, Locus of Control Scales, and Feelings of Inadequacy Scales). The second stage, 2 years later, included the EDI, a demographics form and a clinical interview by a psychiatrist in those students with either current weight, weight history, menstrual status or EDI results suggestive of ED. Diagnosis was made according to DSM-III-R (AN), Russell (BN) and Szmukler (partial syndrome). The participation rate in the interview stage was 58.2%. The remained of the students were not available because they had left school, two of them because of AN. The authors could not report the change in diagnosed EDs, because an interview was only conducted on follow-up. However, the EDI scores, both in the ED group and in the non-ED group, remained stable or worsened, particularly in the ED groups (except for lower scores in perfectionism and maturity fears). Drive for thinness and body dissatisfaction scores on first testing predicted the development of an ED at follow-up. Only two of 15 students with an elevated drive for thinness score remained at school without developing an eating disorder. Most of the AN cases followed up gained some weight, but unfortunately, no exact data are reported. Unlike Szmukler's school study, weight loss and excessive thinness in students was not responded to with a firm hand by the staff.

In their conclusions, the authors are not as optimistic as Szmukler et al. and express their concern about serious eating disturbances in their follow-up sample, which contained 25.7% AN, 2.9% BN and 11.4% partial syndromes. Garner and colleagues do not regard EDs in the ballet dancers as benign disorders and are pessimistic about spontaneous recovery, a view supported by their data on a 6-year follow-up of five students who remained at school in spite of their ED.

King (1989) conducted the only two-stage survey in an adult population in England. The sample was selected from consecutive attenders at local general practitioners' (GP) surgeries. Inclusion was restricted to the age group 16–35, and 534 woman and 186 men took part, representing an impressive 96.3% response rate. Demographic characteristics and social class distribution seem representative; although ethnicity is not mentioned, reported country of birth suggests an underrepresentation of the 20% nonwhite population of South London. Respondents were screened using the EAT-26: all high scorers (n = 76) and a random selection of low scorers (n = 45) were invited for interviews 1–2 weeks later; 90.1% complied with this. The Clinical Interview Schedule (CIS; Goldberg et al. 1970), the Eating Interview (Szmukler 1983) and the Symptom Rating Test (SRT; Kellner and Sheffield 1967, 1973) were used. Six women and one man had BN according to Russell's criteria (prevalence 1.1%); there were no cases of AN but 2.8% of subjects had a partial syndrome (Button and Whitehouse 1981).

At follow-up 12–18 months later all high scorers were again approached by mailed questionnaire (EAT-26, SRT and a questionnaire about symptomatic eating behaviour). No further interviews were conducted. Unfortunately no low scorers were included in postal follow-up, but we should question the tacit assumption that they will not develop an ED.

Four men and 57 women responded, giving a high response rate for mailed surveys (88.4%). The mean EAT scores were lower on follow-up for each diagnostic group except for the bulimic subjects. There was a slight, but nonsignificant increase of SRT scores. With regard to individual cases all six bulimic women remained clinically ill, two having worsened.

The male subject with bulimia had reduced his binge eating and stopped laxative abuse and vomiting although his weight had dropped. Four women who originally had partial syndromes were now exhibiting more severe problems and three of these were now diagnosed bulimic, and four had improved, now scoring below cut-off on the EAT. The remaining subjects showed little change.

If we only consider female bulimics, partial syndromes, obsessive dieters and normal dieters initially and at follow-up and if we assume that this sequence goes from more disturbed to less disturbed categories, a very clear tendency is shown in King's data: almost three quarters of subjects (73.7%) remain in their respective category 12–18 months later; only three females (5.3%) improve, whereas 12 females (21.1%) deteriorate. Two of six bulimics from first assessment worsened and four showed little change, but three new cases developed from those with the partial syndrome. GP records for the EAT high scorers revealed that no subject had been referred for psychiatric treatment of ED in the past nor did anyone ask for specific help with their eating problem, thereby indicating low self-disclosure to the GPs.

In this study a transition of subjects in either direction from normal dieting to BN can occur, but the main result is stable diagnostic categorization in the chosen groups, at least over a time period of 12–18 months. The results seem to support more a static than a dynamic continuum, but movement in either direction along the spectrum of eating pathology was possible. Dieting as a possible pathway to more disturbed behaviour should be emphasized, as this adult sample seems to represent an untreated adult community sample with all degrees of severity except AN. We are waiting for the reinterviews, because this is the only traceable adult sample as of yet!

Patton et al. (1990) extended the cross-sectional two-stage survey of Johnson-Sabine et al. (1988) to a 12-month follow-up to study the importance of the many putative risk factors in the evolution of EDs. Female adolescents aged 14–16 years were chosen to assess subjects near an age of peak incidence for AN. Both initial screening sample size (n = 1010) and initial interview sample size (n = 230) are very large with fairly high response rates (90.0% and 90.2%, respectively). The state secondary schools (avoiding the normal bias towards private schools) seemed broadly representative of London in affluence and social class distribution, but were not randomly selected. A careful analysis of nonrespondents in one of eight schools yielded two cases of AN in treatment, again confirming the significant nonresponse bias in ED screenings. Only nonparticipation in screening, but not drop-out at interview stage or at follow-up seemed to be biased. EAT-26 and General Health Questionnaire (GHQ-28; Goldberg and Hillier 1979) were used as screening tools. The semi-structured interview covered a broad range of clinical features and putative risk factors for EDs derived from the literature; in addition the Social Stress and Support

Interview (SSSI; Jenkins et al. 1981) and some personality scales were used in the interview assessment. The rescreening and the reinterview 12 months later again used the same instruments. Rescreening was permitted only in six of eight schools (n = 735), but response rate in these is again remarkable (93.6%). Reinterview was permitted in all schools, but the participation rate at second interview was somewhat lower (76.5%) compared to the first interview (90.2%).

High EAT scorers constituted the at risk group for interview stage (n = 83). Two control cohorts, one with probable psychiatric morbidity (EAT−/GHQ+) and one without (EAT−/GHQ−) were chosen to control for nonspecific psychiatric factors associated with abnormal eating attitudes. There was a considerable overlap between EAT and GHQ high scorers (78%). Also 73 low scorers on both tests and 74 EAT−/GHQ+ subjects were interviewed at T0 (original interview). A considerable change of EAT score occurred over 12 months. Only 3.4% of the rescreened cohort remained high scorers on both occasions. The restitution rate (changing into low scorers) is 5.9%, but this is nearly equalized by an incidence rate of 4.9% new high-scorers at follow-up.

Diagnostic categories for interview stage were BN or partial syndrome, dieter, or nondieter. At follow-up, the calculated minimum point prevalence rate for EDs (including partial syndromes) was higher than at initial assessment (8.1% compared to 2.77%; 26 compared to 17 cases). Dieting seemed somehow to wane from age 14–16 to age 15–17 (21.3% point prevalence compared to 31.0% at T0). The total prevalence rate for attempts to control weight showed a lesser (nonsignificant) reduction, from 33.8% to 29.4%. Considering the longitudinal evolution of the diagnostic categories, some within-group change was observed.

Clinical cases (BN) can evolve to dieters, but even nondieters can evolve to clinical BN 1 year later. The greatest proportion of change took place in dieters: although the majority of them did not progress to a clinical condition, the 13 (21.3%) who did constituted half the cases at follow-up. So the relative risk for dieters developing an ED 1 year later was calculated 7.9 times that for nondieters. Two thirds of the interview sample remain in their diagnostic category, whilst amelioration (18.2%) or deterioration of symptoms (16.5%) is evenly distributed in their sample. Thus, in contrast to Schleimer's study, the authors found some interchange in diagnostic categories, which would support the dynamic continuum hypothesis.

Most putative risk factors (e.g., social class or sexual activity) were unrelated to any diagnostic category. Some, such as premorbid personality, family weight pathology and above average premorbid weight, were associated with attempts to control weight only. The latter finding is especially important, because overweight and greater discrepancy between actual and desired weight along with higher perceived social stress predicted the later development of dieting in nondieters. General psychiatric morbidity was strongly associated with ED and increased GHQ score was the only distinguishing factor for dieters who progressed to a clinical condition. But the overall explained variance of these putative risk factors is low.

Weeda-Manak (1984) carried out a study to determine the role of anorectic and psychological characteristics in the development of AN. Two samples of women from possible high risk groups for AN were chosen (ballet students, BA, and women with secondary amenorrhoea of at least 3 months duration, AM and compared to asymptomatic female (AF) controls. No data on representativity and response rates in the selected samples are provided. At initial assessment and at follow-up height and weight were measured and the Anorexia Nervosa Inventory for Self-Rating (ANIS, Fichter and Keeser 1980) and the Achievement Motivation Test administered. In addition, a semi-standardized psychiatric interview was conducted. Of the total sample 307 (72%) were traced longitudinally, although the response rates in the three groups varied from 61% (AM) to 72% (AF) to 82% (BA). Further bias is introduced in that only subjects in the AF group were offered financial compensation for their participation in the follow-up study; subjects in the other two groups were not. Comparison of drop-outs with responders showed that in the AM group the drop-outs were significantly lighter than those who participated in the longitudinal follow-up. Also for the asymptomatic group drop-outs were significantly older and shorter than participants. The follow-up duration also varied considerably between the groups: from 6 months (AM) to 9 months (BA) to 14 months (AF).

Unfortunately Weeda-Mannak (1985) does not present absolute changes in scores for the groups over time. Instead a multiple regression analysis was applied to identify the factors at initial assessment which are predictors of weight loss and increased preoccupation with weight and dieting by follow-up. She found that predictors of further weight loss were: for the AF group initial weight; for the AM group ANIS score and for the BA groups menstrual cycle and the psychological factor of drive to achieve. No other predictors could be found in these groups and the amount of variance explained is generally very low. Maybe this is due to the rather arbitrary selection of some predictor variables like achievement motivation and to the interrelationship of others (ANIS), which are predictor and dependent variables. In addition, the mean weight loss (approx. 2–3 kg) does not seem to be clinically meaningful, because such degree of weight variation can be easily found in normal dieting women within 1 year. An analysis of the 3%–10% subjects in these groups who experienced a weight loss of more than 5 kg might have brought more interesting predictors, or an analysis of all subjects, not just those who lost weight. No results on change in diagnostic category, resumption of menses in amenorrhoeic patients and evolution of dieting are presented, i.e. the study as reported is not making use of the interview data, only the self-report.

Marchi and Cohen (1990) traced maladaptive eating patterns in a large random sample of children. Interviews were taken three times (in 1975,

1983 and 1985), giving an admirable follow-up duration. Sampling and tracing characteristics and also representation are good. The longitudinal sample (326 girls and 333 boys) were assessed at mean ages of 6, 14 and 16 years, respectively. As the first interview took place very early (1975), at which time standardized interview schedules were hardly available, the authors applied a semi-structured interview of mothers covering problem meals, pickiness, pica and digestive problems. Follow-up assessments were carried out with the mothers and the children using the Diagnostic Interview Schedule for children (DISC) covering DSM-III-R criteria for EDs. The idea of investigating the longitudinal course of features of EDs may have come up later in the study process (in line with the growing scientific awareness of EDs), and DSM-III-R criteria must have been retrospectively applied. Nevertheless the operationalization of eating behaviour in childhood (age 1–10) seems appropriate, because up to now only patients' retrospective recall of early childhood eating behaviour is available, and this might be affected by the present diagnosis ED.

All variables showed significant stability over the entire time span, although, as expected, stabilities were higher across the 2-year span in adolescence than across the longer 8-year interval. All four of the problems measured at time 1 were predictive of at least one problem 8 to 10 years later, e.g. pica and problem meals in early childhood were a risk for BN or extreme symptoms of BN later on. Interestingly, picky eating proved to be a protective factor for bulimic symptoms later on. Finally, diagnosable levels of AN or BN were presaged by elevated symptoms of these disorders 2 years earlier (at mean age 14), suggesting both an insidious onset and maybe an opportunity for secondary prevention.

Rathner (1991)[1] conducted a study among 517 female adolescents from two state secondary schools (age range 11–15 years) and both high schools (age range 15–19 years) in a small rural town in Italy. As this town is bilingual, only girls whose native language is German were included. The broad age range (11–21 years) was chosen to cover the peak incidence age for both AN and BN. Written informed consent of parents may have reduced the response rate for screening, which was 80.4%. The study population was simply selected because of permission by school authorities, which was refused in Austria. The sample size seems appropriate because of the supposed 1% point prevalence of AN and the supposed lower prevalence of BN in this age bracket. Concerning representativity, one state secondary school was excluded randomly and the screening sample included every second girl of this town. Social class distribution matched the general population, although working adolescents (above age 15) were not included as they were no longer attending school.

The screening stage did not only rely on a single instrument, the EAT, because of its well-known low positive predictive value (see Williams et al.

[1] Detailed data are available from the author on request.

1982), but instead relied on four risk indices: EAT, Anorexia Nervosa Inventory for Self-Rating (ANIS, which covers also more general aspects of eating pathology), measured underweight (<85% MPMW) and secondary amenorrhoea (>6 months). In addition a subrisk group with sub-cut-off test scores was formed to overcome the problem of the rather arbitrary cut-off points of EAT and ANIS. Unfortunately, no self-report instrument for general psychiatric morbidity like the GHQ has been included in the screening. For evaluation of the natural course the interview stage was conducted 1 year later. It included the same instruments as in the screening and a structured interview for ED (Fichter 1985). Third party information was provided by investigation of case records of four referral hospitals in the vicinity to find out whether an ED was missed in the 517 girls.

All risk cases and a random sample of the subrisk (approx. every second person) and the control groups (7.8% of total control group) were invited for interview at follow-up and only few at-risk individuals refused (participation rate of 94.4%; n = 101). The categories for interview stage were AN/BN (DSM-III-R), subclinical AN/BN, overweight dieters, normal dieters and normal. There was no significant difference between the interviewed sample and the screening sample except for defined risk criteria. As an interview assessment was conducted only at follow-up, obviously no change in diagnostic categories can be reported. But the longitudinal development of eating attitudes and behaviour in different risk groups and in relation to the expert rating in the follow-up revealed the following results. Individuals with clinically defined and partial syndromes were taken together. Longitudinally, in all diagnostic groups there was no change of (relative) weight and desired weight except for a somewhat more realistic desired weight in overweight dieters. Figure consciousness as the prerequisite of dieting is extremely stable for 1 year in all groups, overweight dieters at the top of the scale with 96% who felt too fat now. Dieting showed a slight nonsignificant decrease (from 33% to 27%; exactly resembling the figures of Patton et al. 1990) in the whole longitudinal sample and in normals and overweight dieters, but not in normal dieters. The sharp increase of dieting in the ED group is in accordance with the fact that two girls developed AN in the months between.

With regard to the evolution of test scores, an analysis for total longitudinal sample, single (1-year) age brackets and different risk groups yielded almost no significant change in EAT and ANIS factors or total scores. Within the diagnostic groups, the EAT and ANIS did not improve 1 year later, except insufficiency (ANIS) for all groups except normal girls. The ED groups also improved in this factor – perhaps the control of eating gave them a feeling of personal worth. But the latter group showed higher scores in negative effects of food and bulimia (ANIS) at follow-up.

The core result is the extreme stability of both disturbed and undisturbed eating attitudes and behaviour 1 year later. Overweight dieters showed most abnormal eating attitudes, even higher than the ED group. It seems that minor reductions of dieting frequency, a somewhat healthier weight (ED group) and a more realistic desired weight (overweight dieters) can be achieved, but generally subjects start in one diagnostic category and remain there. These findings need to be tested in the reinterview of this sample in order to make more conclusive statements on the interchange between this diagnostic groups and the relationship between clinical and subclinical forms.

A summary of the 8 longitudinal two-stage surveys is shown in Tables 3 and 4. All but one study were conducted in adolescents at risk (except Marchi's and Cohen's random sample from census tracts in schools because of easier sampling and two studies in particular risk groups, ballet dancing students). Only one study (King 1989) investigated adults using GP records. Whereas Marchi and Cohen adopted a two-stage interview approach, the rest most commonly used the EAT as the only screening tool. Only rarely were additional criteria for selection of cases at risk applied. Follow-up duration ranged from 6 months to 10 years, and interview response rate varied from 58%–94%. Only in most recent studies have structured interview schedules been applied and only three studies with a large sample used two longitudinal interviews. Some studies adopted a longitudinal follow-up by mailed questionnaires, whereas others used screening at initial and follow-up stages but an interview at follow-up only. Diagnostic categories seem broadly adequate. Despite these inconsistencies some preliminary trends can be drawn: there seems to be a general stability of categories, e.g. Schleimer's finding that most subjects are still in the dieting, subclinical or clinical category 10 years later. Other studies found some shift between different categories. The overall impression is that between two thirds and three quarters of subjects remain in their category, at least in the short term of 1 or 2 years. Also only minor improvements in disturbed attitudes and behaviour are shown. These tendencies are not contradicted by Szmukler's finding of considerable improvement of cases in a very small high risk sample. Despite the stability, dieting is probably not an innocuous practice, as some findings show dieting as a possible pathway to ED. The search for risk factors seems unrewarding as yet and meaningful results are scarce. Beside these trends the question of evolution of normal and disturbed eating remain open, but further longitudinal community studies of this kind may clarify this issue and the relationship of clinical and subclinical forms.

Conclusions

Longitudinal data on the course of EDs are needed to investigate new case onset (incidence), recurrence or remission rates. That means: following up nonclinical conditions at initial interview to calculate incidence rates, following up those with an ED to calculate remission rates and following up

Table 3. Review of longitudinal two-stage surveys: populations and sampling characteristics

Authors	Location	Subjects	Sample finding procedure	Screening sample size (T0)	Interview sample size (T0)	Age (years) at T0 mean (SD) (range)	Participation rate[a] (%)	Representativity
Schleimer 1983	Urban Sweden	AF	Upper secondary, trade and vocational schools	1241 F 1129 M	111 F	17.5 (14–28)	Nonparticipation matter of chance/85.4	Very high; interview sample mainly upper social classes
Weeda-Mannak 1984	Urban NL	AF	Secondary schools and university	237 AF	Same as screening	AF 19.5	n.r.	n.r.
		AM BA (all female)	Gynecologists Ballet schools	82 AM 105 BA		AM 24.3 BA 19.4 (min. 16 years)		
Szmukler et al. 1985	Metropolitan UK	AF	Ballet school	100	25	15.6 (1.6) 13–20	About 50/100	High standard, ballet school
Garner et al. 1987	Metropolitan Canada	AF	Ballet school	55	–	13.0 (0.6) (11–14)	100/–	High standard, ballet school
King 1989	Metropolitan UK	Adults	GP attenders	534 F 186 M	96 F 13 M	24.9 F 25.2 M (16–35)	96.3/90.1	Social class representative; overrepresentation of caucasians

Patton et al. 1990	Metropolitan UK	AF	State secondary schools	1010	230	14–16	90.0/90.2	Schools broadly representative for London, but not randomly selected; social class representative
Marchi and Cohen 1990	Metropolitan USA	Mothers and children	Random sample from census tracts	–	976	about 6 (1–10)	–/86.0	Yes
Rathner 1991	Rural Italy	AF	State secondary and high schools	517	–	14.8 (2.2) (11.4–20.4)	80.8/–	Social class corresponds to general population; every second girl included

AF, adolescent females; AM, amenorrhoea patients; BA, ballet students; GP, general practitioner; n.r., not reported; F, female; M, male.
[a] With regard to selected screening sample/to interview stage.

Table 4. Review of longitudinal two-stage surveys: logitudinal sampling, methods and results

Authors	Longitudinal sample (n)	Follow-up duration	Participation rate at follow-up[a] (%)	Screening methods	Selection cniteria for 2nd stage	Interview schedules	Diagnostic criteria	Results
Schleimer 1983	Reinterview 97	10 years	87.4 (74.6 compared to selected interview sample T0)	Nylander questionnaire	Anorectic symptoms, weight loss, dieting or psychiatric referral	Clinical psychiatric, semi-structured	Dahlem Conference (AN) Fries (anorectic behaviour)	Dieting, anorectic behaviour and AN remain distinct categories 10 years later
Weeda-Mannak 1984	AF 171 AM 50 BA 86 Reinterview	14 months 6 months 9 months	72.2 61.0 81.9	Weight and height measured, ANIS, Achievement Motivation Test	All subjects interviewed	Clinical psychiatric, semi-structured	Modified DSM-III (AN)	Different predictors of weight loss and preoccupation in differ groups; except covariation of weight and ANIS longitudinal data n.r.
Szmukler et al. 1985	Reinterview 7	1 year	n.r.	EAT-40, self-constructed questionnaire	Interviews with high-risk and few low-risk subjects	Clinical psychiatric, semi-structured	Possible cases approximated DSM-III (AN)	Considerable improvement of cases
Garner et al. 1987	Interview 32	2.7 years (mean) (min 2 years)	58.2	EDI, FAM, LC, Feelings of Inadequacy Test	Weight, weight history menstrual status, EDI high scorers (score unspecified)	Clinical psychiatric	DSM-III (AN) Russell (BN) Szmukler (PS)	Fluctuation between diagnoses; outcome not favourable

King 1989	(Mailed) 61 only high scorers	12–18 months	89.1 F 80.0 M	EAT-26, demographic questionnaire	High scorers + 7% of low scorers	CIS, Eating Interview, SRT	Russell (BN), partial syndrome	3/4 remain in their group, deterioration more likely than amelioration; EAT decreased, SRT increased
Patton et al. 1990	Rescreening 735 Reinterview 176	1 year	Rescreening 93.6 Reinterview 76.5	EAT-26, GHQ-28	EAT and GHQ high scorers, some low scorers	Clinical psychiatric, semi-structured, SSSI	Russell (AN/BN), partial syndrome	Restitution rate of EAT set off by incidence rate, cases in remission nearly doubled by new cases; 2/3 remain in their diagnostic category
Marchi and Cohen 1990	326 F, 333 M and mothers; reinterview	10 years	68.2	–	–	DISC, lay interviewers	DSM-III-R	Stability of all variables over time; picky eating protective for later bulimia
Rathner 1990	Interview 101	1 years	94.4	EAT-40, ANIS, weight, height, questionnaire	EAT and ANIS high scorers, weight, amenorrhoea; 46% medium and 8% low scorers	SIANEX, clinical psychiatric	DSM-III-R, subclinical AN/BN	Stability of disturbed and normal eating attitudes and behaviour

AN, anorexia nervosa; AF, adolescent females; AM, amenorrhoea patients; BA, ballet students; ANIS, Anorexia Nervosa Inventory for Self-Rating; EAT, Eating Attitudes Test; EDI, Eating Disorders Inventory; FAM, Family Assessment Measure; GHQ, General Health Questionnaire; LC, Locus of Control Scales; BN, bulimia nervosa; PS, partial syndrome; CIS, Clinical Interview Schedule; SRT, Symptom Rating Test; SSSI, Social Stress and Support Interview; DISC, Diagnostic Interview for Children; SIANEX, Standardized Interview Anorexia Nervosa Expert Assessment; M, male; F, female; n.r. not reported.

these with disorders in the past to calculate recurrence rates. A longitudinal study also makes it possible to study in detail the interval service use rates in mental health, medical, and other human service sectors by people who fall in the distinct categories at the time of the first interview. Obviously the reasons that two-stage surveys are relatively rare are the methodological difficulties and the labour intensity. Compared to the developing methodology and procedures in general and child psychiatric epidemiology in the last decade (e.g., Costello 1989; Bland et al. 1988; Eaton et al. 1984) there is a delay in adoption of these developments in the field of ED.

The tentative conclusion from the available research on the natural course of EDs is that there is considerable stability of disturbed attitudes and behaviour at least in the short term. As this happens in the high risk period of developing an ED, it challenges our efforts in secondary and tertiary prevention. With regard to eventual preventive significance of epidemiological surveys it would seem that one can detect cases effectively with an adequate two-stage survey, but a considerable proportion of these cases would never seek treatment. Perhaps future epidemiological studies should not express any clinical interest or interest in referring for further treatment as this would not be accepted by subjects – see high denial rates of previous psychiatric contacts (King 1989) or nonacceptance of treatment offered (Rathner 1991). Instead studies should emphasize the scientific interest (and the favour people do to us) for investigating the course of today's common problems!

With regard to primary prevention the importance of the sociocultural ideal of slimness is a common finding in epidemiological surveys; dieting does not seem to be so innocuous and perhaps treatment of this 'normal' phenomena should be considered more seriously (Polivy and Herman 1987). A shift is needed from the individual high-risk approach to the population or public health approach (Rose 1989). If it is true that the value of the average (the ideal of slimness in Western industrialized countries) determines the prevalence of extreme values (EDs), then, rather than being inclined to rescue operations like treatment, a radical but perhaps more effective approach to prevention should be socioeconomic, politic and behavioural.

Recommendations

This review of the available state-of-the-art longitudinal studies reveals that each study – including our own – is hampered by at least one methodological flaw. Most apparently only two studies with a larger sample have conducted interviews at two time points. Instead of going into detail with further critique, it may be more beneficial to develop some provisional guidelines for studies of the natural history of ED in the community which should facilitate further research.

The Perfect Study Design

1. Defined aims of study: No more cross-sectional epidemiological studies seem necessary, but two-stage prospective and longitudinal epidemiological surveys are needed to increase our knowledge of natural history.
2. Clear-cut diagnostic criteria (DSM-III-R, ICD-10) must be adopted for meaningful clinical comparison between studies to be made.
3. Operational definitions of partial/subclinical forms of EDs, of normal dieting (in normal weight and overweight individuals) and of obesity (including overweight bulimics) should be obligatory. Because of the continuing debate about diagnostic issues, it would be wise to separate subclinical diagnoses into restricting and bulimic forms (Rathner 1990). This enables us to monitor the changes between different diagnostic categories at the clinical and subclinical level. Obesity should be included, because overweight subjects may show the most disturbed eating attitudes, as being most deviant from current sociocultural ideals.
4. Screening stages should use accepted, reliable and valid instruments measuring specific and general psychopathology of ED also and measuring general psychiatric status. Objective measure (weight, height) and of hard data (menstruational status, oral contraceptive use) should be included. Because of the low positive predictive values of any known screening tool for EDs more than one self-report test should be used, with a predetermined criterion change imposed. Thus a multi-method approach is necessary.
5. Samples must represent the general population and a shift is necessary from studies relying only upon private schools or – better – state schools to now include working female adolescents. There are considerable difficulties in sampling adults; however, general population adult surveys are needed. Particularly general practice surveys, studies in the work-site or random samples of small communities should be done.
6. Defined cut-off scores for high-risk subjects, but also for subthreshold test scores should be applied. So far only one risk study (Rathner 1990) has used multiple cutting-off points for screening tests. Discordance occurs mostly among the borderline cases. A control group must be included in the interview sample to detect false-negative cases and to establish the sensitivity and specificity (negative and positive predictive value) receiver operating characteristics within that sample.
7. Personal interviews with structured interview schedules for ED and for general psychiatric status should be conducted at both initial and follow-up testing. Re-interviews are mandatory for meaningful conclusions about natural history.

8. Interviewers should not know test scores. They should have been trained previously or be experienced clinicians, but at best both. The use of trained lay interviewers and of standardized interview instruments like the Diagnostic Interview Schedule (Robins et al. 1981) has, of course, advantages for research economy, reliability and replicability in gathering information, but has also disadvantages especially in diseases like EDs, in which denial, shame, guilt and therefore nondisclosure may play a major role.
9. Cross-validation of the results of interview stages by third party information (relatives, GPs or hospital records) is a possible corrective measure for denial, limited self-disclosure and social desirability in ED subjects.
10. Participation rate in these two stages should be as high as possible since anything less than 100% introduces response biases. As a minimum all available information on the nonresponders should be statistically checked against that of responders to estimate response biases. Efforts should be made to establish nonresponse bias by checking doctors/hospital records, etc.
11. Short-term follow-up (1–2 years later) and mid-range follow-up at least 4 years past first interview should be used to detect time and age factors in the evolution.
12. Rates of treatment received (with the usual standard in follow-up or psychotherapy research studies) should be monitored.
13. Investigation of intervening life events with standardized methods would be useful because the tremendous impact of life events on amelioration or deterioration of physical and psychological well-beings is well known.
14. Investigation of the personality axis and the social axis should be performed.
15. Although overall stability is of importance, special scrutiny should be given to the changes within groups (see 3) and between diagnostic groups and possible causative or mediating factors.
16. The problem of statistical significance vs. clinical meaning: elaborate statistical procedures cannot overcome clinically meaningless data. Despite the predominant nomothetic approach in psychological sciences, some idiographic ideas can be included in longitudinal epidemiological studies.

Acknowledgements: For their criticisms of earlier drafts I should like to thank Dr. Bridget Dolan and Dr. Chris Evans (London).

References

Attie I, Brooks-Gunn F (1989) Development of eating problems in adolescent girls: a longitudinal study. Dev Psychol 25:70–79

Bland RC, Newman SC, Orn H (1988) Epidemiology of psychiatric disorders in Edmonton. Acta Psychiatr Scand [Suppl] 338:1–80

Bruch H (1988) Conversations with anorexics. Basic Books, New York

Button EJ, Whitehouse A (1981) Subclinical anorexia nervosa. Psychol Med 11:509–516

Costello EJ (1989) Developments in child psychiatric epidemiology. J Am Acad Child Psychiatry 28:836–841

Crisp AH (1980) Let me be. Academic, London

Derogatis LR, Lipman RS, Rickels K, Uhlenhuth EH, Cori L (1974) The Hopkins Symptom Checklist (HSCL): a self-report symptom inventory. Behav Sci 19:1–14

Dolan B, Lacey HJ, Evans C (1990) Eating behaviour and attitudes to weight and shape in British women from three ethnic groups. Br J Psychiatry 157:523–528

Drewnowski A, Yee DK, Krahn DD (1988) Bulimia in college women: incidence and recovery rates. Am J Psychiatry 145:753–755

Eaton WW, Holzer CE, Von Korff M, Anthony JC, Helzer JE, George L, Burnam MH, Boyd JH, Kessler LG, Locke BZ (1984) The design of the epidemiological catchment area surveys. The control and measurement of error. Arch Gen Psychiatry 41:942–948

Fenley J, Powers PS, Miller J, Rowland M (1990) Untreated anorexia nervosa: a case study of the medical consequences. Gen Hosp Psychiatry 12:264–270

Fichter M (1989) Bulimia nervosa und bulimisches Verhalten. In: Fichter M (ed) Bulimia nervosa. Enke, Stuttgart, pp 1–9

Fichter MM (1985) Strukturiertes Interview zur Anorexia nervosa (SIAN-EX). University of Munich, Internal Technical Report

Fichter MM, Keeser W (1980) Das Anorexia-nervosa-Inventar zur Selbstbeurteilung (ANIS). Arch Psychiatr Nervenkr 228:67–89

Fries (1977) Studies on secondary amenorrhoea, anorectic behavior and body-image perception: Importance for early recognition of anorexia nervosa. In: Vigersky RA (ed) Anorexia nervosa. Raven, New York, pp 163–176

Garner DM, Garfinkel PE (1979) The Eating Attitudes Test: an index of the symptoms of anorexia nervosa. Psychol Med 9:273–279

Garner DM, Olmsted MP, Bohr Y, Garfinkel PE (1982) The Eating Attitudes Test: psychometric features and clinical correlates. Psychol Med 12:871–878

Garner DM, Olmsted MP, Polivy J (1983) Development and validation of a multidimensional eating disorders inventory for anorexia nervosa and bulimia. Int J Eating Dis 2:15–34

Garner DM, Garfinkel PE, Rockert W, Olmsted MP (1987) A prospective study of eating disturbances in the ballet. Psychother Psychosom 48:170–175

Garrow JS, Crisp AH, Jordan HA (1975) Pathology of eating. Group report. In: Silverstone T (ed) Dahlem Konferenzen. Life Sciences Research Report 2, Berlin, pp 405–416

Goldberg DP, Cooper B, Eastwood MR, Kedward HB, Shepherd M (1970) A standardised psychiatric interview for use in community surveys. Br J Prev Soc Med 24:18–23

Goldberg DP, Hillier VF (1979) A scaled version of the General Health Questionnaire. Psychol Med 9:139–145

Hsu LKG (1988) The outcome of anorexia nervosa: a reappraisal. Psychol Med 18:807–812

Huenemann RL, Shapiro LR, Hampton MC, Mitchell BW (1966) A longitudinal study of gross body composition and body conformation and their association with food and activity in a teenage population. Am J Clin Nutr 18:325–338

Jenkins R, Mann AH, Belsey E (1981) The background design and use of a short interview to assess social stress and support in research and clinical settings. Soc Sci Med 153:195–203

Johnson CL, Stuckey MK, Lewis LD, Schwartz D (1982) Bulimia: a descriptive survey of 316 cases. Int J Eating Dis 2:1–16

Johnson C, Tobin DL, Lipkin J (1989) Epidemiologic changes in bulimic behavior among female adolescents over a five-year period. Int J Eating Dis 8:647–655

Johnson-Sabine E, Wood K, Patton G, Mann A, Wakeling A (1988) Abnormal eating attitudes in London schoolgirls – prospective epidemiological study: factors associated with abnormal response on screening questionnaires. Psychol Med 18:615–622

Kellner R, Sheffield BF (1967) Symptom rating test scores in neurotics and normals. Br J Psychiatry 113:525–526

Kellner R, Sheffield BF (1973) A self-rating scale of distress. Psychol Med 3:88–100

King MB (1989) Eating disorders in a general practice population. Prevalence, characteristics and follow-up at 18 months. Psychol Med [Monogr Suppl] 14:1–34

Lacey JH (1985) Time-limited individual and group treatment for bulimia. In: Garner DM, Garfinkel PE (eds) Handbook of psychotherapy for anorexia and bulimia. Guilford, New York, pp 431–457

Mann AH, Wakeling A, Wood K, Monek E, Dobbs R, Szmukler G (1983) Screening for abnormal eating attitudes and psychiatric morbidity in an unselected population of 15-year-old school girls. Psychol Med 13:573–580

Marchi M, Cohen P (1990) Early childhood eating behaviors and adolescent eating disorders. J Am Ac Child Psychiatry 29:112–117

Nylander I (1971) The feeling of being fat and dieting in a school girl population. Acta Sociomed. Scand 1:17–26

Patton GC, Johnson-Sabine E, Wood K, Mann AH, Wakeling A (1990) Abnormal eating attitudes in London schoolgirls – a prospective epidemiological study: outcome at twelve month follow-up. Psychol Med 20:383–394

Polivy J, Herman CP (1987) Diagnosis and treatment of normal eating. J Consult Clin Psychol 55:635–644

Pyle RL (1983–85) The epidemiology of eating disorders. Pediatrician 12:102–109

Rathner G (1989) Häufigkeit der Anorexia nervosa und Bulimia nervosa. In: Söllner W, Wesiack W, Wurm B (eds) Soziopsychosomatik. Springer, Berlin Heidelberg New York, pp 39–54

Rathner G (1991) The natural course of adolescent eating attitudes and behaviour: a prospective epidemiological study (to be published)

Robins LN, Helzer JE, Croughan J, Ratcliff KS (1981) National Institute of Mental Health Diagnostic Interview Schedule. Arch Gen Psychiatry 38:381–389

Rose G (1989) High-risk and population strategies of prevention: ethical considerations. Ann Med (Finnish Medical Society Duodecim) 21:409–413

Rosen JC, Tacy B, Howell D (1990) Life stress, psychological symptoms and weight reducing behavior in adolescent girls: a prospective analysis. Int J Eating Dis 9:17–26

Russel GFM (1977) General management of anorexia nervosa and difficulties in assessing the efficacy of treatment. In: Vigerski R (ed) Anorexia nervosa. Raven, New York, pp 270–289

Rutter M, Sandberg S (1985) Epidemiology of child psychiatric disorder: methodological issues and some substantive findings. Child Psychiatry Hum Dev 15:209–233

Schleimer K (1983) Dieting in teenage schoolgirls. A longitudinal prospective study. Acta Paediatr Scand [Suppl] 312:1–54

Sours (1980) Starving to death in a sea of objects. The anorexia nervosa syndrome. Aronson, New York

Striegel-Moore RH, Silberstein LR, Frensch P, Rodin J (1989) A prospective study of disordered eating among college students. Int J Eating Dis 8:499–509

Szmukler (1983) Weight and food preoccupation in a population of English schoolgirls. In: Bergmann JG (ed) Understanding anorexia nervosa and bulimia. Ross Laboratories, Columbus, pp 21–28
Szmukler G, Eisler I, Gillies C, Hayward ME (1985) The implications of anorexia nervosa in a ballet school. J Psychiatr Res 19:177–181
Vandereycken, Meermann (1984) Anorexia nervosa: a clinician's guide to treatment. de Gruyter, Berlin
Weeda-Mannak WL (1984) Anorexia nervosa: towards an early identification. Thesis Rijksuniversiteit Limburg, Maastricht
Weeda-Mannak WL (1985) The discriminative value of psychological characteristics in anorexia nervosa. Clinical and psychometric comparison between anorexia nervosa patients, ballet dancers and controls. J Psychiatr Res 19:285–290
Williams P, Hand D, Tarnopolsky A (1982) The problem of screening for uncommon disorder – a comment on the Eating Attitudes Test. Psychol Med 12:431–434
Yager J, Landsverk J, Edelstein CK (1987) A 20-month follow-up study of 628 women with eating disorders. I. Course and severity. Am J Psychiatry 144:1172–1177
Yager J, Landesverk J, Edelstein CK, Jarvik M (1988) A 20-month follow-up study of 628 women with eating disorders. II. Course of associated symptoms and related clinical features. Int J Eating Dis 7:503–513
Yager J, Landsverk J, Edelstein CK (1989) Help seeking and satisfaction with care in 641 women with eating disorders. J Nerv Ment Dis 177:632–637

Important Factors in the Process of Recovery According to Patients with Anorexia Nervosa*

G. Noordenbos

Introduction

According to Theander (1985 p. 493) "when assessing the total impact of a disease not only is the final outcome of importance, but also the course of the illness over time: the total duration of the illness; the suffering and discomfort it inflicts on the patient and family; whether the course over time is stable or fluctuating; which complications may be encountered; and whether relapses are apt to occur". Stipulating the precise moment when it can be said that a patient with anorexia nervosa has recovered is difficult, because different opinions are possible. Can we say that a patient with anorexia nervosa has recovered when (s)he eats normally and has a normal weight? Or can we only talk about recovery when the patient has a more positive image of herself and her body, which reduces the likelihood of relapses? According to Theander (1985 p. 496) the following criteria are the most important for a favourable outcome in an eating disorder: "spontaneous and regular menstruation, a stable body weight within 15% of the average body weight for height, normal eating habits without bulimic episodes or undue dieting, and a reasonably normal conception of the body without unrealistic overestimation of the body size". Lafeber (1981 p. 90/91) goes a step further in her description of recovery of anorexia nervosa: "As long as the patient is worried about the symptoms of anorexia nervosa, she risks the possibility of relapsing to slimming behaviour when confronted with disappointing or traumatic situations. Only when this flight reaction is definitively conquered can we say that the patient has recovered".

In order to obtain more information concerning the process of recovery of patients with anorexia nervosa, people who considered that they were well on the way to recovery from anorexia nervosa, or had already recovered, were invited to participate in this research study. Only they knew when their anorexia nervosa started, how long they had denied their eating disorder and how long their anorexia nervosa had lasted.

* Parts of this chapter have been published in the *British Review of Bulimia and Anorexia Nervosa* (1989 and 1990).

Further, only they could inform us about their motivation to ask for help and which ways of treatment really helped them to improve from their eating disorder. In this article I first want to describe the process of recovery as experienced by patients with anorexia nervosa; their development from denying to admitting having eating problems; whether and how they become motivated to seek help and to visit a general practitioner; the activities and referrals by general practitioners; the treatments patients received and the duration of their anorexia nervosa. I shall then describe which factors anorexia patients judged to be important for their improvement and recovery, and in which aspects their lives changed. This article is based on qualitative and quantitative research into the processes of development and recovery from anorexia nervosa. Information about the different stages in the process of recovery is based on responses to the questionnaire, while the interviews gave information about factors which the respondents considered to be important for their recovery.

Methods

The research was carried out at the University of Groningen in the Netherlands from 1983 to 1987. To select respondents for the research, I placed advertisements in local newspapers. I briefly described the characteristics of anorexia nervosa and asked for respondents who had improved or had recovered from anorexia nervosa to participate in the research. They had to fill in a questionnaire regarding eating behaviour, weight history, family characteristics, predisposing factors and questions about their treatment and process of recovery.

In total 129 persons asked for further information about our research, after which 120 respondents filled in the questionnaire. In total 108 questionnaires could be used: 12 questionnaires were not usable, because there were too many questions were left unanswered. The questionnaire was completed by 104 women and 4 men. Thirty-seven respondents were interviewed about their answers. All of them admitted having had anorexia nervosa. Their mean lowest weight was 40.9 kg (SD = 7.6) and their mean percentage of weight loss was 31% (SD = 11.2%). The lowest body mass index was 14.6 (SD = 2.6; Noordenbos 1987b). During the research they had improved or had recovered according to the criteria of improved weight and they were no longer obsessed by dieting and slimming. Possible relapses in the future, however, are not excluded, so these criteria must be considered to be relative.

Hypotheses of Stages in the Recovery Process

In general the recovery process of people with anorexia nervosa comprises the following stages:

1. Passing from denial to admission of the condition
2. Becoming motivated to look for help
3. Taking the step to consult the general practitioner
4. Following one or more treatments
5. Changing in physical, psychological and social aspects

Not every person with anorexia nervosa follows all these stages. Not all of them visit a general practitioner or are treated for anorexia nervosa. But in this research most respondents did follow the above-mentioned stages.

The following hypotheses have been made about the process of recovery:

1. In the literature on anorexia nervosa, much has been written about the denial of being ill and the resistance to looking for help by patients with eating disorders. It is assumed, however, that this is only valid for the first stage of anorexia nervosa, in which the person experiences slimming behaviour positively. The denial of illness does not last during the whole course of the eating disorder, especially when patients are confronted with all the negative effects of their slimming behaviour.
2. It is assumed that when patients are informed about the characteristics of anorexia nervosa, this has a positive effect on their motivation to look for help.
3. Confronted with all the negative effects of their slimming behaviour, patients with anorexia nervosa admit that their physical condition has to be improved.
4. Considering the resistance to change anorectic behaviour, patients with anorexia nervosa hardly ever take the initiative to consult a general practitioner and are usually sent to a doctor by their parents, partners or friends.
5. If the general practitioner does not diagnose anorexia nervosa and prescribes medication or refers patients to a medical specialist for a physical examination, effective treatment is likely to be delayed.
6. Because anorexia nervosa is very complex, it is expected that patients with this eating disorder follow more than one therapy, especially when they first receive a treatment which is exclusively directed at normalizing eating behaviour and bringing about weight gain.
7. To recover from anorexia nervosa, it is not enough to normalize eating behaviour and weight. It is assumed that a positive outcome of the treatment correlates with the degree of attention which is paid to the following aspects:
   a. Increased self-confidence
   b. A more positive feeling regarding one's body
   c. Better relationships with others
   d. Learning to express negative emotions better
   e. Learning to resolve conflicts

f. Learning what is important for oneself, instead of conforming to the expectations of others
g. More insight into the problems underlying one's eating behaviour.

In the following paragraphs I shall describe the factors which were helpful in getting patients to admit that they had eating problems; in becoming motivated to ask for help; in going to the general practitioner; in getting treatment for anorexia nervosa; and in changing eating behaviour, physical condition, and feelings about themselves, their bodies and their social contacts.

Results

Did Patients with Anorexia Nervosa Admit to Having Problems and Did They Decide to Look for Help?

Most patients (72.2%) denied having any problems at the beginning of their disease. But almost 27.8% admitted early in the course of the disease that their way of eating was abnormal. In accordance with our hypothesis, denial of illness is not a permanent characteristic of people with anorexia nervosa, since 95.2% admitted sooner or later that something was wrong. Most of these individuals (81.9%) even had the feeling that they could no longer continue their anorectic behaviour and had to do something to change their situation. Only 18.1% never felt the need to change their behaviour. One third of individuals realized in the 1st year they had anorexia nervosa that they must change the anorectic behaviour, but for 9.9% this turning point did not come until they had been ill for 5 years or more.

What helped these individuals to admit that their eating behaviour was abnormal? Most respondents indicated that reading about the characteristics of anorexia nervosa was the most important factor in admitting to having anorexia nervosa. Most respondents did not hear about anorexia nervosa from a doctor or a specialist, but rather from the media, self-help organizations for anorexia nervosa, or from family or friends. The effect of admitting to having anorexia nervosa was that patients no longer had the alienating idea that they were the only ones who were obsessed with dieting and slimming.

Were Patients Who Admitted They Had Eating Problems Then Motivated to Seek Help in Changing Their Behaviour?

In Table 1 the time at which after the start of the severe slimming behaviour respondents admitted to having anorexia nervosa and in how many motivation for treatment improved are indicated. In addition the

Table 1. Time of admitting to having anorexia nervosa, improvement in motivation to seek help, and duration of anorexia nervosa

Time of recognition	Percentage of respondents (n = 108)	Improvement in motivation (%)	Duration of anorexia nervosa (years)
Within 6 months	17	29	5.5
Between 6 months and 1 year	21	75	4.4
Between 1 and 2 years	14	50	6.5
Between 2 and 3 years	15	50	6.5
After more than 3 years	33	54	12.3

duration of anorexia nervosa in these patients is shown. Except for those patients who admitted having anorexia nervosa within 6 months of starting the severe slimming behaviour, 50% or more of those who admitted to having anorexia nervosa became more motivated to seek help. It could be that 6 months is too short a time to change such behaviour. However, it is interesting to see that those who admitted to having anorexia nervosa within 1 year most often answered that they became motivated to seek help, while after 1 year it may be more difficult to become motivated, because it is more difficult to change behaviour. Even when these individuals felt they could no longer presist in their anorectic behaviour, slightly less than half did not decide to seek help, because they were too afraid and too ashamed to talk about their eating behaviour. One patient related the following:

"I got this feeling only after 5 years. I weighed 40 kg and wanted to stop, but I felt that I could not stop slimming, which made me anxious. I had oedema and much abdominal pain, because I used a lot of laxatives, even when I had eaten only one piece of bread. But I felt very ashamed and I dared not say anything about my eating behaviour".

Admitting to anorexia nervosa early in the condition does not, however, guarantee a shorter duration. Other factors play an important role, too, such as the specific information heard about anorexia nervosa, the length of time spent waiting for help, the treatment received, and so forth.

Do Patients with Anorexia Nervosa Admit That Their Physical Condition Has to Improve When Confronted with All the Negative Effects of Their Slimming Behaviour?

In total 77 respondents (71.3%) answered that their physical condition was very bad while they had anorexia nervosa. Of all respondents 24% answered that their life had been in danger. Only 57% of those who

Table 2. Basis of initiative to consult a doctor (n = 93)

Basis	Percentage
Own initiative	45
In agreement with others	11
Others wished they would go	12
Others insisted they go	32

answered that their physical condition was bad or critical admitted *at that time* that some treatment was necessary, while 43% still did not feel that it was necessary to improve their physical condition. Only *afterwards* did 86% of those who evaluated their physical condition as bad admit that it was necessary to restore their physical condition. In the interviews most stated that it had taken a long time before they admitted that they had endangered their lives. Admitting that their physical condition was bad did not mean that they would then admit that it was necessary to restore their physical condition Hence, waiting with physical treatment until patients become motivated may delay a treatment which only *afterwards* can be evaluated as meaningful.

Did Patients Visit a General Practitioner with Complaints Concerning Anorexia Nervosa?

Did patients with anorexia nervosa consult a doctor of their own accord or were they usually sent to a doctor by someone else? Some 93 respondents (87%) consulted a doctor sooner or later, with complaints which – in retrospect – were connected with the symptoms and effects of anorexia nervosa. In a comparable research study in the Netherlands de Bloois (1987) found that 93% of the 293 respondents with anorexia nervosa consulted a doctor sooner or later. Table 2 shows the breakdown percentages in patient initiative.

It is remarkable that almost half of the respondents took the initiative to consult a doctor. Most of those who did lived away from their parents during the time they had anorexia nervosa and were older than those who went to the doctor because others wished them to go. Those who took the initiative themselves were more reluctant to discuss their eating behaviour and degree of slimming, because they felt very ashamed or fearful that the doctor would tell them they had to eat more and gain weight. Most of them reported complaints which were only indirectly connected with anorexia and bulimia nervosa, such as cold fingers, constipation, pain in the stomach or intestines from using laxatives, etc. Some of them tried to involve the doctor in their slimming by asking for laxatives or by complaining that they always had pain in their stomach or by asking for a diet, which would legitimize their eating behaviour for others.

Table 3. Treatments prescribed by doctors consulted (n = 93)

Treatment	Percentage
None	20.4
Prescribed diet	15.7
Prescribed medication	38.0
Talked with me about my problems	13.9
Referred to a medical specialist	52.8
Referred to a psychiatrist	18.5
Referred to a psychologist	10.2
Told me that I had anorexia nervosa	2.8

Table 4. Number and kind of treatments

Number of treatments (n = 108)	Percentage	Kind of treatment (n = 78)	Percentage
0	28	Medical treatment	51
1	23	Behavioural treatment	19
2	22	Psycho-analysis	9
3	16	Psychotherapy	51
4	8	Family therapy	17
5 or more	3	Self-help	23

Most of those who were sent to a doctor by others lived at home during their anorexia nervosa (61%). They were less or not at all motivated to discuss their eating behavour, tried to manipulate their weight and so forth. In general their mothers accompanied them to the doctor. This made it easier for the doctors to diagnose anorexia nervosa, because the mothers generally reported about eating behaviour, degree of slimming, amenorrhea or other visible symptoms.

Did Doctors Prescribe Medication or Refer These Patients to a Medical Specialist for Physical Examination?

In Table 3 the treatments prescribed by doctors who where consulted by anorectic patients are listed. The 93 respondents who had consulted a doctor mentioned in total 215 activities on the part of their doctors; that is a mean of 2.3 per person. This may be an indication that most patients consulted a doctor more than once with complaints concerning anorexia nervosa. These results are comparable with those of Winstead and Willard (1983): 38% of these women indicated that their doctors prescribed medications and 52.8% of them were referred to a medical specialist. The high number of patients who were treated medically implies that physical aspects were very important to these doctors, or that they needed further medical examination for their diagnosis. Prescribing medications such as vitamins, laxatives, drugs for increasing appetite, or hormones to stimulate menstruation appeared not to be effective according to the respondents and for the most part they delayed a more effective treatment. The same can be said for further specialized investigations, such as X-rays, gastro-intestinal tests, etc.

How Many Treatments Did These Patients Follow?

In Table 4 the number and kind of treatments patients with anorexia nervosa followed are summarized. Of the 108 respondents 72% received treatment for anorexia nervosa, while 28% did not. Most of the treated anorexia nervosa patients received more than one treatment. Generally their first treatment was medical or behavioural, followed by psycho-analysis, psychotherapy, family therapy or self-help. De Bloois (1987) found in her research on 293 patients that the mean number of treatments was 5, while 68% of patients received more than one treatment simultaneously. Some 61% of the patients were admitted to hospital a second time.

What Aspects of Treatment Were Evaluated as Positive for Recovery?

It is remarkable that according to the respondents nearly all the treatments were directed towards different aspects of anorexia nervosa. No treatment was directed solely at weight gain and normalization of behaviour. Relatively, medical treatment emphasized attention to weight gain and normalization of eating behaviour most and other aspects least. This treatment was evaluated the most negatively. Surprisingly behavioural therapy was generally evaluated positively, but in these treatments many aspects were dealt with. After medical treatment, psychoanalysis ranked least positive. Psychotherapy and self-help, in contrast, ranked most positive. These treatments emphasized the other aspects of anorexia and bulimia nervosa more, without completely neglecting weight gain and normalizing eating behaviour. In conclusion, most positively evaluated were those therapies which emphasized many of the following aspects of anorexia nervosa: increasing self-confidence, developing a more positive attitude to one's body, forming better relationships with others, learning to express negative emotions better, learning to resolve conflicts, learning better what is important for oneself instead of conforming to the expectations of others, getting more insight into the problems underlying one's eating behaviour.

Table 5. Important goals for treatment according to the patients (n = 108)

Important goals according to respondent	Respondents who ranked this important (%)
Becoming more satisfied with yourself	98.0
Getting more self-confidence	97.1
Feeling more at home in your body	96.1
Better feelings about your body	95.8
Learning what is important for you	94.8
Becoming more assertive	94.2
Strengthening feelings of identity	93.3
Feeling less guilty after conflicts	92.9
Learning to resolve conflicts	91.6
Evaluating your weight as less important	89.5
Becoming less dependent on others	89.2
Taking more initiative towards others	88.5
Learning to express negative feelings	87.5
Becoming less afraid of losing control	87.8
Solving problems better	86.7
Taking more initiative in your life	86.2
Paying less attention to your weight	84.9
Improving relationships	84.2
Not feeling too heavy any longer	80.9
Learning about reasons for your anorexia	78.1
Becoming less obsessed with slimming	78.0
Learning to relax physically	77.9
Learning to enjoy your meals	77.2
Learning not to escape in slimming when you have problems	70.2

At the end of the questionnaire the respondents were asked the following question: What in your opinion are important goals in a therapy for anorexia nervosa? A list of 24 statements about relevant goals for patients with anorexia nervosa is presented in Table 5 in rank order according to the number of respondents rating the goals as important. The aspects of a positive experience of one's self and one's body were evaluated as the most important goals for therapy, while aspects of eating and weight were evaluated as the least important goals. This confirms the hypothesis that therapies for eating disorders should be directed towards all the problems which patients with anorexia nervosa have, and not only to their eating behaviour and weight.

Some Differences Between Treated and Untreated Patients

Nearly all the information concerning anorexia nervosa is about patients who were treated for anorectic symptoms. Very little is known about patients who did not receive any treatment. It is supposed that the prognosis of untreated patients is bad compared with patients treated for anorexia nervosa (Lafeber 1963; Cremerius 1978). To gather more information about untreated patients with anorexia nervosa we compared a group of 78 patients who were treated for anorexia nervosa with 30 anorectic patients who had never received any special treatment. Patients treated for anorexia nervosa received one or more treatments such as medical treatment, behavioural treatment, psychoanalysis, psychotherapy or family therapy.

The results of this investigation revealed that untreated anorexia nervosa patients differed significantly from treated anorexia patients in the following respects: (1) they were more often male; (2) they lost less of their original weight; (3) they had anorexia nervosa for a shorter period of time; (4) they admitted less often that they had anorexia nervosa; (5) they consulted a doctor less often; (6) they were less often referred to a medical specialist, psychiatrist or psychologist by their doctor. The first conclusion might be that to visit a doctor and to be treated for anorexia nervosa patients need to have had more severe anorectic characteristics for a longer period. However, 73% of the untreated patients had had anorexia nervosa for more than 3 years and sometimes considerably longer. In total 66% of the untreated patients consulted a doctor with symptoms concerning anorexia nervosa. When we take into account that 46% of the untreated patients had a weight loss of more than 25% of their original weight, we can conclude that these patients were either incorrectly diagnosed or not treated while they in fact needed treatment.

In a second analysis I divided the untreated patients into the "real anorectic patients", who lost more than 25% of their original weight, and the "subclinically anorectic patients", who lost less than 25% of their original weight. It is remarkable to note that the level of education of the parents of real anorectic patients was significantly lower than that of the subclinical patients, which might have hampered their possibilities of looking for help. Vandereycken and Pierloot (1983) found in a study at dropout that characteristics such as coming from a lower social class and lower educational level were also associated with more dropout. Concerning the subclinical patients the question arises of whether treatment was necessary. I found, however, that they developed bulimia at an earlier stage than the real anorectic patients and had considerable weight fluctuations significantly more often. The subclinical patients might therefore be characterized as patients with bulimia nervosa. This may explain why their problem was not recognized or not taken seriously enough to start any treatment.

Duration of Anorexia Nervosa

Several anorexia nervosa researchers have concluded that the process of recovery usually takes a long time and is accompanied by many relapses

(Sours 1969; Morgan and Russell 1975; de Bloois 1987). In this study respondents were asked to indicate age at the start and at the "end" of their anorexia nervosa. The mean age at onset was 18 years (SD = 4.4, range 13–33). The age at the end of anorexia nervosa averaged 25.6 years (SD = 5.6, range 14–60). Nearly 80% of the respondents indicated that anorexia ended before they were 30 years old, while 20% answered that they were older then 30 years. However, it is difficult to determine what "end" of anorexia nervosa is because a relapse can never be ruled out. Not only is it difficult for a therapist to say when the anorexia has ended, but this is also the case for the patient herself: "At that moment I thought that I had recovered, I ate more and my weight was restored. But after having had a relapse, I will never say again that I have completely recovered, for that is not true. But now I know that when I have difficulties with eating and try to slim again, there are other problems which I have to solve, and I have learned that slimming is no solution for my problems but a signal that I have to solve my problems".

Of the 108 respondents 85% answered that they had anorexia nervosa during one period in their lives, while 10% reported they had had two outbreaks, and 5% had two or three outbreaks of anorexia nervosa. The mean time between these outbreaks was 2 years, but some respondents had a second outbreak of anorexia nervosa after 15 years. Little systematic information has been gathered about the total duration of anorexia nervosa. It appeared that the mean duration of anorexia nervosa of 108 respondents was 7.5 years, ranging from 1 to 44 years (Noordenbos 1987a). Those patients who answered that their anorexia nervosa had ended more than 1 year previously had a mean duration of anorexia nervosa of 6.4 years (SD = 5.5), while those who answered that they had not completely recovered reported a mean duration of 8.2 years (SD = 7). In her research into the process of recovery of 293 people with anorexia nervosa de Bloois (1987) found similar results: the mean duration of the anorexia nervosa of those who had recovered was 5.5 years (SD = 4.1), ranging from 1 to 23 years, and 9.5 years for those who were not completely recovered (SD = 6.6), ranging from 1 to 43 years.

In general it appears that the shorter the duration of illness and the earlier an effective treatment is received, the better is the recovery prognosis. This does not mean, however, that with a duration of anorexia nervosa of longer then 10 years recovery is ruled out. In total 25 anorexia nervosa patients reported that their anorexia nervosa had lasted 10 years or longer, but half of them answered that they had recovered for more than 1 year. It is difficult to say what the possibilities of recovery are after having anorexia nervosa for more than 10 years. This depends on the physical and psychological condition of the patients, the treatment received and the aftercare. But because some patients may improve or recover even after 10 years, recovery is never excluded (Theander 1985).

State of Improvement or Recovery

Although there are no standard recovery criteria for anorexia nervosa, we used that of patients no longer being obsessed by eating and slimming and not being underweight according to the body mass index (BMI). Further, in the course of 1 year they must not have had a relapse. In this study subjects were asked to indicate their weight at the time of the research, their menstrual periods and the duration of their improvement or recovery. Cremerius (1978) and Lafeber (1963) report that the prognosis for untreated anorexia nervosa patients is worse and that untreated patients are more often chronically ill. Hence the state of recovery of treated and untreated patients was compared.

Actual Weight

At the time of the research the mean weight of the respondents was 54 kg. Their BMI was 19.1. The anorexia group was no longer underweight. When we take into account that the lowest BMI during illness was 14.6, then the weight has been restored to nearly normal. Compared to the mean BMI just before the start of the slimming behaviour, which was 21.1, these individuals are not yet back to their original weight. This gives rise to the question of the weight at which it may be reasonable to conclude that patients with anorexia nervosa have recovered. Comparison of the treated patients with the untreated patients during the research indicates that only 10% of the untreated patients were still under 50 kg, compared with nearly 40% of the treated patients. The mean weight of untreated patients was higher than that of the treated patients, maybe due to bulimic behaviour.

Menstruation

During the research only 23 (22%) of the 104 female respondents had regular menstrual periods; 18 (17%) had irregular periods and 64 (62%) had none at all. There was no significant difference between the treated and untreated patients. Of both groups only 22% menstruated regularly. According to the data in Table 6 we cannot be optimistic about their recovery, for although the weight of treated and untreated patients was not remarkably low, most of the female patients still had amenorrhoea, so they had not completely recovered.

Duration of Recovery

As to the question of how long recovery from anorexia nervosa lasted, in total 69 persons (64%) answered that they were well for 1 year or longer,

Table 6. Restoration of menstruation

Menstruation	Untreated (n = 27)	Treated (n = 77)
Not yet	17 (63%)	47 (60%)
Irregular	4 (15%)	14 (18%)
Regular	6 (22%)	17 (22%)

while 39 (36%) replied that they had improved or were well for a period of less than 1 year. Of the treated patients, 52 (67%) answered that they had been recovering for 1 year or longer before participating in the research, compared with 17 (55%) of the untreated patients. However, one third of the treated patients and nearly half of the untreated patients answered that the duration of their recovery was very short, so that relapse could not be excluded. When we look at the prognosis of untreated patients for recovery from anorexia nervosa a seemingly "spontaneous" recovery of anorexia nervosa appears to be possible. Not all untreated patients with anorexia nervosa are doomed to stay chronically ill, for 21% of the treated patients and 25% of the untreated felt that they had been well for more than 6 years.

The Importance of Useful Criteria for Recovery

When we look exclusively at the percentage of weight loss, the abnormal eating behaviour and amenorrhoea, then the end of anorexia nervosa can be described as the point at which weight is restored, eating behaviour normalized and the women have normal menstrual periods. But when we look at the underlying problems of eating disorders, such as lack of self-confidence, self-worth and a negative experience of one's body, then the criteria mentioned above are incomplete. In this case we can only speak of recovery when persons with anorexia nervosa have feelings of self-confidence, self-worth and positive feelings regarding their body. Lafeber (1981) goes a step further and speaks of real recovery only when former patients definitely have overcome their tendency to escape into abnormal eating patterns when they meet with troubles in their lives, so that relapses are excluded. Up to now considerable attention has been paid to criteria for the diagnosis of eating disorders, but it would be well worth the effort to define useful criteria for recovery, so that research into the process of recovery of patients with eating disorders can become more standardized and comparative.

In looking for useful criteria for recovery we asked our respondents to indicate the most relevant factors in their lives which had changed and which had been important in the recovery process. The following changes were reported most often:

1. To increase self-confidence: "I have much more self-confidence and I think I'm a worthwhile person".
2. To have a stronger feeling of identity: "Before and during my anorexia nervosa I never talked about myself, but always about general things. I had a "pseudo-self" who played a role. Now I can say that I express my own feelings and ideas".
3. To become more assertive: 'I followed a training programme to become more assertive and I learned to more clearly express to other people my own wishes and limits".
4. To become less dependent of others: "If you orient yourself at other people for such a long time, then you no longer know what your own wishes are. Now I act according to my own feelings, and not to the expectations of other persons".
5. To learn better what one's own feelings are: "I always suppressed my own feelings, but in the treatment I got I learned what my own feelings were, and what is more important I learned to trust and express my feelings".
6. To learn to face conflicts: "Before and during my anorexia nervosa I always tried to avoid conflicts, because I was very afraid that people would not accept me any longer if I did not agree with them. But I learned that I can discuss problems and state my opinion without losing my friends".
7. To become aware of one's own needs: "Before and during anorexia nervosa I needed to be strong and independent. Having to eat proved that I was weak. I needed a lot of warmth and attention, but I could not combine that with my ideal to need nobody. Now I can enjoy dealing with the experiencing security without losing my own autonomy".
8. To be less obsessed by food and weight: "I think food will remain a weak point, but I can enjoy eating and I do not think about what I eat each time and what the consequences will be for my weight".
9. Having fewer binges followed by vomiting or purging: 'I tried to fight against my bulimia, but I always failed. At last I accepted that I had binges and vomited. Through permitting myself to have binges each time I had done something which had given me pleasure I discovered that my binges became less necessary, and now I do not need them to reduce my tension anymore".
10. To pay less attention to weight and have a more positive experience of one's body: "I discovered that my outward appearance was not as important as being accepted by others. I learned to be responsible for my body, because when I feel more positive about myself and my body I feel much better and have a better outlook".

Content analysis of the answers to the questions of aspect in which the patients improved revealed a series of factors, which are presented in Table 7. Emotional factors were most often mentioned followed by cog-

Table 7. Factors considered as important for improvement and recovery (n = 108)

Factors	Percentage
Emotional: express better feelings, conflicts and tensions	75
Cognitive: having different ideas and insights	54
Behavioural: becoming more assertive, take more initiatives	32
Eating and weight: weight and food became less important	20

nitive factors. Hence, the criteria of recovery of anorexia nervosa need not be reduced to aspects of behaviour, food intake and weight, but should include psychological, emotional and cognitive factors as important criteria for the recovery from eating disorders.

Conclusions and Recommendations

In this chapter the different stages in the process of recovery of patients with anorexia nervosa have been explored and the aspects which have been evaluated as positive for recovery.

1. In accordance with the stated hypothesis, denial of illness is not a permanent characteristic of patients with anorexia nervosa, since sooner or later nearly all patients admitted that something was wrong. Three quarters of the respondents believed during the first stage of anorexia nervosa that they were the only people who were dieting and slimming so extremely, but heard sooner or later that such behaviour is a clinical condition known as anorexia nervosa. Most respondents did not hear about anorexia nervosa from a doctor or a specialist, but rather from the media, self-help organizations for anorexia nervosa, or from their family or friends. The most important effect of admitting to having anorexia nervosa was that they no longer had the alienating idea that they were the only person obsessed by eating and slimming. More than half of those who admitted to having anorexia nervosa became more motivated to seek help. For that reason it is important for general practitioners and therapists to inform their patients about the diagnosis anorexia nervosa as soon as possible and to inform them about the physical, psychological and behavioural consequences. In her book *The Golden Cage* Bruch (1978) describes how important an adequate description of the diagnosis can be for the patient and that the therapists should be well informed about the characteristics of anorexia nervosa. In addition, according to Purgold (1987), it should be apparent to the sufferer and is very important that the therapist be personally understanding of the patients.

2. Early recognition of anorexia nervosa does not, however, guarantee that help is sought. Other factors play an important role as well, such as information received and duration and severity of the anorexia nervosa. Even of those who felt they could no longer continue in their anorectic behaviour, less than half of them did *not* decide to seek help because they were too afraid to talk about their eating behaviour. To help patients overcome their fear, it is desirable that therapists understand the feelings of fear, guilt and shame, and not speak in a blaming way about eating behaviour, because this increases the chance that the patient will withdraw from further contact and treatment. Much, of course, depends on the expectations the patients have of the treatment. When they suppose that treatment only means that eating behaviour has to be improved, strong feelings of fear develop and patients are not motivated to seek help. Consequently it is very important to give patients adequate information about the treatment, which must not only concentrate on her weight, but also on developing a better self-image, becoming more assertive and developing a positive body image.
3. In total 71.3% of the respondents answered that their physical condition had been very bad and 24% of them answered that their life had been in danger. But only 57% of them admitted *at that time* that some treatment was necessary, while 43% still did not feel this need. To wait for improvement of physical condition until patients are motivated may retard a treatment which apparently only *afterwards* can be valued as important.
4. Most respondents with anorexia nervosa do not take the initiative to go to a general practitioner themselves. Those who do are very reluctant to disclose their eating behaviour and way of slimming, because they feel very ashamed and fear that the doctor will tell them they have to eat more and gain weight. Taking into account how difficult it is for people with anorexia nervosa to inform the general practitioner about their eating problems, it might be advisable for general practitioners to ask in a sympathetic way about eating problems if they suspect that the patient has anorexia or bulimia nervosa. This may be the case in the following situations:
   a. Low weight or strong fluctuations in weight
   b. Complaints about constipation or request for laxatives
   c. Feelings of being cold and tired due to weight loss
   d. Pain in the throat or deterioration of teeth caused by frequent vomiting
   e. Irregular menstruation or amenorrhea

   In addition it may be advisable for the general practitioner to ask parents or partners for more information if anorexia nervosa is suspected (see Vandereycken and Meermann 1984).
5. Most of the respondents consulted a doctor more than once. Prescribing medication or further medical investigations often delayed a more effec-

tive treatment. Vandereycken and Meermann (1984) hold the view that patients with anorexia nervosa are often exposed to many physical examinations before the right diagnosis is made. Only if the diagnosis of anorexia nervosa is seriously doubted do they request further physical examinations directed to some alternative diagnosis. It is not only desirable that general practitioners learn to diagnose anorexia nervosa better at an earlier stage; they also need better guidelines for referrals to more specialized forms of treatment.

6. Some 72% of the respondents received treatment for anorexia nervosa, while 28% were never treated. Most of the treated anorexia nervosa patients received more than one treatment: medical treatment or behavioural therapy generally was their first treatment, followed by psychoanalysis, psychotherapy, family therapy or self-help. Because treatment which is directed exclusively to weight restoration often produces relapses and reduces motivation for treatment, it is desirable to consider the psychological and emotional aspects of anorexia nervosa already at the beginning of treatment.
7. It was remarkable that nearly all the treatments were directed towards several aspects of anorexia. No treatment was solely directed at weight gain and normalization of eating behaviour. After medical treatments psychoanalysis was evaluated least positive. Psychotherapy and self-help were evaluated most positive: these treatments placed a lot of emphasis on other aspects of anorexia nervosa, without completely neglecting weight gain and normalizing of eating behaviour. Treatment which concentrates on both to emotional, psychological and social aspects of the patient and also on the improvement of eating behaviour and weight is according to patients the most beneficial for improvement and recovery. Positive experiences of one's self and one's body were considered the most important goals for therapy, while aspects concerning eating and weight were considered the least important goals. For this reason positive experience of one's self and one's body should be the essential goals in the treatment of anorexia nervosa, without neglecting the improvement of weight and eating behaviour.
8. Preliminary analysis of the results of the research revealed that untreated anorectic patients differed significantly from treated patients in the following respects: (1) They were more often male; (2) they lost less of their original weight; (3) they had anorexia nervosa for a shorter period; (4) they admitted that they had anorexia nervosa less often; (5) they consulted a doctor less often; and (6) they were less often referred by their doctor to a medical specialist, psychiatrist or psychologist. However, 73% of the untreated patients had anorexia nervosa for more than 3 years and sometimes for a considerably longer time, and a total of 66% of the untreated patients consulted a doctor with symptoms concerning their anorexia nervosa. Furthermore, 46% of the untreated patients lost more than 25% of their original weight. We can conclude

that these patients were incorrectly diagnosed and were not treated although treatment was needed. General practitioners should learn to diagnose an eating disorder even when the characteristics are not fully developed, because early identification of the onset characteristics of anorexia nervosa is important for secondary prevention and better prognosis.

9. At second analysis the untreated patients were divided into the real anorectic patients, who lost more than 25% of their original weight, and the subclinically anorectic patients, who lost less than 25% of their original weight. The level of education of the parents of real anorectic patients was significantly lower than that of the subclinical patients, which might have hampered possibilities of seeking help. Patients with eating disorders who come from a lower class background need more attention from general practitioners and therapists. This is also the case for patients who develop bulimia nervosa in an earlier stage and whose weight fluctuates more. We might conclude that the need for help should not only be related to the degree of weight loss, but also to eating behaviour (binges, laxatives or vomiting). Being severely obsessed by food and weight, living in isolation and being depressed should be important characteristics for the diagnosis of eating disorders. Moreover, it is desirable that eating disorders be diagnosed at an early stage and not only when severe symptoms are visible. Early identification of the subclinical characteristics of anorexia and bulimia nervosa may prevent further development of these disorders, recovery from which would cost patients years, with or without treatment.

Discussion

The goal os this study was to acquire more insight into the process of recovery in patients with anorexia nervosa. To this purpose a questionnaire was used and completed by 108 respondents. These respondents were found through articles in a number of newspapers. This kind of self-selection has some limitations. Since only those persons were selected who had improved or who were on the way to recovery, we have no information about patients who have not improved from their anorexia nervosa. Further research is needed to gather information about these patients. Moreover, it is possible that the results are too optimistic, for we do not know how many patients are still at risk of relapse in the future. To what extent answers on the questionnaire and the interviews are valid or have been distorted in the memory of the respondents is another open question. Hence it is important to compare the results of this research with other studies using general practitioners and therapists as informants. Such a study could reveal important information about matters such the likelihood of diagnosing anorexia nervosa at an early stage, and the frequency with which patients are informed about their diagnosis, treatments given and goals, such studies could also provide information about practitioners experiences with denial, motivation for treatment and evaluation of treatment by the patients.

In the present research on patient evaluation of different kinds of treatments, it is important to take into account that only a relatively small number of patients followed a special kind of treatment. But the results presented here are very similar to those of de Bloois (1987) who studied the evaluation of treatment by 293 patients with anorexia nervosa. She also concluded that a treatment solely directed at weight increase is evaluated the least positive, while individual psychotherapy and self-help groups are valued very positively for recovery. Both surveys, however, started around 1983, and in the meantime much more information is available on anorexia nervosa, so it is a question of further research to ascertain to what extent the results presented here retain their validity.

References

Bloois M de (1987) Anorexia nervosa: from within. How do patients with anorexia nervosa evaluate their treatment? A research of registered readers of 'Antenne', the journal of the Dutch Foundation for Anorexia nervosa. University of Leiden (in Dutch)

Bruch H (1978) The golden cage. The enigma of anorexia nervosa. Harvard University Press, Cambridge (Mass.)

Cremerius J (1978) Zur Theorie und Praxis der psychosomatischen Medizin. Suhrkamp Frankfurt am Main

Lafeber C (1963) Klinisch psychiatrische studie over de anorexia nervosa. Schrik's Drukkerij, Asten

Lafeber C (1981) Anorexia nervosa. Problems in adequate therapeutic planning. In: van de Loo KJM et al. (eds) Anorexia nervosa. Diagnosis, treatment and research. Dekker and van de Vegt, Nijmegen, pp 85–103 (in Dutch)

Morgan HG, Russell GFM (1975) Value of family background in clinical features as predictors of long term outcome in anorexia nervosa: four year follow-up study of 41 patients'. Psychol Med 5:335–371

Noordenbos G (1987a) A multidisciplinary approach in the treatment of anorexia nervosa, possibilities and limitations. Mon J Ment Health 10:1087–1099 (in Dutch)

Noordenbos G (1987b) Unlimmited slimming behaviour. Research on cultural and gender factors in the development of anorexia nervosa. Dissertation, University of Groningen, Leiden, DSWO-Press (in Dutch)

Purgold J (1987) Who gets better? Who remains ill? Br Rev Bulimia Anorexia Nervosa 1:77–82

Sours A (1969) Anorexia nervosa. In: Caplan G, Lebovici S (eds) Adolescence: psychological perspectives. Basic Books, New York

Theander S (1985) Outcome and prognosis in anorexia nervosa and bulimia: Some results of previous investigations, compared with those of a Swedish long-term study. J Psychiatr Res 19:493–508

Vandereycken W, Meermann R (1984) Anorexia nervosa. A clinician's guide to treatment. De Gruyter, Berlin

Vandereycken W, Pierloot R (1983) Drop-out during in-patient treatment of anorexia nervosa. A clinical study of 133 patients. Br J Med Psychol 56:145–156

Winstead D, Willard SG (1983) Bulimia, diagnostic clues. South Med J 76:313–315

Social Support and Long-Term Course of Anorexia Nervosa*

R. Manz, H.-C. Deter, and W. Herzog

The Importance of Social Support

In anorexia nervosa there is a very large body of literature which discusses the importance of family structures in the development and management of the syndrome (see Vandereycken et al. 1989). Several kinds of social structures in a client's primary group (his/her family) can be subsumed under the concept of social support. However, little effort has been made to use the social support concept for research in anorexia nervosa (literature research using DIMDI and PSYCHLIT has discovered only three articles dealing with anorexia nervosa and the social support concept published within the last 5 years). Research on this concept could lead to a better understanding of the course of anorexia nervosa: First, because the importance of family of origin structures for patients decreases over time, other social relations – especially close intimate relations – become more important; second, especially in anorexia nervosa, social support may be a useful way to overcome either social isolation or dependency and therefore provides a basis for personal development and future health.

Although many studies on social support have been conducted (e.g., Cohen and Syme 1985; Henderson et al. 1981) little attention has been given to the theoretical development of the concept. Weiss (1982) postulated several types of social relationships which fulfill the social needs described by Henderson et al. (1980 p. 725):

1. Attachment – from close affectionate relationships
2. Social integration – from membership of a network of persons having shared interests and values
3. The opportunity to nurture – nurturing others (usually children) gives some incentive for continuing in the face of adversity
4. Reassurance of personal worth – the promotion of self-esteem comes both from those at home and from colleagues
5. A sense of reliable alliance – mainly from kin
6. Obtaining help and guidance when difficulties have to be resolved – from informal advisors

* We gratefully acknowledge support from the BMFT (the German *Bundesministeium für Forschung und Technologie*).

According to Weiss the satisfaction of these personal social needs can be regarded as social support. Social networks can be regarded as defined numbers of members (nodes) and specific relations between these members (Wellman 1981), and support is derived from the interactions for which the network provides a basis. There are two kinds of relations – direct and indirect (Mueller 1980), but as Wellman (1981) outlined, not every primary relation is supportive, and so the concept of social networks represents the object and the concept of social support the subject of studies on social support.

Several authors have made a distinction between the main and the buffering effect of social support (e.g., House 1981; Cohen and Wills 1985; Thoits 1982). A buffering effect can only be expected if a person is threatened by some kind of stress (life events). Here the effect of social support is to help maintain psychological well-being. A problem in many studies on social support is that these two concepts are confounded (Cohen and Syme 1985). Critical life events can often also be interpreted as changes in a person's support system. To clarify the relations between stress, social support, and well-being it is necessary to distinguish these concepts by means of their operationalization (Monroe and Steiner 1986). The main effect is based on the assumption that social support is a ubiquitous phenomenon of social being and a precondition for psychological well-being.

Social support should no longer be regarded solely as an independent variable. It also appears to be an outcome variable (Manz et al. 1987; Manz and Schepank 1989). That means that social support represents a wide range of a person's social functioning. Social functioning can thus be interpreted as a part of social and mental health. In our study we focussed on the social and psychological aspects of interpersonal relations. The self-rating network questionnaire (see below) measures important aspects of a person's social environment and social integration. The interview procedure (confidant rating) focusses on the psychological dimensions of social interaction with significant others. However, close relationships not only function to help maintain psychological well-being, they can also be a source of chronic interpersonal stress. Many pathological relation structures in anorectics described by Bruch (1973), Selvini-Palazzoli (1974), and Minuchin et al. (1978) – such as helplessness, inmeshment, and rigidity – can be regarded as nonsupportive interaction. Hence close social relationships seem to be an important factor in the ongoing process of the anorectic disorder (Engel 1987).

Methods and Results

As part of a follow-up study described elsewhere in this book (see Herzog et al., this volume), our research group used two different assessment

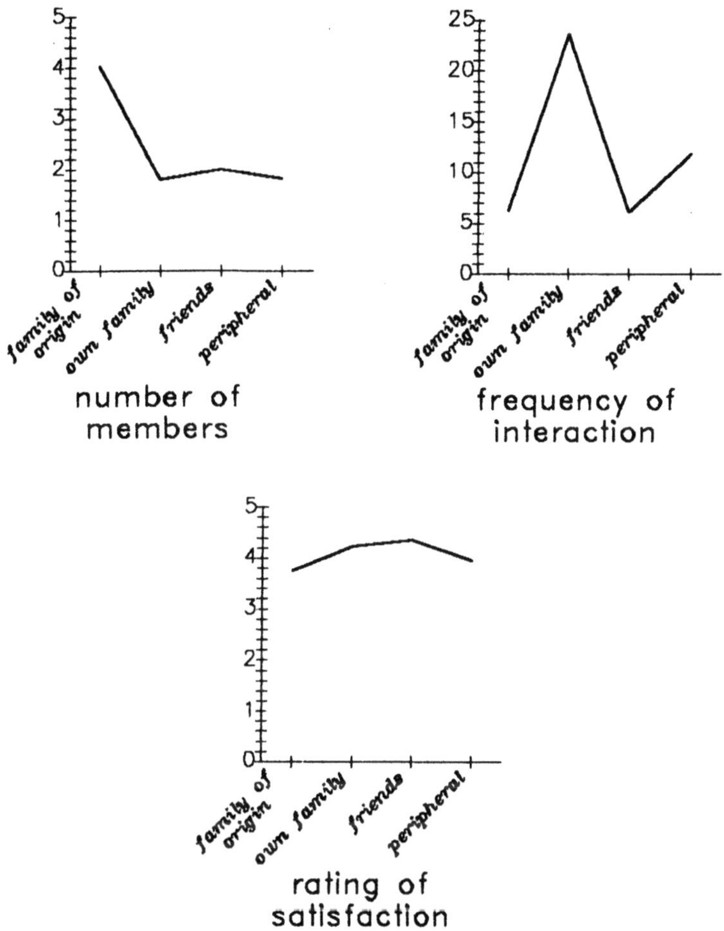

Fig. 1. Mean scores on Network Questionnaire for 58 former anorexia patients

methods which underline the importance of the operationalization of the social support concept. Both instruments were applied to anorexia nervosa patients as well as to patients suffering from Crohn's disease and psychosomatic inpatients. The self-rating questionnaire was also applied to "healthy" people. In addition we used two questionnaires widely applied in anorexia nervosa patients: the Eating Disorder Inventory (EDI; Garner et al. 1983) and the Eating Attitudes Test (EAT-26; Garner et al. 1982; German translations: Meermann and Vandereycken 1987).

Self-rating Network Questionnaire

The self-rating network questionnaire (NQ) designed by the first author registers the number of persons who are contacted by a patient during the

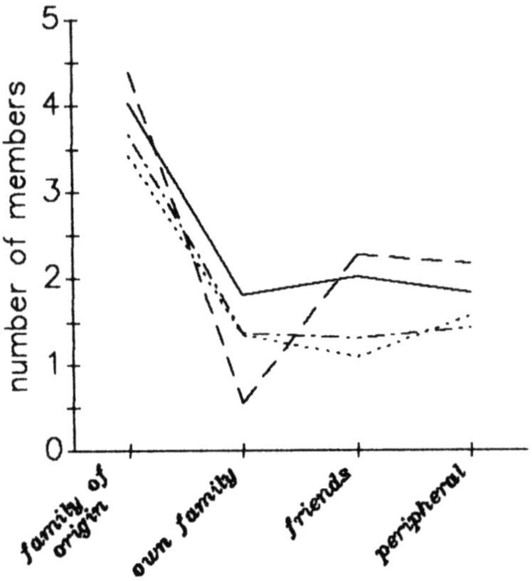

Fig. 2. Average number of network members for former anorexia patients (n = 84, *solid line*), psychosomatic inpatients (n = 117, *dash-dot line*), Healthy students (n = 53, *dashed line*) and patients with Crohn's disease (n = 42, *dotted line*)

preceding month, as well as frequency and satisfaction of these contacts. The indices are divided into four aspects of the social environment: contact to family of origin, own family, friends and peripheral contacts (e.g., colleagues; Fig. 1). In comparison to family of origin and friends, former anorectics appear to have much more contact to their own family and to peripheral relations. It is important, however, to compare the average number of network members with data from other groups. Figure 2 shows such a comparison with findings in patients suffering from Crohn's disease, "healthy" students and psychosomatic inpatients.

In comparison to these groups the former anorectics appear to have no lack of social relations as regards the structure of these relations. They appear, like the healthy controls, to have more friends than psychosomatic inpatients and patients suffering from Crohn's disease. Differences between the groups seem to be related to age and therefore to role differences (younger people are more likely to be single): the healthy group has a mean age of 26 years and the inpatients of 38 years; the other groups range between them.

Confidant Rating

Using a half-standardized interview, the confidant rating (CR), also designed by the first author (Manz and Schepank 1989; Manz 1990), we rated

Table 1. Cross-tabulation of confidant 1 by confidant 2 for 84 former anorexic patients

Confidant 1	Confidant 2							
	No friend		Nonsupportive friend		Supportive friend		Total	
	(*n*)	(%)	(*n*)	(%)	(*n*)	(%)	(*n*)	(%)
No partner	7	(8.33)	0	(0)	13	(15.48)	20	(23.81)
Nonsupportive partner	5	(5.95)	3	(3.75)	9	(10.71)	17	(20.24)
Supportive partner	16	(19.05)	2	(2.38)	29	(34.52)	47	(55.95)
Total	28	(33.33)	5	(5.95)	51	(60.71)	84	(100)

the existence of confiding others in a person's social environment (i.e., partner, spouse, very good friend) and the quality of the social interaction by means of several dimensions (e.g., emotionality, instrumentality, sexuality, etc.).

The CR method provides data for the quality of the relationship to a partner (confidant 1) and a good friend (confidant 2). Quality is categorized by the interviewer into no relationship, nonsupportive relationship, and supportive relationship (several interrater studies showed good interrater reliability of $K = .88$; $p \leqslant 0.015$ for dichotomized data: supportive versus nonsupportive or no relation).

Table 1 shows that 24% of the sample have no partner. In 20% of the cases there exists a nonsupportive relationship which may be a source of chronic emotional stress. However, the majority (56%) has a supportive, and therefore stress-reducing confidant 1. Thus approximately 15% of the sample either has no confiding relationship or a stressful relationship. These people are in danger of developing psychological or physical symptoms.

In comparison to an epidemiological random sample (Manz 1990) and patients with Crohn's disease (Deter et al. 1991) the former anorectics suffer from a lack of confiding relationships (see Fig. 3).

As shown in Fig. 3 25% of the epidemiological sample live without a partner or with a nonsupportive partner. In contrast 44% of the former anorectics and 35% of patients suffering from Crohn's disease do not have a supportive partner. There is a marked difference between the anorectic group and the epidemiological sample ($p \leqslant .001$). Although patients with Crohn's disease also suffer from a lack of confiding relationships there is no significant difference in comparison to either the anorexic or the epidemiological sample.

To test the hypothesis that the absence of a confiding relationship leads to increased symptom status we carried out two-way analyses of

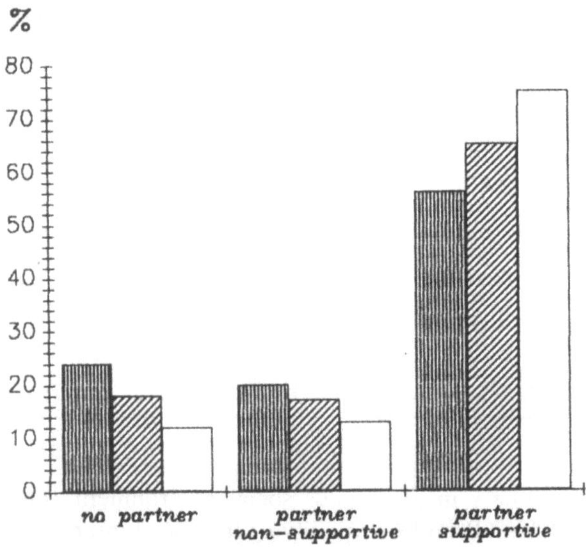

Fig. 3. Quality of partner relationship for former anorexia patients (n = 84, *striped bars*) in comparison to an epidemiological sample (n = 200, *open bars*) and patients suffering from Crohn's disease (n = 42, *hatched bars*)

variance. The ratings for confidant 1 (partner) and confidant 2 (friend) were dichotomized (supportive vs. no or nonsupportive relationship). Dependent variables were subscales of the impairment score (IS; Schepank 1987), a clinical rating system measuring the degrees of psychogenic impairment on somatic, psychic and social adaptation dimensions. Further we used the anorexia nervosa symptom score (ANSS; see Deter, this volume), a multidimensional system using somatic (for example, blood parameters), as well as defined social (for example, days of illness during the last year) and psychological (for example, depression, anxiety) symptoms which are specific for anorexia nervosa.

For the social and the psychological dimensions there is always a significant main effect for confidant 1 and (with only a few exceptions) for confidant 2, whereby the effects of the former appear statistically to be much stronger than for the latter. People living without a partner or with a nonsupportive partner show higher impairment scores then people living in a confiding relationship. On the one hand, these results underline the importance of close social relationships for a person's well-being; on the other they underline the limitations of the support concept. Hence the CR is influenced by social as well as psychological symptoms. So – in contrast to physical symptoms – the former are confounded with the CR and there are no differences for somatic scores between people with good and poor support. However, we investigated a follow-up sample, where fewer

Table 2. Average scores on EDI and EAT-26 subscales in 59[a] former anorexia patients

Subscales	Confidant 1		Confidant 2	
	None or nonsupportive	Supportive	None or nonsupportive	Supportive
EDI 1 Drive for thinness	5.16	1.82**	3.50	2.68
EDI 2 Bulimia	1.76	0.76	1.30	0.95
EDI 3 Body dissatisfaction	7.56	4.27*	6.68	3.53
EDI 4 Ineffectiveness	6.80	3.24*	4.84	4.70
EDI 5 Perfectionism	5.72	4.01	4.95	4.70
EDI 6 Interpersonal distrust	4.40	3.21	4.47	3.50
EDI 7 Interoceptive awareness	4.24	1.97	3.05	2.88
EDI 8 Maturity fears	4.28	2.35*	4.05	2.75
EAT 1 Dieting	8.12	4.23*	6.00	5.79
EAT 2 Bulimia and food preoccupation	2.32	0.83	1.47	1.45
EAT 3 Oral control	4.12	1.88*	2.84	2.83

F-test (Scheffe test for group differences, two-way ANOVA); *p ≤ .05; **p. ≤ .01.
[a] Sample size was changed because of missing data on EDI and EAT-26.

somatic symptoms are to be expected than in an acute anorectic group. Therefore ceiling effects may be the reason for missing group differences.

In a second step we also analyzed typical anorectic features using the EDI and EAT-26. Here also two-way analyses of variance were carried out. There are significant main effects for confidant 1 in several subscales of EDI and EAT-26; however, there was no such effect for confidant 2 and no interactive effect (see Table 2).

For the EDI subscales "drive for thinness" and "body dissatisfaction," and EAT subscales "dieting" and "oral control," which represent typical features of anorexia, there are significant group differences. People living without a partner or with a nonsupportive partner show markedly higher symptom scores than people with a supportive close relationship. There are also significant differences for "ineffectiveness" and "maturity fears," which represent more general psychopathological dimensions not specific to anorexia. These results lead to the conclusion that a supportive partnership seems to be a protection against typical anorectic behavior and experience. In contrast to general psychopathological impairment there is no association between confidant 2 and typical anorectic symptoms represented by EDI or EAT-26. These results are validated by the self-rating questionnaire (NQ). As the NQ provides data for social integration in several sectors of the social environment, we are able to compare the importance of more intimate and less intimate social relations for symptom status in our sample. Table 3 shows the correlations between the NQ scales described above and general psychopathological as well as typical features of anorexia.

Although Table 3 includes a large number of correlation coefficients and the significance level has to be adjusted ($p_{adj} = .0002$), it is remarkable that satisfaction with contacts to the patient's own family (often only the partner) shows significant correlations to almost all general and specific symptom dimensions considered in this study. In contrast to this, the other social environment parameters show only a few significant correlations. These results again underline the importance of the quality of the relationship to the patient's own family, above all the partner, for the health status of our sample.

Quality of Social Network and Impairment

As mentioned above, structural parameters of the social network per se seem not to be associated with the impairments studied. These parameters together may build a more elaborated relation structure which, as hypothesized, should be associated with impairment. There must be a change in perception from quantity (number of different structure parameters) to the quality of the social network (i.e., quality of a specific structure). The social network structure shows the relative importance of the various network sectors for a person's social welfare and therefore his or her social integration. Less socially integrated people, for example, may have contacts to only one or two social sectors.

In order to examine in what types of social structures the former anorexia patients live and how their social reality is associated with impairment, we carried out a cluster analysis of the NQ ratings of number of social members. Using Ward's method (Ward 1963) which generally leads to balanced cluster proportions, we found three clusters:

- Cluster 1 consists of 20 people with a large number of contacts to family of origin, friends and peripheral persons but very few contacts to own family.
- Cluster 2 consists of 18 people with relatively few contacts to family of origin and very few contacts to own family, friends and peripheral persons.
- Cluster 3 includes 20 people with a moderate number of contacts to family of origin, friends and peripheral persons, but with a large number of contacts to the own family (see Fig. 4).

With the exception of own family, cluster 1 appears to be well integrated. People of this cluster have many friends and peripheral contacts and also a

Table 3. Correlations between NQ scales and general psychopathological as well as specific anorexic impairment (IS, ANSS, EDI, EAT-26) for 58 former anorexic patients

Impairment	Members				Frequency				Satisfaction			
	FO	OF	FR	PER	FO	OF	FR	PER	FO	OF	FR	PER
IS-Ph	−.104	−.179	−.095	−.126	−.039	−.124	−.177	−.039	−.182	−.500***	−.008	−.001
IS-Ps	−.205	−.404*	−.191	−.322*	−.127	−.331	−.043	−.033	−.337*	−.461**	−.201	−.231
IS-So	−.198	−.240	−.270	−.362**	−.163	−.318	−.099	−.028	−.226	−.615***	−.232	−.380*
ANSS-Ph	0.031	0.010	−.029	−.042	−.298	−.350	−.061	−.299	−.185	−.418*	−.104	−.150
ANSS-Ps	−.011	−.121	−.100	−.159	0.308	−.298	−.013	−.012	−.156	−.573***	−.151	−.301
ANSS-So	−.187	−.420*	−.193	−.304*	0.256	−.485**	−.055	−.080	−.215	−.350*	−.151	−.068
EDI 1	0.078	0.020	−.030	−.113	0.216	−.047	−.131	−.230	−.158	−.529**	0.085	−.285
EDI 2	−.022	−.124	−.176	−.177	0.293	−.227	−.199	0.134	−.260	−.381*	0.043	0.239
EDI 3	0.259	0.098	0.079	−.009	0.077	0.003	0.029	−.226	−.234	−.361*	0.026	0.206
EDI 4	0.019	−.082	−.009	−.195	0.284	−.308	−.155	0.133	−.277	−.411*	−.124	−.382*
EDI 5	0.120	−.083	0.220	0.019	0.008	−.078	−.331	−.131	−.303	−.391*	−.120	−.180
EDI 6	−.007	−.112	−.143	−.148	0.153	−.305	−.169	0.248	−.312	−.507**	−.127	−.218
EDI 7	0.083	−.062	0.029	−.112	0.232	−.250	−.163	0.098	−.227	−.305	−.035	−.056
EDI 8	−.125	−.001	−.345	−.391*	0.150	0.056	−.046	0.252	−.280	−.194	−.537***	−.354
EAT 1	0.080	0.051	−.005	−.080	0.187	−.051	−.216	−.151	−.147	−.493**	−.025	−.292
EAT 2	−.120	−.096	−.166	−.221	0.479**	−.257	−.220	0.128	0.076	−.366	0.108	−.016
EAT 3	−.014	−.111	−.010	−.144	0.255	−.289	−.138	0.002	−.047	−.441*	0.149	−.144

FO, Family of origin; OF, own family; FR, friends; PER, peripheral; IS, impairment score; Ph, physical; Ps, psychological; So, social adaptation; ANSS, Anorexia Nervosa Symptom Score; Ph, physical; Ps, psychological; So, social; EDI, Eating Disorder Inventory (see Table 2); EAT-26, Eating Attitude Test (see Table 2); NQ, network questionnaire.

$^{*}p \leqslant .01$; $^{**}p \leqslant .001$; $^{***}p \leqslant .0001$.

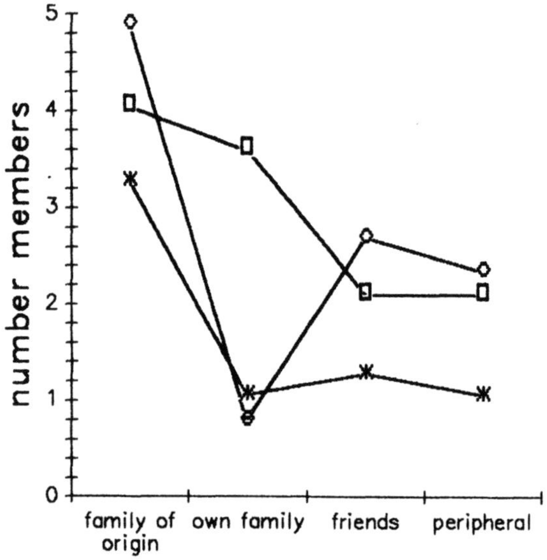

Fig. 4. Number of network members defining the three clusters (mean values). –○– cluster 1 (n = 20); –*– cluster 2 (n = 18); –□– cluster 3 (n = 20)

large number of contacts to family of origin. Only 45% of them have a supportive partner (confidant 1) but 90% have a supportive friend (confidant 2).

Cluster 2 is characterized by few contacts to own family, friends and peripheral persons. They also have relatively few contacts to family of origin. These people seem not to be socially integrated except within their family of origin. Only 50% have a supportive confidant 1 and 50% have a supportive confidant 2.

Cluster 3 shows the best integration of all. These subjects have balanced contacts to all social sectors and in contrast to cluster 1 and cluster 2 they usually have their own family and children. And 70% of them have a supportive confidant 1 and 60% a supportive confidant 2.

As shown in Table 4 85% of cluster 3 are married. In contrast to this only 30% of cluster 1 and 16.7% of cluster 2 live with a spouse. For social class and education we carried out nonparametric ANOVAs (H test) to examine differences between the clusters. Cluster 1 shows a significantly higher level of education than cluster 3, whereby cluster 2 ranges between them. However, there are no differences between the three groups regarding social class. The most important differences between these groups are not general sociodemographic characteristics but individual parameters of social behavior such as marital status and social relations to different parts of the social environment. These differences seem to depend on personal motivation and abilities to build up an own social network. Whereas

Table 4. Sociodemographic characteristics for the three NQ clusters in 58 former anorexia patients

Demographic variables	Cluster 1 (n = 20) (%)	Cluster 2 (n = 18) (%)	Cluster 3 (n = 20) (%)
Marital status			
single	55.0	66.7	10.0
engaged	10.0	5.6	–
married	30.0	16.7	85.0
divorced	5.0	–	–
widowed	–	5.6	5.0
other	–	5.6	–
Social class			
lower	–	27.8	5.0
lower middle	35.0	22.2	25.0
middle	20.0	16.7	30.0
upper middle	30.0	16.7	30.0
upper	15.0	16.7	20.0
Level of education[a]			
0	–	–	–
1	–	5.6	–
2	5.0	27.8	30.0
3	25.5	11.1	30.0
4	–	5.6	10.0
5	70.0	50.0	30.0

[a] This is not comparable to British or American school systems.

Table 5. Group differences for three NQ clusters on several impairment parameters (means and Scheffe test for contrasts) in 58 former anorexia patients

Impairment	Cluster 1 (n = 20)	Cluster 2 (n = 18)	Cluster 3 (n = 20)	p value	Groups different ($p \leqslant .05$)
IS-Ph	0.9	1.06	1.06	n.s.	–
IS-Ps	1.7	2.11	1.10	.002	(3-2)
IS-So	0.9	1.89	0.80	.001	(3,1-2)
ANSS-Ph	22.61	28.92	18.26	n.s.	–
ANSS-Ps	37.78	56.17	27.78	.003	(3-2)
ANSS-So	28.75	38.89	11.34	.000	(1,2-3)
% IBW	95.70	84.22	94.50	.020	(2-1)

IS, impairment score; Ph, physical; Ps, psychological; So, social adaptation; ANSS, Anorexia Nervosa Symptom Score; Ph, physical; Ps psychological; So, social; % IBW, relative Broca weight; NQ, network questionnaire.

people of cluster 3 usually live in their own family with the spouse as the most supportive person, members of cluster 1 have a large number of very close friends. Both groups are also well integrated into the wider social environment. In comparison, cluster 2 suffers from a lack of intimate relationships and of social integration into the wider social environment. These effects are not related to either age or general sociodemographic or socio-economic resources such as education and social class.

What are the associations between these social realities and type of impairment? To answer this question we carried out one-way ANOVAs with some of the impairment parameters described above (see Table 5).

Table 5 shows some differences between the social groups defined by NQ. Although there are no group differences for the physical subscales of impairment score (IS) and ANSS, cluster 2 shows a markedly lower body weight than cluster 1 and obviously (even if not statistically significant) than cluster 3. There were no differences for the self-rating procedures EDI and EAT-26.

These results are not easy to interpret. On the one hand there are group differences for psychological and of course social impairment which indicate that these groups differ in terms of general psychopathological impairment. This result can also be expected for other symptom groups as they do not correspond to specific anorectic behavior. On the other hand there are differences in relative body weight, indicating that the members of cluster 2 continue to suffer from anorectic impairment. It is possible that these groups represent different stages of anorexia nervosa and possibly different ways of overcoming it. The social networks of the three clusters seem to represent an outcome variable of this adaptation process.

Discussion

As often described (see, e.g., Bruch 1973), anorectics suffer from a strong ambivalence between autonomy and dependency. People of cluster 3 might solve this conflict in the direction of dependency by founding their own family and living together with a supportive partner. Members of cluster 1 might solve this conflict in the direction of autonomy. They do not have their own family but they do have close relationships to good friends. In this constellation a person can have close relationships without general mutual obligation. Cluster 2 appears to be relatively socially isolated and its members seem to make no effort to solve this basic conflict. Analyses of the quality of intimate social relationships using the confidant rating yield similar results. The existence of an intimate confidant 1 is associated with less impairment. Our analyses lead to the conclusion that specific social relation structures work as the basis for qualitative social support. The social relations structures represented by NQ clusters include different types of intimate social relations to either spouse or good friends or no intimate social relationship at all. They each represent different aspects of the social needs which have been described and therefore represent different types of relations which are more or less intimate and dependent.

Our efforts to examine the meaning of several social support dimensions produced results which are in line with the theoretical assumptions described above. Structural support dimensions such as quantity of network members and frequency of contacts (even for distinct sectors of the social network) appear not to be associated with psychological and physical impairment in an anorectic sample. On the other hand, the quality of social interaction with a close partner shows a strong association with symptom status. Here, expert as well as self-rating procedures have produced identical results. Various dimensions of the social network together lead to specific social structures which then provide the basis for social support. The quality of close relationships represents one of the most important factors in maintaining psychological well-being. At the same time close relationships are considered an outcome criterion in the clinical rating systems described.

References

Bruch H (1973) Eating disorders. Obesity, anorexia nervosa, and the person within. Basic Books, New York

Cohen S, Syme SL (1985) Issues in the study and application of social support. In: Cohen S, Syme SL (eds) Social support and health. Academic, Orlando

Cohen S, Wills TA (1985) Stress, social support, and the buffering hypothesis. Psychol Bull 98:310–357

Deter HC, Manz R, Becker M, Gladisch R (1991) Soziale Unterstützung in ihrer Beziehung zu körperlichen und psychischen Befunden bei psychosomatischen Patienten (am Beispiel des Morbus Crohn). Z Psychosom Med 37:45–59

Engel K (1987) Prognostische Faktoren der Anorexia nervosa – eine empirische Arbeit zur Aufstellung einer hierarchischen Ordnung der bedeutsamen Faktoren. Z Klin Psychol Psychopathol Psychother 35:313–320

Garner DM, Olmsted MP, Bohr Y (1982) The Eating Attitudes Test: psychometric features and clinical correlates. Psychol Med 12:871–878

Garner DM, Olmsted MP, Polivy J (1983) Development and validation of a multidimensional eating disorder inventory for anorexia nervosa and bulimia. Int J Eating Dis 2(2):15–34

Henderson S, Duncan-Jones P, Byrne D, Scott R (1980) Measuring social relationships: the interview schedule for social interaction. Psychol Med 10:723–734

Henderson S, Byrne DG, Duncan-Jones P (1981) Neurosis and the social environment. Academic, New York

House JS (1981) Work stress and social support. Addison-Wesley, Menlo Park (CA)

Manz R (1990) Social support. In: Schepank H (ed) Verläufe. Seelische Gesundheit und psychogene Erkrankungen heute. Springer, Berlin Heidelberg New York

Manz R, Schepank H (1989) Soziale Unterstützung, belastende Lebensereignisse und psychogene Erkrankung. In: Angermeyer MC, Klusmann D (eds) Soziales Netzwerk: ein neues Konzept für die Psychiatrie. Springer, Berlin Heidelberg New York

Manz R, Valentin E, Schepank H (1987) Soziale Unterstützung und psychogene Erkrankung. Ergebnisse aus einer epidemiologischen Feldstudie. Z Psychosom Med 33:162–170

Meermann R, Vandereycken W (1987) Therapie der Magersucht und Bulimia nervosa. De Gruyter, Berlin

Minuchin S, Rosman B, Baker L (1978) Psychosomatic families: anorexia nervosa in context. Harvard University Press, Cambridge (Mass)

Monroe SM, Steiner JB (1986) Social support and psychopathology: interactions with preexisting disorders, stress, and personality. J Abnorm Psychol 95:29–39

Mueller DP (1980) Social networks: a promising direction for the research on the relationship of the social environment to psychiatric disorder. Soc Sci Med 14:147–161

Schepank H (1987) Epidemiology of psychogenic disorders: the Mannheim Study – results of a field survey in the Federal Republic of Germany. Springer, Berlin Heidelberg New York

Selvini-Palazzoli M (1974) Self-starvation: from the intrapsychic to the transpersonal approach to anorexia nervosa. Chaucer, London

Thoits PA (1982) Conceptual, methodological, and theoretical problems in studying social support as a buffer against life stress. J Health Soc Behav 23:145–159

Vandereycken W, Kog E, Vanderlinden J (1989) The family approach to eating disorders. PMA, New York

Ward JH (1963) Hierarchical grouping to optimize an objective function. J Am Stat Assoc 58:236–244

Wellman B (1981) Applying network analysis to the study of support. In: Gottlieb BH (ed) Social networks and social support. Sage, Beverly Hills

Weiss RS (1982) Relationship of social support and wellbeing. In: Schulberg HC, Killilea M (eds) The modern practice of community mental health. Jossey Bass, San Francisco

Ego Strength, Object Relations, and Life Events as Outcome Predictors in Restricting Anorexia Nervosa and Normal-Weight Bulimia Nervosa*

S. Sohlberg and C. Norring

There is still much to be discovered regarding factors that predict outcome in eating disorders (Szmukler and Russell 1986; Sohlberg et al. 1989). The present project addressed this issue and was conceived as a traditional follow-up study with some important modifications. By "traditional" we mean selection of a group of patients who have applied for help during a set time period, the follow-up of them after several years, and no tight controls on treatment. The latter feature is often necessary either because the study is retrospective or because efforts required to control treatment type, quality and duration are prohibitive (Gowers et al. 1989). Our design was prospective but treatment decisions were made on clinical grounds and we settled for post hoc comparison of results among those who did and did not receive formal therapy.

The personality functioning of the patients was not investigated in previous studies and this is a possible reason for the difficulties in predicting their follow-up status. By the term "personality functioning" we don't mean personality traits or styles (e.g., impulsivity vs rigidity and inhibition), but a dimension spanning a range from the optimally adaptive personality to the least adaptive. As elaborated by the psychoanalysts Hartmann, Jacobson, Bellak and Kernberg from the origins in Freud's later writings, we understand the adaptive personality to be rich in nuances yet cohesive, characterized by subtle and flexible coping styles that permit the person to handle life challenges without losing contact with reality. The less adaptive one is undifferentiated and unintegrated, rigidly relying on grossly distorting defense mechanisms that may be effective in dealing with select problems but at a high cost in terms of long-term overall adaptation. Understanding of the latter personality may be increased by comparison with a child's functioning, even though its presence in an adult may be explained by several causes beyond stunted development. The first modifi-

* This study was supported by grants from the Bank of Sweden Tercentenary Foundation. We thank Börje Rosmark, M.Sc., who made half of the interviews and shared responsibility for project initiation. During the preparation of this manuscript the first author (SS) was a Visiting Research Scholar at UCLA Neuropsychiatric Institute with financial support from the Sasakawa Young Leaders' Fellowship Foundation.

cation in this project of the traditional approach was thus to include a thorough assessment of personality as conceptualized in this "ego psychological" sense (Bellak et al. 1973).

However, it also seems reasonable that some part of outcome variance is truly unpredictable as long as we study factors in the patient or her environment only at the moment of initial presentation. Specifically, major negative life events that occur during the course of the disorder could have an effect on recovery and relapse. Stress factors have attracted attention in eating disorders research (Strober 1984; Mitchell et al. 1985; Greenberg 1986; Cattanach and Rodin 1988; Johnson et al. 1989), but no firm data on their association with outcome exist yet. A decision to study the occurrence of such events was thus the second modification introduced in the design. Various statistical controls were planned in the event that correlations with outcome were obtained in order to probe whether these associations might actually reflect causal relationships.

Finally, rather than doing a one-shot follow-up, it was decided to make repeated ones (Theander 1985). Clinically there is a need to make forecasts also for briefer time ranges than the criterion of at least 4 years as proposed in previous studies (Morgan and Russell 1975). From a scientific standpoint repeated follow-ups permit the identification of patterns and within-sample replication of results.

We have previously reported presentation and 1st- and 2nd-year data for the project sample. In summary, patients were strikingly dissimilar in personality functioning despite resemblance in symptoms (Norring et al. 1989; cf. Hirshberg 1989). Patients with less adaptive ego functions had a poorer overall adjustment at follow-up, and needed 2 years to improve in their eating disorder as much as those with more adequate ones did in 1 (Norring and Sohlberg 1991). The number of untreated patients was small, but analysis of 1st-year results showed ego functioning to be predictive both among them and the others, with no significant difference (Sohlberg and Norring 1989). Stressful events made for a statistically significant but modest difference in 1st-year recovery rates (Sohlberg and Norring 1991).

Excluding those meeting DSM-III-R (APA 1987) criteria for both anorexia nervosa (AN) and bulimia nervosa (BN), as well as those with atypical syndromes as judged by that manual, we will now analyze classical restricting anorectic and normal weight bulimic individuals. Starting in our own clinical experience as well as that of others, we have long maintained that pathological fixation on body weight and shape is the focal symptom in AN and BN (Guiora 1967; Ehrensing and Weitzman 1970; Holmgren et al. 1983). Its importance overshadows that of the dissimilarities between the syndromes and is a sufficient diagnostic feature even when weight loss or binge frequency is not dramatic. This view rests in particular on the finding that patients may repeatedly shift between restriction and bingeing with vomiting, gaining, and losing weight, while retaining throughout a desperate attitude that the most important thing in the world is not to be fat (Holmgren et al. 1983).

In keeping with this view researchers comparing restricting anorectic with normal weight bulimic individuals have sometimes found them surprisingly similar (e.g., Humphrey 1986). However, differences in various background variables have also repeatedly been reported even between patients with unquestionable and uncomplicated syndromes (Rybicki and Lepkowsky 1987). It seemed warranted therefore to investigate whether the results we reported earlier can be replicated when atypical patients are excluded and analysis restricted to those with AN or BN.

Subordinate to that goal, we tested whether the level of object relations is particularly important for outcome. In a previous study we found that, controlling for the difference in overall ego strength, the one significant difference between borderlines and neurotic patients was more prominent problems with object relations in the former group (Norring et al. 1989). Thus, they not only tended to show more evidence of unconscious schemata with blurred distinctions between own and others' feelings, thoughts, and emotions, limited capacity to meet own needs without the presence of important others, and shifts between black-and-white "good" and "bad" views of self and others. This area of dysfunction was also more conspicuous than their problems with reality testing, defense mechanisms and so forth. We now wanted to see whether focus on this particular ego function translates into superior prediction of outcome compared with focus on the overall level of ego strength.

Outcome was operationalized by interview assessments of preoccupation with weight and shape, which we consider to be the most important variable, and by scores on the well-known self-rated Eating Disorder Inventory (EDI; Garner and Olmsted 1984). The issues addressed were: (1) Does ego strength predict these variables among anorectics and bulimics? (2) Do life events predict them? (3) Are object relations more important as a predictor than the overall level of ego strength in either or both of the two groups?

Method

Patients

Fifty-three referred or self-referred patients on their first visit to a university hospital psychiatric clinic were seen during the data collection period. Forty-eight (46 women, 2 men) were originally included in the project as being preoccupied with weight and/or shape and having a DSM-III-R eating disorder according to two independent interviewers. Full confidentiality was guaranteed and all were informed that refusal to participate would not reduce their chances to receive whatever help they needed from the hospital. The average body mass index (BMI: kg/m^2) of two normal control groups of comparable age were used as estimates of expected weight (see Drewnowski and Garn 1987). The present study deals only with those 22 patients who had BN and the 11 with restricting AN. The bulimic subjects were 22.6 ± 4.9 years old (M ± SD), had been ill for 6.6 ± 4.3 years, had a BMI of 21.8 ± 2.4, and binged 6 ± 1.9 times a week. Twenty were vomiters or abused laxatives (91%), and 17 had been treated for their disorder before entry into our project (77%). The anorectic subjects were 24.6 ± 9.3 years old, had been ill for 7.3 ± 10.0 years, had a BMI of 15.4 ± 1.2, and by definition no one binged. Six were vomiters or abused laxatives (55%), and 8 had been treated for their disorder before (73%). The groups came from similar social backgrounds (3.3 ± 1.6 vs 3.6 ± 1.2; 1 = high 6 = low; Swedish Statistics, SCB 1982). Using *t*-tests and chi-square no differences were significant apart from those expected to be so by definition (binge frequency, weight, and vomiting/laxative abuse).

All had a clinically significant preoccupation with weight and/or body shape. Some of the questions used to elicit evidence of such preoccupation were: Is it of particular importance for you to stay at a certain weight level; do you want to reduce; do you set new goals for weight loss if you succeed in attaining one previously set; are you dissatisfied with any particular part of your body; do you feel fat; do you compensate in any manner for the intake of food. Weight/shape preoccupation was deemed present if judges considered that the responses reflected an attitude overriding vital needs (nutrition, friendship, sex) and did not permit the patient to have a single normal-sized meal without being severely anxious or having to compensate in some manner (for example, "paying" in advance by cleaning her apartment or afterwards by vomiting, running etc.).

Treatments included psychoanalytic therapy, eating regimens, antidepressants, and combinations of these. Two anorectic (22%) and three bulimic (16%) individuals were untreated the 1st year, and three anorectic (38%, data unavailable in one case) and ten bulimic (58%) individuals the 2nd year.

Procedure at Presentation

A diagnostic interview and a 3- to 4-h personality assessment interview (Ego Function Assessment, EFA; Bellak et al. 1973) were included in the protocol at presentation. The EFA interview concerned the patient's functioning during the past years with a focus on the most typical levels. This instrument is a set of interview questions and 13-point rating scales which define 12 ego functions, 3 superego aspects and 4 id aspects. All scales cover a wide range from psychotic functioning to that which is optimally adaptive and flexible. Theoretical ranges for severity of ego dysfunction are: psychotic = 1–6, borderline = 4–8, neurotic = 6–10, and normal = 8–13. We developed a semi-structured interview using the EFA questions as appropriate and assessed the 12 ego functions and 1 superego aspect. Only "ego strength" (an average of the individual functions) and "object relations" are used in this study. The functions are: reality testing, judgment, sense of reality (identity and self-boundaries), regulation and control of affects and impulses, object relations, thought processes, adaptive regression (capacity for controlled regressions and creative use of these experiences to further adaptation), defensive functioning, stimulus barrier (filtering of and coping with nonhuman stimuli), autonomous functioning (functional level of psychomotor, intellectual, and acquired capacities), synthetic functioning (capacity for and activity towards integrating potentially discrepant feelings, ideas, and behaviors), and mastery/competence (capacity to actively master life challenges, and own feelings about this capacity). Superego adaptation refers to degree of realism and adaptiveness in feelings of guilt, shame, self-criticism etc. The superego is traditionally defined as a partially independent structure, but was here conceptualized as an aspect of the ego. Interviews were made by either of two clinical psychologists with extensive experience in eating disorders. Control ratings of the audiotaped interviews by the first rater for the second and vice versa showed good agreement (r = .70–.82 across functions, no level or range differences). Final scores were based on a consensus discussion.

Procedure at Follow-ups

At follow-up patients were asked to complete self-report instruments at home and shortly thereafter to take part in an interview, either at the clinic or by telephone. Previous analyses in the project have indicated that the telephone interviews produced valid data (details in Sohlberg and Norring 1989). The questionnaires included the EDI and a short version of the Psychiatric Epidemiology Research Interview (PERI) Life Events Scale (Dohrenwendt et al. 1978). The EDI is a 64-item, psychometrically sound instrument (Norring and Sohlberg 1988; Welch et al. 1990) comprising eight scales of relevance for eating disorders. In the present study the eight scales were summed into two variables: EDI symptoms (subscales Drive for Thinness, Bulimia, and Body Dissatisfaction), and EDI Attitudes (subscales Ineffectiveness, Perfectionism, Interpersonal Distrust, Interoceptive Awareness, and Maturity Fears). This distinction has both theoretical (Garner and Olmsted 1984) and empirical (Norring 1991) support.

From the 102 PERI Life Events we composed a list of 20 events that are major, negative events "certainly independent of both respondents' physical health and psychological condition" according to Dohrenwendt et al. (1978): p. 216; e.g. Have you during the last year found out that you cannot have children; has during the last year a close friend of yours died; has during the last year someone close to you (defined as close friend or relative, partner, or child) suffered a serious illness or injury). In an attempt to counter some of the problems with self-reports of life events (Monroe 1982), the subsequent interview contained a control of the reports, which in a few cases resulted in the subtraction or addition of events.

The interview was made by the person who made the initial EFA for that particular patient and focused on eating disorder and other psychiatric symptoms, as well as social functioning, subjective well-being, and treatment history during the past year. For the present study we judged whether a preoccupation with weight and shape as described above was present. Reliability checks on twenty interviews at each follow-up by a control rater blinded for patients' initial status indicated that agreement on the presence of a weight/shape preoccupation was satisfactory at $K = .69$ after 1 year and .60 after 2 years (agreement in 85% and 80% of cases; Norring and Sohlberg 1991). Later addition of twenty more tapes the 2nd year and substitution of the control rater's assessments for the original ones resulted in negligible changes in the results.

Results

Nine of the 11 anorectic individuals participated at the first follow-up (T1). Seven of the nine and two who were dropouts at T1 took part at the next follow-up for a total of nine again at T2 (one of whom did not take part in the interview). Nineteen of the 22 bulimic individuals participated at T1; 16 of the 19 and 3 who were dropouts at T1 took part at T2 for a total, again, of 19. Mean follow-up time to T1 was 13 months (range: 12–17) and from T1 to T2 12 months (range: 9–14) with small and nonsignificant differences between the groups.

Seven anorectics (78%) had a weight/shape preoccupation at T1 versus 12 of the bulimics (63%). A year later the proportions were reversed, with 5 anorectics being ill (63%) compared with 14 bulimics (74%). None of these differences was significant (X^2, $p > .72$). Average scores on the EDI were somewhat higher for bulimics, but due to the considerable within-group dissimilarites these differences were not significant either (*t*-tests, $p > .21$):

EDI Symptoms at T1, AN 20.8 ± 18.5 versus BN 28.6 ± 15.0;
at T2, AN 19.6 ± 18.1 vs BN 26.5 ± 10.8;
EDI Attitudes at T1, AN 31.9 ± 19.3 vs BN 33.2 ± 22.1;
at T2, AN 21.9 ± 15.2 vs BN 30.1 ± 16.8.

We searched then for variables that might predict outcome in the long term. Using dummy coding we set presence of weight/shape preoccupation

Table 1. Percent outcome variance explained by Ego Strength, Object Relations, and Life Events

		Weight/shape preoccupation		EDI symptoms		EDI attitudes	
		Year 1	Year 2	Year 1	Year 2	Year 1	Year 2
AN n = 9[a]	Ego Strength	39*	*18*	7	1	15	34
	Object Relations	43*	*20*	6	1	9	39*
	Life Events	14	4	16	0	*3*	*5*
BN n = 19	Ego Strength	36***	16*	19*	6	44***	37***
	Object Relations	30**	13	20*	2	41***	30**
	Life Events	17*	9	*2*	8	2	1

(*Italicized* figures imply that a better outcome was associated with less Ego Strength, lower levels of Object Relations, and more Life Events, respectively).
*$p < 0.10$; **$p < 0.05$; ***$p < 0.01$.
[a] Weight/shape preoccupation data available for $n = 8$ at year 2.

to 1 and its absence to 0, and so we were able to use Pearson's correlation formula with this variable as well as for the quantitative EDI variables. This is a convenient and informative technique, but since it may be unreliable with respect to the one-tailed significance tests that theory would allow here (Harris 1975), we report two-tailed probability figures. Table 1 illustrates outcome variance (i.e., squared Pearson coefficients) that Ego Strength, Object Relations and Life Events explained in the two groups.

Considering Ego Strength first, only the association with weight/shape preoccupation at T1 approached significance among the anorectic subjects. The 39% variance is enough to imply that there is only an 8% chance that anorectics at lower levels of ego strength have the same chance for recovery as those at higher levels. Several more trends and significances were obtained in the bulimic sample. The significances imply that bulimics at lower levels of ego strength, similar to the tendency among anorectics, have less chance of being free from weight/shape preoccupation after 1 year. They also show higher levels of eating disorder attitudes both after 1 and 2 years.

Looking only at the magnitude of the associations the most conspicuous difference between the syndromes was found in the EDI at T1. Ego strength was more relevant for predicting 1st-year outcome among the bulimic than the anorectics subjects (19% and 44% versus 7% and 15%). At one point, relationships were actually in reverse of expectation: anorexics in this sample were more likely to be judged well after 2 years if they had *less* ego strength at initial presentation (numbers *italicized* in the Table 1).

Stressful Life Events during the course appeared to be less related to status at follow-up than initial Ego Strength, the correlation with 1st-year weight/shape preoccupation among bulimics being the only one actually approaching significance. A notable difference between the two syndromes was found in EDI symptoms after the 1st year. These were more closely related to Life Events for anorectic than for bulimic individuals (16% variance, indicating more symptoms for those with more events, vs 2%, indicating fewer symptoms).

Our third question concerned the prognostic importance of object relations as contrasted with overall ego strength. As evident in Table 1, all conspicuous changes are associated with the move from one outcome measure and point of follow-up to another, rather than with the use of object relations or ego strength as predictor. This does not depend on whether anorectics or bulimics are studied. Consider e.g. anorectics' object relations predictions for T1 and T2 weight/shape preoccupation and T1 results for EDI symptoms. From T1 to T2 there is a 43 − 20 = 23 drop in percent variance. Going from T1 weight/shape preoccupation to T1 EDI symptoms the drop is 43 − 6 = 37%. In contrast, the difference between using object relations and ego strength as predictor for T1 weight/shape preoccupation is only 43 − 39 = 4%. This pattern largely applies to the entire Table 1.

Discussion

Our original observation that ego strength predicts 1st-year recovery (Sohlberg and Norring 1989) was replicated for those with BN. Bulimics with lower levels of ego strength also had significantly more EDI eating disorder attitudes at both follow-ups.

These findings are clearly consistent with clinical observation (Johnson 1991) and also with previous suggestions that patients with concurrent DSM-III Axis II-disorders fare less well (Merrill et al. 1987). Note, however, that the low levels of ego strength called "borderline" in the EFA system are compatible with other kinds of personality disorder (PD) besides the borderline PD, the prevalence of which in bulimics is still disputed (Pope et al. 1987; Piran et al. 1988). Thus, our findings generalize more to the psychoanalytic notion of "borderline" than to the implications of that particular disorder. The small number of anorectics makes reliance on significance tests difficult, but if one were to base interpretation on magnitude of correlations and consistent patterns rather than on statistical significance, it is clear that results were generally similar in AN and BN. Coefficients linking ego strength to 1st-year EDI scores were notably lower among the anorectics than among the bulimics, but the smaller range of scores among anorectics implies that restriction of range may provide a purely mathematical explanation of this difference (ego strength range 4.73 versus 5.81; EDI attitudes range 51 vs 68).

More interesting, the next-to-nothing 1% risk for 2nd year weight/shape preoccupation that ego strength represented in the original sample (Norring and Sohlberg, 1991) concealed reverse-direction relationships among anorectic and bulimic subjects. Thus, ego strength accounted for 16% variance in the theoretically expected direction among the bulimics but 18% in the reverse direction among the anorectics.

The result among the bulimic individuals only came close to significance, but suggests that our earlier conclusion that those with lower levels of ego functioning have equal chances for a recovery within 2 years may not hold for those with BN. With respect to the anorectics the result may well be due to chance since it deviates from the 1st-year data and the group was small. It is also necessary to consider that attempts to assess personality run the risk of being confounded by the effects of malnutrition per se on functioning. It was our impression during the pilot phase of this project that changes in patients assessed before and after at least a month of refeeding were small in comparison with the entire range of functioning encountered. While patients may have become more talkative and less single mindedly preoccupied with weight and food, they did not appear radically different, and the information they gave about their typical ways of thinking, feeling and behaving was the same. Nevertheless, the most severely emaciated anorectic subjects in this study also had very low levels of ego strength. We cannot rule out the possibility that there were some among them whose ego strength scores did not accurately represent their true level.

Our first study demonstrated that when overall ego strength differences were accounted for, object relations remained a more conspicuous problem for borderline patients in comparison with those at neurotic levels (Norring et al. 1989). In addition it may be true that rather than being just one of the ego functions, object relations is the central one on which the development of the others depends (Horner 1984). If so, this aspect would require focal attention also at neurotic levels and when improvement in other functions is the ultimate goal of therapy. Nonetheless, the present data demonstrate quite convincingly that for prognostic purposes object relations do not occupy a privileged position vis-à-vis overall ego strength. Both among anorectic and bulimic subjects the results were strikingly similar whether object relations or overall ego strength was used as the predictor.

Concerning Life Events, the one correlation close to significance was the one for weight/shape preoccupation the 1st year among bulimics. It therefore remains to be conclusively demonstrated that adverse events are correlated with outcome in these syndromes. This is not a question of sample size; the coefficients obtained for Life Events on 1st-year weight/shape preoccupation were actually lower than in the original sample (Sohlberg and Norring 1991). Though restriction of range may explain this for the anorectics (two events maximum recorded) this is not the case for the bulimics (four events maximum, same as in original sample). Chance may be the most likely explanation, but it is theoretically conceivable that the course of typical AN and BN is less influenced by stress than in the case of atypical syndromes.

A discrepancy between the groups was that life events explained 16% variance in 1st-year EDI symptoms among the anorectics compared with 2% in the reverse direction for bulimics. Though this may also be a chance occurrence, restriction of range is not a convincing explanation in this instance since the range of scores was greater among the bulimics.

In summary, we found no support for the possibility that object relations are more important for purposes of forecasting than overall ego strength. Neither was the influence of life events strong enough to allow the conclusion that this factor is associated with outcome in unquestionable and uncomplicated AN or BN. However, EDI attitudes in BN 1 and 2 years after initial presentation are predicted by ego functioning. Ego functioning also predicts weight/shape preoccupation at follow-up in this syndrome. Importantly and in contrast to our earlier conclusion concerning all types of weight/shape preoccupied patients, this may hold not only for the 1st year but for the 2nd as well. Analysis of 3rd year follow-up data will enable us to confirm, refine, or refute these conclusions.

References

American Psychiatric Association (1987) Diagnostic and statistical manual of mental disorders, 3rd edn: revised (DSM-III-R). APA, Washington DC

Bellak L, Hurvich M, Gediman H (1973) Ego functions in schizophrenics, neurotics, and normals. Wiley, New York

Cattanach L, Rodin J (1988) Psychosocial components of the stress process in bulimia. Int J Eating Dis 7:75–88

Dohrenwendt BS, Krasnoff L, Askenasy A, Dohrenwendt BP (1978) Exemplification of a method for scaling life events: the PERI life events scale. J Health Soc Behav 19:205–229

Drewnowski A, Garn S (1987) Concerning the use of weight tables to categorize patients with eating disorders. Int J Eating Dis 6:639–646

Ehrensing R, Weitzman E (1970) The mother-daughter relationship in anorexia nervosa. Psychosom Med 32:201–208

Garner D, Olmsted M (1984) Manual for Eating Disorder Inventory. Psychological Assessment Resources, Odessa

Gowers S, Norton K, Yeldham D, Bowyer C, Levett G, Heavey A, Bhat A, Crisp A (1989) The St. George's prospective treatment study of anorexia nervosa: a discussion of methodological problems. Int J Eating Dis 8:445–454

Greenberg B (1986) Predictors of binge eating in bulimics and nonbulimics. Int J Eating Dis 5:269–284

Guiora A (1967) Dysorexia – a psychopathological study of anorexia nervosa and bulimia. Am J Psychiatry 124:391–393

Harris R (1975) A primer of multivariate statistics. Academic, New York

Hirshberg L (1989) Rorschach images of symbiosis and separation in eating-disordered and in borderline and nonborderline subjects. Psychoanal Psychol 6:475–493

Holmgren S, Humble K, Norring C, Roos B-E, Rosmark B, Sohlberg S (1983) The anorectic bulimic conflict – an alternative diagnostic approach to anorexia nervosa and bulimia. Int J Eating Dis 2:3–14

Horner A (1984) Object relations and the developing ego in therapy. Jason Aronson, New York

Humphrey L (1986) Structural analysis of parent-child relationship in eating disorders. J Abnorm Psychol 95:395–402

Johnson C (1991) Treatment of eating disordered patients with borderline and false self/narcissistic features. In: Johnson C (ed) Psychodynamic treatment of anorexia nervosa and bulimia. Guilford, New York, pp 165–193

Johnson C, Tobin D, Enright A (1989) Prevalence and clinical characteristics of borderline patients in an eating-disordered population. J Clin Psychiatry 50:9–15

Merrill CA, Mines RA, Starkey R (1987) The premature drop-out in the group treatment of bulimia. Int J Eating Dis 6:293–300

Mitchell J, Davis L, Goff G (1985) The process of relapse in bulimia. Int J Eating Dis 4:457–463

Monroe S (1982) Assessment of life events: retrospective vs concurrent strategies. Arch Gen Psychiatry 39:606–610

Morgan H, Russell G (1975) Value of family background and clinical features as predictors of long-term outcome in anorexia nervosa. Psychol Med 5:355–371

Norring C (1990) The Eating Disorder Inventory: its relation to diagnostic dimensions and follow-up status. Int J Eating Dis 9(6):685–694

Norring C, Sohlberg S (1988) Eating Disorder Inventory in Sweden: description, cross-cultural comparison, and clinical utility. Acta Psychiatrica Scand 78:567–575

Norring C, Sohlberg S (1991) Ego functioning in eating disorders: prediction of outcome after one and two years. Int J Eating Dis 10(1):1–13

Norring C, Sohlberg S, Rosmark B, Humble K, Holmgren S, Nordqvist C (1989) Ego functioning in eating disorders: description and relation to diagnostic classification. Int J Eating Dis 8:607–621

Piran N, Kennedy S, Garfinkel P, Whynot C (1988) Eating disorders, affective illness, and borderline personality disorder (letter). J Clin Psychiatry 49(3):125

Pope H, Frankenburg F, Hudson J, Jonas J, Yurgelun-Todd D (1987) Is bulimia associated with borderline personality disorder? A controlled study. J Clin Psychiatry 48(5):181–184

Rybicki D, Lepkowsky C (1987) Assessment of anorexia nervosa and bulimia using the Millon Clinical Multiaxial Inventory. BASH Mon Mag 6:194–196

Sohlberg S, Norring C (1989) Ego functioning predicts first-year status in adult patients with anorexia nervosa and bulimia nervosa. Acta Psychiatrica Scand 80:325–333

Sohlberg S, Norring C (1991) Influence of life events and social support on the course of anorexia nervosa/bulimia nervosa (to be published)

Sohlberg S, Norring C, Holmgren S, Rosmark B (1989) Impulsivity and long-term prognosis of psychiatric patients with anorexia nervosa/bulimia nervosa. J Nerv Ment Dis 177:249–258

Strober M (1984) Stressful life events associated with bulimia in anorexia nervosa: empirical findings and theoretical speculations. Int J Eating Dis 3:3–16

Swedish Statistics (1982) Swedish Socioeconomic Classification. Reports on statistical coordination 4, SCB, Örebro

Szmukler G, Russell G (1986) Outcome and prognosis of anorexia nervosa. In: Brownell K, Foreyt J (eds) Handbook of eating disorders. Basic Books, New York

Theander S (1985) Outcome and prognosis in anorexia nervosa and bulimia. J Psychiatr Res 19:493–508

Welch G, Hall A, Norring C (1990) The factor structure of the Eating Disorder Inventory in a patient setting. Int J Eating Dis 9:79–86

What Happened to the "Fasting Girls"? A Follow-up in Retrospect

R. van Deth and W. Vandereycken

As late as 1873, morbid self-starvation was identified as a distinct clinical entity and recognized as such by the medical establishment. First termed hysterical anorexia, it soon became known as anorexia nervosa. The British physician W.W. Gull and the French neuropsychiatrist E.C. Lasègue must be awarded "joint parenthood" for the first description that opened the eyes of their colleagues to this peculiar disorder (Vandereycken and Van Deth 1989). This is far from saying that self-starvation was unknown in former days. Our ancestors surely were familiar with self-starvation, but they did not consider it primarily a pathological phenomenon.

Before the establishment of anorexia nervosa as a clinical entity in the late nineteenth century two main periods in the history of self-starvation can be distinguished (Van Deth and Vandereycken 1989). The first phase is chiefly concerned with extreme fasting as part of penitential or ascetic practice of many pious Christians. Although this kind of self-starvation occurred incidentally even in our century, it was a common part of the pattern of piety of deeply religious women especially in the late Middle Ages. The second phase started with the appearance of the so-called "miraculous maids" or "fasting girls." In the history of self-starvation this phenomenon is the connecting link between medieval fasting within a Christian tradition and modern anorexia nervosa within a medical context.

From the sixteenth century on, these fasting girls – just like giant babies, sea monsters, and other monstrosities – belonged to the sensational news of the time. Until the late nineteenth century they achieved local or even national notoriety, because they ate very little or even nothing at all and nevertheless stayed alive, sometimes even in excellent health! Just like their illustrious medieval predecessors these "natural wonders" would have lasted for months or even years. Though we are poorly informed about the opinions of ordinary people in those days, for example, through popular media like pamphlets and newspapers, a fasting girl appears to have been often considered a miracle, a sign of God's presence on earth, and (later) an astonishing natural phenomenon. Sometimes thousands of people, including kings and other dignitaries, paid a visit to such fasting girl and offered money and goods. Did this rather peculiar phenomenon really disappear or was it replaced by a medicalized version of its appearance: anorexia nervosa?

Speculations on the Nature of Prolonged Fasts

In the High Middle-Ages protracted abstinence from food could give rise to either devote veneration or suspicions of possession: it was either God or the devil who enabled the person fasting to endure the abstinence from food. In early modern Europe a clear line of demarcation between the merely pathological and the miraculous was still lacking. However, the Scientific Revolution and its concomitant process of secularization gradually minimized the belief in the supernatural causes of this unusually protracted fasting. Furthermore, rising protestantism and the meanwhile formalized reservedness of the Roman Catholic Church towards new miracles did not leave much space for potential fasting saints. When the connection between fasting and religion became more and more loose, the medical profession increasingly stressed the pathological nature of prolonged abstinence. Although adherents of miraculous and naturalistic explanations competed until the late nineteenth century, ultimately by the turn of the century self-starvation was generally considered an affliction with natural causes.

This gradual demystification of self-starvation is reflected in several medical treatises from different periods. In treatises on prolonged abstinence, physicians were especially concerned with the question of how long a person could actually live without eating. Lacking reliable empirical data on the topic, the answers varied. Contradicting accounts of shipwrecked persons or locked miners and experiments on hibernating animals only caused confusion. Though already in the sixteenth century physicians expressed doubts about the occurrence of the "inedia prodigiosa" or "prodigious abstinence" (Harvet 1597), scientists seriously discussed, even into the nineteenth century, the possibility of fasting for years.

In the premodern era, many doctors presumed protracted abstinence from food to have a supernatural cause. The Swiss physician Jacob Zwinger, for instance, discussed in the early seventeenth century several natural causes of prolonged fasting and concluded that these fasts are against every natural law (Zwinger, in Fabricius Hildanus 1656). He rejected contemporary theories that one might become accustomed to not eating or that extreme fasters are to be compared to hibernating animals. In his opinion a person could not actually live more than 4 or 5 days without food. However, instead of seriously doubting contemporary accounts of extended fasts, he considered them for this very reason miracles of a divine origin. Moreover, if these fasts are natural, why should we consider Christ's fast in the desert miraculous?

These kinds of metaphysical explanations were gradually substituted by more earthly, though not far less speculative, theories. In a Latin dissertation from 1737 on the blind Swiss fasting girl Christina Kratzer, Johann Jacob Ritter discussed several explanations of this 3-year fast. One of them concerns Christina's amenorrea. If a woman feeds her fetus with her own blood, why would Christina not have been able to feed herself with her menstrual blood (Ritter 1737)? Such suppositions are found in great variety at the time. In a well-documented treatise on prolonged fasts one doctor mentioned no less than sixteen explanations (Reies Franco 1661).

One of the very favorite explanations of the "inedia prodigiosa" was based on the assumption that human beings could live on "feeding particles" in the air. Almost every author addressing the problem of extended fasting devoted a passage to this theory. An early example is a treatise from 1551 by the Italian philosopher Simone Portio on the 2-year-old fast of an unnamed miraculous maid. However, the air-as-food theory was rooted in antiquity. The Roman chronicler Pliny, for example, described a people on the Ganges, called the Astoni, said to have had no mouth and who lived by smell alone. In early modern Europe these stories were retold and amplified with own observations. Returning from exotic countries voyagers made mention of people living solely on the smell of fruits and spice. An often rehearsed tale was about people near Moscow, the Lucomoria, who would hibernate for months (see Citois 1603). These stories, combined with observations on the mystery of hibernating animals, formed the most important sources of inspiration for air-as-food speculations.

Even the late eighteenth-century revolutionary scientific discoveries of oxygen and its role in the process of combustion did not put an end to it. In 1800 the German physician Ludwig Schmidtmann rejected every suspicion of deceit in a 2-year fast of Anna Maria Kienker from Osnabrück. In his opinion she resorbed from the atmosphere "elementary particles" which were sufficient for the nutrition of her body. And in respect to the 3-year fast of the Italian Anna Garbero it was suggested in the 1820s that she extracted chemicals from other people by way of the air. That would explain why her sister, who slept with her, had lost weight . . . (Luciani 1890). Similar explanations can even be traced in our century. Peculiar enough, in 1948 the well-known psychiatrist and former pupil of Freud, Carl-Gustav Jung, suggested the possibility that in prolonged fasts nourishment might be effected by the passage of living molecules of albumen from one body to another (Jung 1977).

However, in our days the medicalization of self-starvation has been accepted for decades. In the latter part of the nineteenth century fasting girls mostly were diagnosed and treated as hysterics. Already a century before, the famous Swiss physiologist Albrecht von Haller regarded many fasting girls as hysterics (Von Haller 1757–66) and his view was shared by several early nineteenth-century physicians (e.g., Rostan 1832; Bourgeois 1855). Even after anorexia nervosa had been introduced in the medical literature this view on "hysterical fasting" remained unchanged for decades (e.g., Campbell 1878; Hammond 1879; Holländer 1921; Morgenthaler 1926).

It was common knowledge among physicians of those days that hysterics often abolished all desire for food and were able to endure lengthened periods of abstinence. Besides, hysterics were notorious for their tendency to deceive and to attract attention, which explained their unbelievable assertions with regard to the length of their fasts. Even their eating in secret, formerly interpreted as clear evidence of deceit, was now considered a symptom of a disease for which the patient could not possibly be held responsible. Consequently, one doctor explained in 1878: "In the present state of our knowledge of the vagaries, the simulations of diseases, and the moral depravity noticed in patients suffering under the hysteric state, I think that no more 'fasting girls' should occur" (Campbell 1878 p. 255). It was not until about 1960 that these stories about fasting girls were related to anorexia nervosa by medical historians (Diekmeier 1959; Von Baeyer 1959; Bliss and Branch 1960; Schadewaldt 1965).

Varying Tasks of the Medical Profession

In cases of self-starvation the task of physicians for centuries did not simply consist of diagnosing and curing the disease. Indeed, from antiquity on, less severe eating disorders such as lack of appetite and aversion to food have been common symptoms in a variety of psychiatric afflictions (Vandereycken et al. 1990). Even long fasts might have been considered pathological if these were attended with symptoms of known diseases like tuberculosis. That may explain why physicians confronted with abstinence from food often firstly tried to find natural causes and to apply medical therapies. However, if this appeared to be of no avail and the abstinence holded on, their task shifted: instead of their regular medical work, they had to take part in sharp discussions on the possibility that a miracle was going on.

From bitter experience, people at the time were well aware that refusal to eat might originate from motives of financial gain or craving for attention. Several accounts of the deceit by fasting pseudosaints and miraculous maids were a serious warning to credulity. On the other hand, everyone was familiar with (and believed in) the prolonged fasting of Moses, Elijah, Jesus, several medieval fasting saints and certain well-attested cases of fasting girls. Small wonder that there was sharp controversy on the topic. Often in charge of the local town council, physicians tried to put an end to rising rumors by making an extensive investigation into the case. Just like medieval clergymen in cases of potential fasting saints, it was now the task of physicians to settle the question whether a miracle or a fraud was taking place. In agreement with the basic principles of the experimental method food intake was controlled by isolating the fasting girl completely from the outside world and watching her continuously for several days. For the fasting girl there was no escape: refusal to cooperate would soon be interpreted as fear of detection.

The results of these watches were quite divergent, varying from a happy end to deep sorrow. Sometimes the girl proved to be a real case of "prodigious abstinence," which assured her and her surroundings of fame and fortune. The German fasting girl Margaretha Weiss, for instance, was said to have fasted in the middle of the sixteenth century for at least 3 years. Several investigations could only but acknowledge this peculiar fact (Diekmeier 1959). But in view of the alleged length of many fasts and applied rigorous methods of observation, it does not come as a surprise that not every case ended as happily. In several instances the fasting girl was unmasked as fraudulent, which could result in imprisonment or even capital punishment. Eva Vliegen, for example, another German fasting girl, ended in prison, when in 1628, after an alleged fast of 30 years, butter, cheese, honey and other edible goods were found in her room (Van Wassenaer 1629). In some cases the fasting girl was unmasked only after several investigations. Ann Moore, the "pretending Fasting Woman from Tutbury," had been placed under surveillance in 1808 by Robert Taylor, a member of the Royal College of Surgeons, who could not establish any deceit. However, 5 years later, Reverend Legh Richmond discovered that she had been fed secretly by her daughter (Richmond 1813). Finally, in a few cases deceit was only discovered posthumously. After no less than 30 years of fasting and one thorough medical investigation in 1826, the Dutch fasting girl Engeltje van der Vlies was ultimately unmasked. After her death in 1853, autopsy revealed scraps of pap in her stomach. She appeared to have been fed for years by an accomplice through a shutter in her cupboard-bed (Wachter and Huijgens 1854; Krul 1894).

In the course of the 19th century, the task of physicians as medical judges gradually disappeared as it became accepted that prolonged fasting was a sign of morbidity and that only some form of deception could explain the alleged length of the fasts. The phenomenon was more and more ridiculed in the medical literature and physicians often displayed a hostile attitude towards fasting girls (Morgan 1977). As soon as miraculous fasting had been annexed by the medical profession and the first signs of a new culture of slenderness emerged, fasting girls disappeared from the scene and were replaced by hysterical or anorectic patients. From then on, discussions on the nature of self-starvation (except for hunger artists and hunger strikers) became completely part of the medical discourse. Methods changed likewise: observations and judicial prosecution of alleged fasting girls were supplanted by medical diagnoses and treatment of anorexia.

Were Fasting Girls Disguised Anorectics?

It is apparent that the interpretations and ways of handling self-starvation have changed considerably over time. The central question is whether these various interpretations and methods have been applied to the same phenomenon. In other words, were fasting girls actually disguised anorectics? There has not been a unanimous answer to this question. Some authors (e.g., Skrabanek 1983; Silverman 1986) plainly equate fasting girls with anorexia patients, while others stress the differences between the two phenomena (e.g., Brumberg 1988; Habermas 1990). Yet, as the regular pre-1850 medical literature is almost completely lacking in discussions of anorexia-like disorders, it seems quite reasonable to presume that accounts of fasting girls may have hidden cases of anorexia nervosa. Applying modern criteria for anorexia nervosa seems the obvious way to explore this presumption.

"Diagnosing" this "historical psychopathology" in retrospect is fraught with difficulty. In addition to the fragmentary character of many historical sources, a major problem concerns the diagnostic criteria one should use for this purpose. As outlined elsewhere (Van Deth and Vandereycken 1991), current criteria of anorexia nervosa are inadequate and have to be modified within two restrictions. Firstly, our adapted criteria should still enable us to differentiate between possible cases of anorexia nervosa and other forms of inanition. Secondly, physicians from previous centuries should have been able to observe and report every single criterion. Following these principles, it is preferable to omit modern, more covert criteria like weight concerns and disturbed body image. Although presentday clinicians consider these features as essential symptoms of anorexia nervosa, they may be predominantly culture bound or too closely related to a modern psychological approach. Instead, it is advisable to focus on more overt, physical and behavioral features of the disorder. Consequently, to label disorders in medical-historical reports retrospectively as "anorexia nervosa," we propose that the following criteria be met:

1. A dramatic weight loss, evident to friends, relatives and medical attendants. Symptoms like hypothermia, constipation and amenorrea may be interpreted as indicators of prolonged starvation.
2. Striking physical (hyper)activity, unusual in medical afflictions with a comparable degree of emaciation or malnutrition.
3. Stubborn defiance, often expressed in reluctance to acknowledge the need of help, and/or resistance to entreaties from the surroundings (doctors, relatives, friends) to start eating again.
4. Evidence of a peculiar eating pattern: abstinence from food in combination with eating in secret, periods of over-eating and/or self-induced vomiting.
5. No other data suggestive of physical illnesses ("wasting" diseases such as tuberculosis, tumors etc.) or psychiatric disorders (especially psychoses and major depression) that would account for the weight loss.

Having established these particular criteria for retrospective diagnostic studies, we analyzed a number of printed primary sources on 25 fasting girls from the sixteenth through the nineteenth century. The first column of

Table 1. Historical accounts of fasting girls and their relation to specific criteria of anorexia nervosa

Source	Fasting girl	Year of birth	Age of onset	Minimal period of fasting (years)	Evidence for prolonged starvation	Hyper-activity	Stubborn defiance	Eating secretly	Data suggestive of other illness
1. Bucoldianus (1542)	Margaretha Weiss	1529	10	3	h,c	±	–	–	Headache; stomachache; contracted limbs
2. Kolb et al. (1585)	Katharina Binder	1558	20	7	a,c	–	–	–	Fits; loss of hearing and "understanding"; headache
3. Wier (1577)	Barbara Kremers	1563	10	1	w,c	±	+	+	Temporary problems in walking
4. Anonymous (1611); Van Wassenaer (1629); Fabricius Hildanus (1656)	Eva Vliegen	1575	19	34	w,a,c	–	–	+	Inability to swallow; visions
5. Fabricius Hildanus (1656)	Puella Coloniensis	1581	7	7	w,c	±	–	–	
6. Lentulus (1604); Anonymous (1607)	Apollonia Schreier	1584	71	6	w,a,c	–	–	–	Swelling; stomachache; convulsions; oppression
7. Citois (1603)	Jeanne Balam	1588	11	3	w,h,c	+	–	–	Fever; loss of speech; paralysis; inability to swallow
8. Hobbes (1668); Reynolds (1669)	Martha Taylor	1649	18	1	w,c	–	–	–	Inability to swallow; "her gutts lye out" (?); lower extremity paralysis

9.	Steill (1744)	Janet Young	1665	18	50	a,c	–	–	–	Vomiting of blood
10.	Stalpart van der Wiel (1686)	Aaltje Kuyper	1667	15	0.5	c	–	–	–	Inability to swallow; fits; closed eyelids; oppression
11.	Anonymous (1722)	Anonymous	1700	20	2		–	–	–	Periods of "lethargy"
12.	Lossau (1729)	Maria Jehnfels	1707	9	12	w,a,c	–	+	–	"Lethargo continuo"; pus in lungs; fever; consumption
13.	Nielen (1775)	Maria van Dijk	1719	48	7	w,c	–	–	+	Menorrhagia; swellings; stomach and lumbar pain; spitting and evacuation of blood and worms
14.	Hoessle (1780)	Anna Maria Zettlerin	1734	20	26	w,a,c	–	–	–	Spitting of blood; indurations in stomach; cramped feet
15.	Varennes (1817)	Marie-Joseph Dahl	1740	33	10	w	–	+	–	Motionless; lockjaw; "physically and morally insensible"
16.	Lardillon (1786)	Christine Michelot	1741	10	4	c	–	–	–	Headache; convulsions; "delirium"; loss of speech; paralysis
17.	Taylor (1808); Henderson (1813); Richmond (1813)	Ann Moore	1761	45	7	w,c	–	+	+	Paralysis; hysterical fits; inability to swallow; headache
18.	Van Stiprian Luisçius and Vorstman 1827; Wachter and Huijgens 1854	Engeltje van der Vlies	1787	27	39	a,c	–	+	+	Oppression; fits; pain in chest; gastroenteritis; "strictura oesophagi"
19.	Schuurman (1837)	Eva Mulder	1823	12	1,5	w,c	–	–	–	Fever

Table 1 (*continued*)

Source	Fasting girl	Year of birth	Age of onset	Minimal period of fasting (years)	Evidence for prolonged starvation	Hyper-activity	Stubborn defiance	Eating secretly	Data suggestive of other illness
20. Whiting (1859)	Betsy Hays	1831	26	2	c	–	–	–	Convulsions; spasms; choking sound in throat; cramped feet
21. Brant (1878)	Martha White	1855	17	5	w,c	–	–	–	Inability to swallow; loss of sight; convulsions; stomachache; Self-multilation dyspnea; contracted limbs; insensibility
22. Hammond (1879)	Sarah Jacob	1857	10	2	c	–	–	–	Fits; loss of consciousness; inability to swallow; rigidity of muscles; opisthotonus; spitting of blood and froth; globus hystericus
23. Sephton (1876)	Ellen Sudworth	1859	12	5	w	–	–	–	Headache; closed eyes; loss of speech; blood from between eyelids and mouth (?)

24. Dougal (1881, 1882)	Christine Marshall	1867	13	1,5	w,a,c	–	–	–	Inability to swallow; cough; pain in chest; stomachache; vomiting of glairy, bloody fluid; headache
25. Anonymous (1881)	Mrs. Lockwood	?	?	7		–	–	–	"Incurably afflicted"; "state of coma"; fainting; headache

w, weight loss; a, amenorrea; h, hypothermia; c, constipation.

Table 1 lists these sources. With a few exceptions (case numbers 8, 9; partly 17; partly unknown 4, 6; unknown 11, 25), all reviewed accounts were drafted by medical practitioners. The next four columns show the name of the fasting girl, her year of birth, age of onset of fasting, and reported minimal period of food abstinence in years (the age of onset is the age at which the girl is reported to have started eating less). To avoid additional problems in interpretation we excluded historical cases of "fasting boys," as well as those from deeply religious women. Indeed, some of the listed fasting girls were pious Christians. A few were deemed by some to be inspired by God or Satan (e.g., cases 1, 4, 12) or considered to have developed clairvoyant faculties (case 4). But none of them fasted explicitly for ascetic or penitential reasons as fasting saints did. Most of the selected girls were well known in their time, though some were not (cases 9, 10, 25).

The last five columns reflect the criteria mentioned above for a positive diagnosis of anorexia nervosa. The subcriteria of periods of overeating and self-induced vomiting were omitted from Table 1, because both went totally unmentioned in the reviewed accounts! A positive or a negative sign is used if data meet criteria or not. If available information was inconsistent or ambiguous we used a ±. If particular criteria were not mentioned in the reviewed sources, we deemed them to have been absent in reality. The latter assumption seems plausible, because physicians at the time carefully scrutinized the girls' bodies and behavior, and they were familiar with each symptom mentioned in our criteria. Besides, they surely would have reported them if they had occurred, for each criterion was of particular interest to them in view of the assessment of the miraculous character of the fast.

It appears from Table 1 that ten girls started fasting between the ages of 13 and 20. This bears some resemblance to anorexia nervosa, a disorder that particularly befalls girls in their teens. As is presumed in modern anorectics, the abstinence of fasting girls may have been a psychological reaction to the turbulence of puberty and adolescence or is suggestive of psychosexual problems (Habermas 1990). However, notwithstanding the suggestive age, from our analysis not much evidence appears in support of this presumption. Although perhaps partly due to the fact that physicians at the time were reluctant to make inquiries into the sexual aspirations of their female patients, our sources are silent on possible psychosexual problems. Instead, the reviewed sources mention more prosaic occasions such as a fall from a horse (case 9), the death of a relative (13) or washing out the linen of a person affected with scrofulous ulcers (17). Only in a few cases there is some suggestion of psychosexual problems. One girl (12) asked whether she had a friend reacted "with a sad face" and started to beat. But contrary to modern psychoanalytically inspired theories on anorexia nervosa, two other cases (6, 15) obviously did not try to forestall adult womanhood and its implications of changing sex roles. Instead, by fasting they desperately tried to reach a particular aim: their lover! Moreover, the majority of the girls in our study were just too young or too old to be bothered by pubertal problems: nine were younger than 13 and five older than 20 when their problems were noticed (age unknown in case 25).

The alleged length of fasting in years, in some cases even extending three decades, is rather surprising. However, we should keep in mind that most fasting maidens only gradually reduced their food intake and the accounts seldom report complete abstinence from food for years, not even in more progressed stages. Instead, most fasting girls did drink and eat small portions. Considering the recorded evidence of prolonged starvation, in about half of the cases we may presume that these girls really must have subsisted on extraordinarily small amounts of food. Doctors rather constantly diagnosed evidence of prolonged starvation over time, especially signs such as constipation and somewhat less frequently amenorrhea and weight loss. Probably most of them got used to a state of chronic malnutrition and just ate enough in secret to sustain life. Nevertheless, this covert eating might fit in the clinical picture of anorexia nervosa, which often attends with tricks and artifices in relation to food.

It is striking that mortality appears to be minimal in these fasting girls. Most of them fasted for years and only died (for example, cases 12, 13, 14, 15) or were unmasked (cases 4, 17, 18) after considerable periods of time. If they were indeed disguised anorectics, the condition must have been of the chronic type in many of these individuals. Data suggestive of prolonged starvation are quite insufficient to conclude a diagnosis of anorexia nervosa. They are only epiphenomena of severe starvation of whatever origin and do not necessarily point to anorexia nervosa. Concerning the other, more crucial criteria, only in a few cases were stubborn defiance (3, 12, 15, 17, 18) and eating in secret (3, 4, 13, 17, 18) reported. In the cases (3, 17, 18) where both criteria were observed, stubborn defiance might have originated from fear of being discovered eating in secret. The fact that most of these individuals not oppose medical authority is not surprising because such a refusal inevitably would make them suspect.

Only in the case of Jeanne Balam (7) is a similarity to the remarkable overactivity of present-day anorectics suggested. In her case, the French physician Citois (1603 p. 11) noticed, that "shee travaile about the house, goe to the market for victuals; sweepe the house; spinne at her wheele; reele off her quill; and gives herselfe (as any other) to all serviceable offices in a familie, & seemes as if shee were not defective, in any part of sence, or moving of her bodie." Prolonged fasting did not impede the normal level of activity in three other girls (1, 3, 5). Though there was no mention of hyperactivity, both fasting girls played, walked up and down, and did all kind of things that were done by children of their age. Some girls did suffer from insomnia (8, 17, 24), but most of them were bedridden. Of course, this apparent lack of overactivity may be attributed to the progressed stage of starvation. But at an earlier stage of the fasts and in cases in which evidence of starvation is scarce (10, 11, 25) overactivity was not noticed either.

Reviewing the whole series, only one patient (3) meets our criteria rather well. This girl, however, started to eat immediately after she had been caught eating in secret. Even her problems in walking disappeared after she had been treated with an ordinary oil. Remembering that none of the accounts mentioned periods of overeating or self-induced vomiting, our study cannot establish any convincing resemblance in "symptoms" between fasting girls and anorectic patients. On the contrary, our results suggest the fasting girls to have suffered from diseases other than anorexia nervosa. Data in the last column of Table 1 corroborate this impression. Strikingly, over time (especially from the nineteenth century on) there are an increasing number of data suggestive of other illnesses, in particular conversion hysteria. The recorded fits, temporary paralysis, problems in walking, hearing, seeing, speech, etc., which often accompanied the prolonged fasts, remind us of nineteenth-century accounts of hysteria.

Two Different Worlds

Designation of the behavior of the fasting girls as hysterical is not without problems. It does not really provide an explanation and it is often used to explain away too much (Kroll and De Ganck 1986). Besides, it is rather difficult to retrospectively distinguish hysteria-like symptoms caused by organic processes from those arising from the mind (Shorter 1986). Indeed, symptoms observed in some cases (12, 14, 18, 24), such as pus in lungs, pain in chest, cough, vomiting of blood, worms, and menorrhagia may indicate grave somatic diseases. However, the striking recovery in some (3, 10) and the temporary character of the symptoms in many cases at least suggest a functional basis or point to the possible influence of social and psychological factors.

To reconstruct parts of the social context of the fasting girls we have to consider the main contemporary opinions about the phenomenon. Basically, the attitude of the medical profession towards fasting girls was often characterized by mistrust, irritation, or hostility. This appears to be a constant factor over the centuries, for even today remnants of this attitude may be observed. Some present-day clinicians confronted with anorectic patients display a similar attitude towards those "obstinate creatures," whose behavior is no more than a "foolish caprice of fashion." This attitude seems to be rooted both in a sense of powerlessness towards self-starvation as well as in a kind of disapproval of those who harm their sanity so "willfully" (Vandereycken and Meermann 1984).

Furthermore, fasting girls, just like modern anorectics, were confronted with several, often conflicting theories on the causes of prolonged abstinence. Extended fasts had been doubted as much as applauded in those days and the title of "fasting girl" was never obtained easily. To make sure of the support of "believers" and to convince skeptics, fasting girls had to persist in refusing almost every piece of food for a substantial period of time. Eating in secret, whether or not with the aid of confederates, might have been a solution, though quite risky because of rigorous penalties when discovered. This begs the question of why they would exert themselves to such extremes? Partly, because they derived considerable secondary gain through these fasts, both materially and psychologically.

The material benefits are the most obvious: in view of the poverty of many fasting girls the offered money and gifts may have been quite welcome. For instance, before the British fasting girl Ann Moore (17) began to attract the attention of the public in the previous century, she had been "labouring under the greatest distresses" and had not even sufficient clothes to cover her bed. But, as a contemporary stated, "... she has turned the exhibition of her person to such account, as to be able, in the course of the present year, to place the sum of £.400 in the stocks" (Henderson 1813 p. 77). Ann Moore was not the only one who exploited her fast. Just like hunger artists around the turn of the century, fasting girls like Puella Coloniensis (5) and Maria van Dijk (13) were even exhibited publicly for payment in taverns. Most fasting girls, however, were more discreet. For example, the Welsh fasting girl, Sarah Jacob (22), had a receptacle lying on her breast into which visitors could put silver. In general, the families (most fasting girls stayed with their parents), but also the local community took advantage of the visitors commercially. Paradoxic as it may seem, by abstaining from food fasting girls and the people in their surroundings were able to earn their bread. This may explain why often fasting girls were only initially put under pressure from those around them to start eating again.

Psychologically they might have profited by the enormous attention which their fasting often excited. For otherwise anonymous girls from rather isolated areas this must have been an attractive perspective indeed. Furthermore, by starving themselves women were able to manipulate not only their bodies, but also their social environments. Their rising status from poor child to famous fasting girl may have procured them some sense of power and autonomy. Often the people around them had an interest in preserving the miracle and for this reason may have been quite willing to dance to the tune of the fasting girl. This idea of self-starvation as a means of obtaining some form of power may be an important clue to the question of why it is that predominantly women have taken refuge in self-starvation throughout history.

Yet, it is an oversimplification to picture miraculous maids just as unmitigated individuals who saw their way clear to escape from their restraining and humble circumstances. That the matter is more complex is apparent from the fact that an unknown number of fasting girls never became celebrities and remained in poverty. Louise Gussie of Anglefort en Bugey, for instance, took no food from January 1770 to August 1773. Her doctor did not suspect any imposture, because she lived in a poor, out-of-the-way hovel, "where such a trick would not bring in six sous by way of alms in the whole of a twelvemonth" (de la Chapelle 1774, cited in Thurston 1952 p. 360). In the case of another fasting woman, named Whistler, it was only her gravestone which reminded people that she had lived 7 years without food and drink (Lossau 1729).

But even those who had been reputed for prolonged fasts did not always simply abuse the credulity of their contemporaries. Rather, they shared the same contemporary opinions about prolonged fasts and considered their behavior in terms of available models of interpretation. In this light, it is not precluded at all that many of them were convinced of the miraculous character of their own fasts! Indeed, at the onset a variety of physical afflictions and/or psychological reasons may have given rise to the refusal to eat. Then the particular social context at the time preserved and intensified their abstinence from food and eventually may have convinced the fasters themselves of the miraculous event. A miracle which had to be a little sustained by the secret intake of some food.

Inversely, through their almost incredibly prolonged fasts they corroborated contemporary metaphysical speculations on the subject. The increasing number of data suggestive of other illnesses in our study may be partly attributed to the growing awareness of doctors that these fasts essentially were pathological. As medicalization of the phenomenon progressed physicians sought and found an increasing variety of symptoms. On the other hand, fasting girls may have adapted their behavior to these changing contemporary frames of reference. This might explain why the motive of the fasting girls' abstinence became less and less religious in nature and increasingly more secular. In a way, fasting girls spoke the language of their time and as such were understood by their contemporaries.

The exhibitory atmosphere in which this took place can hardly be compared with the completely medicalized climate in which modern anorectic individuals live. In the former reverence, admiration and astonishment prevailed, in the latter pity, dismay and care. Small wonder that miraculous maids, unlike modern anorectics, did not conceal that "something peculiar" was going on. As also evidenced in the apparent lack of overactivity, they did not deny their condition or state of physical exhaustion like modern anorectics at all. On the contrary, to convince those in their surroundings of the miracle they had to demonstrate its characteristics. The eighteenth century Swiss fasting girl Josephine Durand, for example, displayed her difficulty in taking nourishment in full: At the request of visitors she took half a spoonful of plain water, which was regurgitated instantly and caused a sort of paroxysm for a quarter of an hour (Thurston 1952). In contrast to modern anorexia nervosa (and in line with classic hysteria) miraculous fasting had a theatrical character. Whereas anorectics tend to seek isolation even within the family, fasting girls needed an audience usually outside of the family circle.

Finally, although we lack statistical evidence, a striking difference in incidence of the two phenomena is also indicated. Miraculous maids presumably must have been rarities, for it seems quite implausible that fasting girls could induce speculations of miracles and reach fame and fortune if at the time prolonged refusal to eat did occur on a large scale or almost "epidemically" as is reported today for anorexia nervosa! This suggests that the incidence of noneating behavior varies considerably over time and points to the importance of sociocultural factors.

Conclusion

No doubt a phenomenon like self-starvation is confined by its biological boundaries. Yet, within these boundaries the exact character of self-starvation reveals striking historical variation. Just like anorexia nervosa, the phenomenon of fasting girls is firmly wedded to the specific cultural context that defines and shapes its expression and facilitates its recognition and incidence. This does not lead, however, to the simple conclusion that one and the same phenomenon has been differently interpreted over time: transcendentally in former days, medically today. Rather, like several other social phenomena in an historical context (see, e.g., Gergen and Gergen 1984), it questions the assumption of the stability of human ("normal" and "abnormal") behavior over time and the belief in a basic and unchanging human nature. Regardless of some superficial similarity in the pure somatic features of starvation of both phenomena, the distinct historical conditions include substantial differences in content. Though former "prodigious abstinence" and modern anorexia nervosa may be both labeled as forms of self-starvation, they are dissimilar as far as motivation, function and meaning are concerned. For the very reason that these forms appear to be essentially socially constructed and historically contingent, there are worlds of differences between them.

References

Anonymous (1607) Warhafftige Beschreibung von einer Jungfrawen die jetzt vergangen Michaelis sechs gantzer Jahr keiner Speiss noch Tranck genossen auch nicht arbeit oder schaffet sondern stäts an einem Orth ligt . . . Bern

Anonymous (1611) Pourtraict a vif d'Eva Vliegen de Meurs laquelle en 14 ans ny mangés ny beu, agée a present de 36 ans. Jansen, Zutphen

Anonymous (1722) Von einem Mägdgen, so bey 25. Monat lang gefastet, und bald sehr schwer, bald sehr leichte ist. Sammlung Natur Medizin Class IV Artic 13:101–105

Anonymous (1881) A fasting woman in Ipswich. Med Press Circ 31:36–37

Bliss EL, Branch CHH (1960) Anorexia nervosa: its history, psychology and biology. Hoeber, New York

Bourgeois X (1855) De l'inanition et de ses rapports avec la thérapeutique, l'hygiène et la médecine légale. Ann Soc MedChir Bruges 16 (2nd series-III): 181–207, 229–261, 277–298
Brumberg JJ (1988) Fasting girls: the emergence of anorexia nervosa as a modern disease. Harvard University Press, Cambridge (Mass)
Bucoldianus G (1542) Von dem Meydlin welchs on essen unnd trincken lebt/eyn kurtze erzelung. Vogtherr and Schiesser, Speyer
Campbell JA (1878) Feeding versus fasting. Br Med J 1:254–255
Citois F (1603) A true and admirable historie, of a mayden of Confolens, in the province of Poictiers: that for the space of three yeeres and more hath lived, and yet doth, without receiving either meate or drinke. Roberts, London
Diekmeier L (1959) Krankheitsbild und künstlerische Darstellungen des "Wundermädchens von Speyer" Margaretha Weiss. Kinderarztl Prax 27:107–115
Dougal J (1881) Report on the fasting girl at Chapelton. Lancet 1:755–756
Dougal J (1882) A fasting girl. Br Med J 1:631–632
Fabricius Hildanus G (1656) Aanmerkingen rakende de geneesende heelkonst. Bestaande in zes deelen, yder deel in hondert geschiedenissen. Leers, Rotterdam
Gergen KJ, Gergen MM (1984) Historical social psychology. Erlbaum, London Hillsdale
Grant F (1878) The Market Harborough fasting girl. Br Med J 1:152
Habermas T (1990) Heisshunger. Historische Bedingungen der Bulimia nervosa. Fischer, Frankfurt
Hammond W (1879) Fasting girls; their physiology and pathology. Putnam's, New York
Harvet I (1597) Discours par lequel est monstré contre le second paradoxe de la premiere decade de M. Laur. Ioubert, qu'il n'y a aucune raison que, quelques uns puissent vivre sans manger, durant plusieurs jours et années. Portau, Niort
Henderson A (1813) Observations on the case of Ann Moore, called the Fasting-Woman of Tutbury. Edinburgh Med Surg J 9:74–83
Hobbes T (1845) Letter to Mr. Beale. In: Molesworth W (ed) The English works of Thomas Hobbes of Malmesbury, vol VII. Longman, London, pp 463–464
Hoessle JG (1780) Krankengeschichte der Anna Maria Zettlerin, welche 10. Jahre lang ohne Speis und Trank lebte. Deren Leichenöffnung und natürliche Erklärung dieses so seltenen Zufalls. Von Jenisch, Augsburg
Holländer E (1921) Wunder, Wundergeburt und Wundergestalt in Einblattdrucken des 15. bis 18. Jahrhunderts. Kulturhistorische Studie. Enke, Stuttgart
Jung CG (1977) The symbolic life. Miscellaneous writings. Routledge and Kegan Paul, London
Kolb CA et al. (1585) Eine warhaffte Histori welcher massen zu Schmidtweyler in dess durchleuchtigsten hochgebornen Fürsten und Herrn Herrn Johann Casimirs . . . ein Mägdlin siben Jar lang weder gessen noch getruncken und doch von Gott wunderbarlicher weiss bei Leben erhalten worden. Strassburg
Kroll J, De Ganck R (1986) The adolescence of a thirteenth-century visionary nun. Psychol Med 16:745–756
Krul R (1894) Drie singulieren. Frans van Dusseldorp, Bestje van Meurs en Engeltje van der Vlies. Tijdspiegel 3:394–412
Lardillon (1786) Une fille passe beaucoup de temps sans manger. Collection Academique Composée des Mémoires 12:352–354
Lentulus P (1604) Historia admiranda de prodigiosa Apolloniae Schreierae, virginis in agro Bernensi, inedia. Le Preux, Bern
Lossau CJ (1729) Wahrhaffte und ausführliche Beschreibung eines besondern und merckwürdigen Casus Inediae, welcher sich im Jahr 1728. mit eines Gärtners Tochter aus Steinbeck im Hollsteinischen bey Hamburg, nahmens Maria Jehnfells zugetragen . . . Brandt, Hamburg

Luciani L (1890) Das Hungern. Studien und Experimente am Menschen. Voss, Hamburg
Morgan HG (1977) Fasting girls and our attitudes to them. Br Med J 2:1652–1655
Morgenthaler W (1926) Eine Hysterika zu Beginn des 17. Jahrhunderts. Arch Gesch Med 8:196–201
Nielen PM (1775) Verhandeling over de oorzaaken van het bestaan des leevens zonder spysen, of zelfs zonder spijs en drank, en over de oorzaaken, kentekenen, onderscheid, voorzegging en geneezing van de spijswalging, of afkeer tegen het gebruik van voedzels. Ten Bosch, Utrecht
Portio S (1551) De puella germanica, quae fere biennium vixerat sine cibo potuque. Ad Paulum III, Pontificem Maximum. Torrentinus, Florence
Reies Franco G a (1661) Elysius iucundarum quaestionum campus . . . Vivien, Brussels
Reynolds J (1669) A discourse upon prodigious abstinence; occasioned by the twelve months fasting of Martha Taylor; the famed Derbyshire Damsel: proving that, without any miracle, the texture of human bodies may be so altered, that life may be long continued without the supplies of meat and drink. With an account of the heart, and how far it is interested in the business of fermentation. London
Richmond L (1813) A statement of facts, relative to the supposed abstinence of Ann Moore, of Tutbury, Staffordshire: and a narrative of the circumstances which led to the recent detection of the imposture: to which is subjoined an appendix, containing medical and other papers, illustrative of the statement . . . Croft, London
Ritter JJ (1737) De impossibilitate et possibilitate abstinentiae longae a cibo et potu; occasione puellae inediam longam fingentis. Decker, Basel
Rostan L (1832) Abstinence. In: Dictionnaire de médecine ou répertoire générale des sciences médicales, tome 1. Béchet, Paris, pp 283–308
Schadewaldt H (1965) Medizingeschichtliche Betrachtungen zum Anorexie-Problem. In: Meyer JE, Feldmann H (eds), Anorexia nervosa. Thieme, Stuttgart, pp 1–14
Schmidtmann LJ (1800) Merkwürdige Geschichte eines jungen Mädchens im Hochstifte Osnabrück was bereits achtzehn Monate ohne Speisen und Getränke lebt, nebst physiologischen und pathologischen Betrachtungen darüber. Hahn, Hannover
Schuurman JB (1837) Berigt van eenen langdurigen en volkomenen afkeer van alle spijs. Practisch Tijdschrift Geneeskunde 16:289–299
Sephton R (1876) The fasting girl in Lancashire. Br Med J 1:329
Shorter E (1986) Paralysis: the rise and fall of a "hysterical" symptom. J Soc Hist 19:549–582
Silverman JA (1986) Anorexia nervosa in seventeenth century England as viewed by physician, philosopher, and pedagogue. An essay. Int J Eating Dis 5:847–853
Skrabanek P (1983) Notes towards the history of anorexia nervosa. Janus 70(1–2):109–128
Stalpart van der Wiel C (1686) Eerste deel van het tweede hondertgetal der zeldzame aanmerkingen, soo in de genees- als heel- en snykonst, meest by eygen ondervinding van tijt tot tijt vergadert. Geselle, The Hague
Steill T (1744) An extraordinary abstinence during fifty years. Med Essays Observations 5(2):477–480
Taylor R (1808) A singular case of abstinence. Med Phys J (Lond) 20:402–405
Thurston H (1952) The physical phenomena of mysticism. Burns and Oates, London
Vandereycken W, Meermann R (1984) Anorexia nervosa: a clinician's guide to treatment. De Gruyter, Berlin
Vandereycken W, van Deth R (1989) Who was the first to describe anorexia nervosa: Gull or Lasègue? Psychol Med 19:837–845
Vandereycken W, van Deth R, Meermann R (1990) Hungerkünstler, Fastenwunder, Magersucht. Eine Kulturgeschichte der Essstörungen. Biermann, Zülpich

Van Deth R, Vandereycken W (1989) Reflections on the history of self-starvation. BASH Mag 8(4):92–94
Van Deth R, Vandereycken W (1991) Was nervous consumption a precursor of anorexia nervosa? J Hist Med 46:3–19
Van Stipriaan Luisçius A, Vorstman JG (1827) Algemeen verslag aan de provintiale commissie van geneeskundig onderzoek en toevoorzigt . . . De Groot, Delft
Van Wassenaer N (1629) Historisch verhael aller ghedenckwaerdiger gheschiedenissen die in Europa, als Duytslandt, Vranckrijck etc., van den beginne des jaers 1621 voorgevallen sijn. Jansz, Amsterdam
Varennes de (1817) On an unmarried woman who, for nearly eleven years, had taken no solid food. Med Phys J (Lond) 38:341–342
Von Baeyer W (1959) Metabletica – Bemerkungen zum gleichnamigen Werke von J.H. van den Berg, zugleich zum Problem der Pubertätsmagersucht. Nervenarzt 30:81–85
Von Haller A (1757–66) Elementa physiologiae corporis humani, vol 6. Bosquet, Lausanne
Wachter GH, Huijgens W (1854) Verslag van de lijkschouwing bij Engeltje van der Vlies. Ned Weekblad Geneeskundigen. February 25:81–82
Whiting LE (1859) The woman who lives without eating. Boston Med Surg J 60:109–111
Wier J (1577) De lamiis liber: item de commentitiis jejuniis. Oporinus, Basel

Treatment of Anorexia Nervosa: An Interview

E. Petzold

The long-term course of eating disturbances has been viewed from various perspectives in this book. This chapter is devoted to reflecting on the changes in the approach to treatment in the past two decades, addressing questions such as: What has actually changed in the treatment of eating disorders? At which stages of development is the change noticeable? Are there typical dynamics in anorexia? What is particularly important in the treatment of anorexia today? Which therapeutic problems still have not been solved?

These questions were posed to E. Petzold, who views the problem eclectically despite being familiar with other approaches to treatment (for results of the long-term studies of the Heidelberg group see Deter, Herzog, and Manz et al., this volume).

E.P.: Since the mid-1960s I have been working at the Medical Clinic of the University of Heidelberg, first as a student, then as a doctoral candidate, later as intern, ward physician and assistant medical director and finally for the past 10 years as director of the clinical psychosomatics section.

From the beginning to the middle of the 1960s anorexia nervosa was treated by a combination of somatic treatment and psychotherapy at the Medical Clinic. In practice this meant the patients were kept in bed and when they became restless received medication – chlorpromazine, for example. They received a nasogastric tube which often stayed in place for a number of days. The patients were tube fed with a high-calorie formula (slowly increasing from 2000 to sometimes 4000 calories) and were summoned to the office of the consultant, a psychoanalyst, for 1-h psychotherapy sessions a few times per week. There was no such thing as a conference among the consultant and the staff, of which as an intern I was a member, in those days. Looking back, there were no facilities, be it space or time, provided for such meetings. The consultant was bound to his oath of secrecy – and the pedestal on which this term stood was very high indeed. With psychoanalysis Freud has given us access to the subconscious, a part of us which is worthy of protection, but doing so often conflicts with the necessity of understanding a patient when more than one individual is involved in the treatment: No one knows what the other is doing or thinking. This inevitably leads to contradictions and of course to conflicts, and it "breeds disease." Unfortunately, such an "infection" can be contracted by the institution itself and make working conditions unbearable.

Let it be said here that various approaches to treatment were being investigated in those days and the resulting conflict potential was not used therapeutically. In anorexia, splitting is not only a potential threat to the family, but also to the therapeutic team.

W.H.: But this comment does not just refer to the process of splitting in combined somatic and psychic treatment . . .

E.P.: No, it also applies to the cooperation between general practitioners or psychotherapists in private practice with the clinic – or in another area – for the relationship between individual versus family therapy.

General care in particular is of utmost importance and should not be thought of as a contrast to psychotherapy, but rather as a supplement. I don't mean general care only for the patients when the often more complicated treatments aimed at change fail, but also for family members seeking advice and help. The general practitioner can let things be: He has a maintenance job; no more is expected, but no less, either. For this job it is useful to see the patient regularly, without overly controlling her. Overcontrol is in my opinion one of the most significant pathogenic factors in anorexia nervosa families. Anorexia itself is often an expression of protest against overcontrol. Of course the attending physician will check the patient's weight and state of nutrition and make it clear to the patient and her family that, following from the mortality rate, the situation is very serious. In cases of life-threatening weight loss the physician will have the patient admitted to the hospital. This usually provides the family with some relief, but in our experience, is not much more than a "breather" and a time in which the family must be made aware of the facts. But, to get back on the track, the family doctor has a firm place in the therapeutic setting.

W.H.: But now to the family, that is to say, the tendency to separate individual from family therapy.

E.P.: I am reminded of a discussion which took place during my psychiatry days at the clinic of the University of Basel with the late psychoanalyst D. Beck, who met with such a tragic death. He was strongly in favor of separating the patient from her family during the treatment period, making reference to the issue of self-esteem – including body perception. He maintained that this was the primary issue and not the relationship. The counterquestion of how and from whom the distorted perception of body came to be was left unanswered.

My stand on the issue developed in the opposite direction: The family should be included in treatment. At the end of the 1960s individual treatment was emphasized for anorexia nervosa. The family was kept in the background. I felt that this was a dilemma.

The work of the recently deceased cultural anthropologist Norbert Elias (for example, 1987) on the "I – we" balance, which has developed in the direction of "I" in the long process of civilization, casts some light on the dilemma today. The "I" stands for a person's individuality. The "we" for the group to which one belongs – the family, the tribe. At the end of the Middle Ages "we" was the leading element. The single person, the individual, did not count nearly as much in courts of law as he or she does today. It seems that in anorexia nervosa the patient and her family are victims of this process of civilization, as if they were less capable of individuation. The symbiotic power of the mother-daughter relationship comes to mind here. Their mutual dependence seems often to be such that one cannot exist without the other. The father-daughter relationship can be similarly close in anorexia patients, especially when the father is a *pater maternelle*, that is, a motherly person with a high potential for caring but also for controlling.

I like to call this dilemma which derives from these distortions in the civilization process the Antigone phenomenon.[1]

Antigone stood decidedly for the traditions and customs of her father Oedipus. When Antigone insisted her brother, a fallen warrior, be buried according to old customs, she came into conflict with her uncle, Creon, the new ruler of Thebes, who had rigorously enforced new norms. She had to die for this – some will know, she was immured and died of hunger. Ismene, her sister, submitted to the new state and survived.

I am not concerned here with the political dimensions of this, but rather with the question of loyalties . . .

W.H.: Now there is a constructivist approach to family therapy . . .

E.P.: All psychotherapy is (re)construction. Radical constructivism is lacking, though, in my opinion, the dimension of depth, that is, the reference to one's past, to a personal relationship. G. Weber and H. Stierlin discuss this constructivist approach in their book In Liebe entzweit (Separated in Loving) I think that G. Weber's standpoint of decided committment to radical constructivism represents a great loss. It seems to me to be definitely useful to keep an eye on the traditional psychotherapeutic means of access – the individual – and not to needlessly abandon the tension which exists between the system and the individual. Stierlin also speaks of "related individuation," of "growing within the relationship." In the treatment of anorexia nervosa I emphasize the family dimension, too, especially in younger patients. I do not begin with an individual approach until the blocks in the family relationships have been lifted or at least definitely identified.

[1] Tolstrup (this volume, p. 228) has used the Antigone metaphor in a different sense.

W.H.: You support, then, an eclectic approach which takes into account somatic and general medical aspects and which emphasizes family counseling in psychotherapy?

E.P.: Yes, the concept "this as well as that" developed in the mid-1970s from the "either – or" attitude. In those days one had the option of medical treatment or psychotherapy. The "this as well as that" concept attempted to link these perspectives.

The "extended this as well as that" concept developed from a review of 14 cases in 1974 which included 3 deaths. In those 3 cases the families had not been included in therapy. In the next stage, a review of 44 patients in 1977, it became clear just how obviously the symptoms of the condition can be seen as a function of the relationship.

W.H.: What determines this relationship in your opinion? Certainly a (family dynamic) specificity postulate has not been propounded?

E.P.: No, but we did find some stereotypes in such families:

1. The pattern of relationships in these families is marked by monotony.
2. The symbiotic relationship between mother and daughter is upheld to an extreme.
3. The role model stereotypes are not pathognomic. There are dominant fathers, dominant mothers, and also inhibited, depressive, timid fathers and mothers.
4. The system appears to be closed. The relationships are stagnant. They seem to have been fixed and unable to bear change.
5. Emotions are seldom adequately expressed. The inner person does not emerge. Any feelings with an inkling of libido are rejected with all consequence, even more so than aggression.
6. Any violation of the family code leads in most cases to even stronger bonding of the individual to the family system.
7. Feelings of impotence and disappointment mixed with anger and wishing the other would die create massed resignation, and isolation and retreat seem to be the only way out. This is where family therapy begins.
8. Anorexia nervosa is a symptom of the protest of the healthiest member of a system against the rigidity and fixation of the system as a whole.

As you see, these observations have a certain similarity to those of Minuchin, Jackson, or Wynne.

W.H.: And just how did this process of change in therapy look?

E.P.: Let me continue with the therapeutic stages. From 1975 to 1978 the instrument of family therapy was investigated more intensively. We recorded family sessions and listened to them, which of course at first was at the initiative of individual therapists. Then the sessions were transcribed and verbatim protocols exchanged among the therapists. The protocols were assessed according to the Gottschalk-Gläser method. At the same time epidemiologic investigations were underway for the entire region around Heidelberg. After this came the first specific follow-ups and numerous team members were trained in family therapy.

The development of the methods of family therapy – between about 1979 and 1984 – began with a working group including G. Weber (of the Institute for Family Therapy and Basic Psychoanalytical Research H. Stierlin) and psychiatrists and psychotherapists in the area. There were meetings with family therapists from Milan (L. Boscolo and G. Cecchin). At the same time we initiated observation studies using one-way screens and videotaped family sessions. We developed techniques of therapeutic hypnosis and concentrated then on the link between clinical-psychosomatic work and inpatient psychotherapy emphasizing the family therapeutic approach. This was further developed by anchoring the family dimension into inpatient psychotherapy, part of which is family tree work and family sculpture.

From 1985–1990, in terms of therapy, we were occupied with the absurdity of one Carl Whitacker (see Petzold et al. 1990, 1991), Frank Farrelly's provocative therapy, the rediscovery of spontaneity, sexuality, sense of humor, appropriate protection of self, and setting of personal goals. In family research we worked with the grid method (Hartmann 1986), the family board (Bergmann et al. 1988), then concentrated, however, on application of the symlog technique (Kröger et al. 1987) of qualitative and quantitative family description.

W.H.: Which questions are still open in your opinion?

E.P.: The splitting of good and bad objects is immanent. This still represents a chance and a danger in the family and also in the extended family, i.e., in the therapist-family system.

The grandiose idea that one can do more than someone else can is also still fascinating and dangerous. The experience of insufficiency follows necessarily in the wake of omnipotence fantasies, resignation being the predictable result of such an illusion. Thoughts of therapeutic omnipotence weaken the development of the person seeking help.

The ethical dimension has been neglected. This is where challenges arise which we have failed to meet so far, where, for example, we have proved ourselves incapable of grieving together. A. Mitscherlich's book from (1967) is called to mind: He presents some of the tasks of an open society. I prefer to speak of remembering work. By that I mean the transformation of "experienced" into "accepted" past. Only when this has transpired can we speak of an "own" past, and only then does an own identity develop.

What do I mean, though, by ethical dimension? In this sense ethical dimension means the dialogue between personal value systems. The

Antigone phenomenon and the reference to the long-term, century-long process of civilization (N. Elias) are examples of this. Still, I do not know why I am so preoccupied with personal values in this particular illness. I sense the enormous amount of energy these patients have (I won't speak here of the enormous amounts of energy spent by therapists). Viktor von Weizsäcker (1988) has said, "Life which hasn't been lived is the force which drives it on." I have sensed life's force in anorexia nervosa patients more than in any other condition. The energy from these patients and their families has encouraged me. Some of these people were able to change, some not. But dealing with both groups has changed me. Certainly also in learning to be more relaxed and to acknowledge no change as a compromise which is just possible in respect of differing systems of values. These different value systems come from the varying family backgrounds of father and mother. Ivan Boszormenyi-Nagy's concept of invisible loyalties is also significant here (Boszormenyi-Nagy and Spash 1973).

A relaxed attitude can have practical consequences, though, and did. For example, the brother of an anorexia patient asked me to commit his sister to a psychiatric clinic after 10 years of being ill because she had been admitted to the hospital severely underweight and the family's potential for hope had been exhausted. I refused on the grounds that she should be preferably allowed to lead her, albeit restricted, life as an anorectic than be institutionalized. The patient seemed to have stabilized enough to return to her studies. When she did die a few months later, of course the question arose of whether the decision was right to allow the girl her personal freedom instead of making the decision for her.

This leads to the question of the ethical dimension of using authority in the therapeutic process (Petzold et al. 1990). With his family lunch Minuchin pointed out the danger of this. Family confrontation therapy which has been carefully conceived deeply penetrates the ethical dimension of treatment. I do regret that the questions surrounding this are not topics of general and scientific discussion.

W.H.: And your conclusion?

E.P.: Let me make an interim status report:

1. Family therapy may prevent the patient from separating from the family of origin, and hence separating the generations, as is necessary – first symbolically, then in effect – such that after successfully taking leave no feeling of loneliness, not to mention inner isolation arises. The object of the illness is to protect solidified family value systems (e.g., father's, mother's, etc.), which would otherwise be lost. The object of treatment is to help the family to access these value systems and put them up for a fair exchange. Not until a core of good inner representations of father and mother (from the paternal and maternal families) has been established can the patient successfully separate, and survive.

2. In an acute crisis the interaction conflict specific to the dynamics of the relationship is in the forefront. At such time the other group, the therapeutic team, whether on the ward, in family therapy sessions, or as represented by the institution of treatment itself, functions as a mirror. Not until this problem area, which appears initially bipolar, is clarified is further individuation possible. Not until these borders have been reached and an exchange has taken place can the patient draw on the strength which is actually and probably abundantly available to develop her own identity.

References and Further Reading

Becker S (1987) Das SYMLOG-Rating-Verfahren eingesetzt zur Erfassung von Interaktionen in Familien mit einer anorektischen Tochter. Thesis, University of Heidelberg.

Bergmann G, Kröger F, Petzold E (1986) Allgemeine klinische Psychosomatik – Weiterentwicklung eines Stationsmodells. Gruppenpsychother Gruppendynamik 21:224–235

Bergmann G, Klotz S, Petzold E (1988) Familienarbeit und stationäre Psychotherapie – Verlaufsbeobachtung und Kontrolle mit dem Familienbrett. In: Schüffel W (ed) Sich gesund fühlen im Jahr 2000. Springer, Berlin Heidelberg New York, pp 544–549

Boszormenyi-Nagy J, Spask GM (1973) Invisible loyalities. Harper, New York

Breiden U (1981) Verlaufsuntersuchungen zur klinischen Behandlung von 69 Patienten mit Anorexia nervosa. Thesis, University of Heidelberg

Deter HC, Petzold E (1979) Klinische Befunde und Verlaufsuntersuchungen bei Patienten mit Anorexia nervosa. Therapiewoche 29:6320–6331

Deter HC, Petzold E, Hengst R, Breiden U, Lanzinger G (1983) Katamnestische Ergebnisse einer klinisch-psychosomatischen Behandlung von 103 Patienten mit Anorexia nervosa aus internistischer Sicht unter besonderer Berücksichtigung der Mortalität. Z Innere Med 10:3–12

Deter HC, Petzold E, Hehl FJ (1989) Differenzierung der Langzeitwirkungen einer stationären psychosomatischen Therapie von Anorexia nervosa Patienten. Z Psychosom Med Psychoanal 35:68–91

Ehlers M, Franzmeier G, Weimann C, Petzold E (1982) Ein Blick in die innere Welt. Adjuvantische Therapie, part 13. Musik Med 1:47–56

Elias N (1987) Die Gesellschaft der Individuen. Suhrkamp, Frankfurt

Hartmann A (1986) Epistemologie der Familie. Eine Untersuchung anhand eines dreigliedrigen empirischen Systems. Thesis, University of Heidelberg

Herzog W, Petzold E, Kröger F (1988) Die Elterngruppe als flankierende Maßnahme bei der Behandlung von Anorexia-nervosa-Patienten. In: Deter HC (ed) Gruppen mit körperlich Kranken. Eine Therapie auf verschiedenen Ebenen. Springer, Berlin Heidelberg New York, pp 281–286

Hofner B (1980) Die Magersuchtsfamilie. Vergleichende Darstellungen bisheriger Forschungsergebnisse, Fragebogenuntersuchung und inhaltlich analytische Auswertung von Familientherapieprotokollen. Thesis, University of Heidelberg

Kröger F, Petzold E, Ferner H (1984a) Skulpturgruppenarbeit. Gruppenpsychother Gruppendynamik 19:361–379

Kröger F, Petzold E, Ferner H (1984b) Familientherapie in der klinischen Psychosomatik – Skulpturgruppenarbeit. Festschrift in honor of Prof. Luban-Plozza Forum Galenus 10. Mannheim. Gruppenther Gruppendynamik 19, 4:361–379

Kröger F, Bergmann G, Petzold E (1986) Clinical psychosomatic medicine: individual's circumstance of admission and a systemic view of symptoms: can the two be linked? In: Lancey JH, Sturgeon DH (eds) Proceedings of the 15th European Conference of Psychosomatic Research John Libbey, London

Kröger F, Drinkmann A, Becker S, Petzold E (1988a) SYMLOG in der Familiendiagnostik: Beobachter-Rating von 13 "Anorexie-Familien" in Familiengespräch. Z System Therapie 6(4):297–302

Kröger F, Drinkmann A, Schneider J, Petzold E (1988b) SYMLOG – Ein Instrument zur Selbst- und Fremdbeschreibung von Familien: Erste Erfahrungen und Validierungen. In: Schüffel W (ed) Sich gesund fühlen im Jahr 2000. Springer, Berlin Heidelberg New York, pp 526–532

Kröger F, Drinkmann A, Herzog W, Petzold E (1991) Family diagnostics. Object representation in families with eating disorders Int J Small Group Res 22(1):99–114

Mitscherlich A, Mitscherlich M (1967) Die Unfähigkeit zu trauern. Reclam, Leipzig

Petzold E (1981a) Familienkonfrontationstherapie bei Anorexia nervosa (2nd edn). Vandenhoeck and Ruprecht, Göttingen

Petzold E (1981b) Macht und Ohnmacht im System Familie. Periskop-Boehringer, Igelheim am Rhein

Petzold E (1983). Anorexia nervosa. In: Luban-Plozza B, Mattern HJ, Wesiak W (eds) Der Zugang zum psychosomatischen Denken. Hilfe für den niedergelassenen Arzt. Springer, Berlin Heidelberg New York

Petzold E (1985) Familienkonfrontationstherapie in der Klinik. Krankenhausarzt 58:70–72 (Reprint issue 1)

Petzold E, Bergmann G (1986) Konfrontation im Systemischen. Die Familiendimension in einem klinischen System. Praxis der Psychotherapie. Psychosomatik 31:87–95

Petzold E, Deter HC (1979) Familiendynamische Gesichtspunkte bei der Langzeitbehandlung von Patienten mit Anorexia nervosa. Therapiewoche 29:6332–6338

Petzold E, Herzog W (1991) The black eye. In: Klippstein H (ed) Ericksonian hypnotherapeutic group induction. Brunner and Mazel, New York

Petzold E, Vollrath P, Ferner H, Reindell A (1976a) Familienkonfrontation – Beitrag zur Therapie der Anorexia-nervosa-Patienten. Therapiewoche 26, 7:970–977

Petzold E, Vollrath P, Ferner H, Reindell A (1976b) Erfahrungen mit Familienkonfrontationstherapie bei Anorexia-nervosa-Patienten. Schweiz Z Psychosom Med 6:207–212

Petzold E, Kröger F, Ferner H, Hartmann-Kotteck-Schröder L, Bergmann G (1990) Die Angst, das eigene Gesicht zu verlieren. Balintarbeit und Familienkonfrontationstherapie. Die Balintgruppe 5

Petzold E, Kröger F, Deter HC, Herzog W (1991) Familienkonfrontationstherapie bei Anorexia nervosa. Z System Familie 4:158–167

Weber G, Stierlin HV (1989) In Liebe entzweit. Rowohlt, Hamburg

Weizsäcker V von (1988) Fälle und Probleme. Suhrkamp, Frankfurt, 248 (gesammelte Schriften, vol 9)

Outlook

W. Herzog, H.-C. Deter, and W. Vandereycken

Having finished our editorial work, it is time to look back at what has been written in this book and to compare it with the expectations we had when we started. We are quite satisfied with the end product thanks to the collaboration of a large group of experienced researchers and clinicians. Several of the authors have presented really new data; others have taken a very personal view of the subject. What might have appeared to be another boring collection of repetitive scientific papers on a small fragment of modern health care turned out to be a kaleidoscopic panorama of eating disorders seen through the eyes of professionals who still are fascinated and challenged at the same time by the experiences anorectic and bulimic patients were able to share with them. Beside purely methodological issues, the previous pages include reports and discussions on the most divergent aspects of eating disorders, ranging from historical and sociocultural viewpoints, to biological and medical approaches, to psychological and psychotherapeutic issues. As such it presents the truly "biopsychosocial" perspective which is required in modern health care.

We realize, however, that this book has its shortcomings, too. If we are slightly frustrated at the end of our editorial work, it is mainly because of our failure to present follow-up studies which all meet the standards set forth in the first chapter on "ideal" investigations. It was, however, a utopian wish since all studies had been planned and carried out long before the idea for this publication was conceived. So, as researchers have to learn from trial and error, other editors, too, may profit from our experience. It would be a great advantage if investigators would agree in the future on some basic guidelines and rules for follow-up research and would publish their findings accordingly. Therefore, as a final contribution to this book, we have outlined a description of follow-up studies and hope that other authors and editors will find this example useful.

Outline for Description of Follow-up Study

1. Introduction: background and aim of the study (hypothesis)
2. Method

2.1. Subjects: at first assessment (a) for whom follow-up data are available and (b) for all dropouts (specify at which stage they dropped out)
- Demographic characteristics (age, social class, marital status, educational level, occupation)
- Clinical features (DSM III-R diagnosis, age at onset or duration of illness, previous treatment)
- Pretreatment selection (what type of patient is referred and what type of selection took place before entering the study)

2.2. Treatment
- Description of the characteristics (in/outpatient, duration, therapeutic interventions) including the usual aftercare

2.3. Follow-up
- Design (prospective, repeated measurements, duration from first assessment)
- Procedure: how patients were invited and assessed, and by whom
- Specify evaluation instruments (use standardized assessment methods and a multidimensional approach)

2.4. Statistics
- Description of the procedures and methods applied

3. Results
- Actual body weight (highest/lowest weight over last 6 months)
- Menstruation (over last 6 months)
- Eating pattern (over last 6 months)
- Attitudes towards body shape and food
- Comorbidity (somatic, psychopathologic)
- Family aspects (family of origin, own family)
- Psychosocial functioning
- Process of recovery/illness

4. Discussion
- Methodological problems of present study
- Comparison with other studies
- Implications of findings (suggestions for further research)

Subject Index

MIX
Papier aus verantwortungsvollen Quellen
Paper from responsible sources
FSC® C105338

If you have any concerns about our products,
you can contact us on
ProductSafety@springernature.com

In case Publisher is established outside the EU,
the EU authorized representative is:
Springer Nature Customer Service Center GmbH
Europaplatz 3, 69115 Heidelberg, Germany

Printed by Libri Plureos GmbH
in Hamburg, Germany